This book is due for return on or before the last date shown below.

Consumers, Tinkerers, Rebels

Making Europe: Technology and Transformations, 1850–2000

Series editors: Johan Schot (Eindhoven University of Technology, the Netherlands) and Phil Scranton (Rutgers University, USA)

Book series overview:

Consumers, Tinkerers, Rebels: The People Who Shaped Europe
by Ruth Oldenziel (Eindhoven University of Technology, the Netherlands) and Mikael Hård (Darmstadt University of Technology, Germany)

Building Europe on Expertise: Innovators, Organizers, Networkers
by Martin Kohlrausch (KU Leuven, Belgium) and Helmuth Trischler (Deutsches Museum, Germany)

Europe's Infrastructure Transition: Economy, War, Nature
by Per Høgselius (Royal Institute of Technology (KTH), Sweden), Arne Kaijser (Royal Institute of Technology (KTH), Sweden) and Erik van der Vleuten (Eindhoven University of Technology, the Netherlands)

Writing the Rules for Europe: Experts, Cartels, and International Organizations
by Wolfram Kaiser (University of Portsmouth, United Kingdom) and Johan Schot
(Eindhoven University of Technology, the Netherlands)

Communicating Europe: Technologies, Information, Events
by Andreas Fickers (University of Luxembourg, Luxembourg) and Pascal Griset
(Paris-Sorbonne University, France)

Europeans Globalizing: Mapping, Exploiting, Exchanging
by Maria Paula Diogo (New University of Lisbon, Portugal) and Dirk van Laak (University of Giessen, Germany)

Initiator: Foundation for the History of Technology (Eindhoven University of Technology, the Netherlands)

The Foundation for the History of Technology (SHT) seeks to develop and communicate knowledge that increases our understanding of the critical role that technology plays in the history of the modern world. Established in 1988 in the Netherlands, SHT initiates and supports scholarly research in the history of technology. This includes large-scale national and international research programs, as well as numerous individual projects, many of which are in collaboration with Eindhoven University of Technology. SHT also coordinates Tensions of Europe (TOE), an international research network of more than 250 scholars from across Europe and beyond who are studying the role of technology as an agent of change in European history. For more information visit www.histech.nl.

Consumers, Tinkerers, Rebels

The People Who Shaped Europe

Ruth Oldenziel

Mikael Hård

First published 2013 by
PALGRAVE MACMILLAN

Palgrave Macmillan in the UK is an imprint of Macmillan Publishers Limited, registered in England, company number 785998, of Houndmills, Basingstoke, Hampshire RG21 6XS.

Palgrave Macmillan in the US is a division of St Martin's Press LLC, 175 Fifth Avenue, New York, NY 10010.

Palgrave Macmillan is the global academic imprint of the above companies and has companies and representatives throughout the world.

Palgrave® and Macmillan® are registered trademarks in the United States, the United Kingdom, Europe and other countries

ISBN: 978–0–230–30801–5

This book is printed on paper suitable for recycling and made from fully managed and sustained forest sources. Logging, pulping and manufacturing processes are expected to conform to the environmental regulations of the country of origin.

A catalogue record for this book is available from the British Library.

A catalog record for this book is available from the Library of Congress.

Contents

Making Europe:
An Introduction to the Series

In a typical conversation about twentieth-century European history, the subject of war will almost certainly arise—whether it is the Great War, the Second World War, or the Cold War. Similarly, historians who write about contemporary European history often view war as the twentieth century's iconic event. In fact, many scholars rely on Europe's political history, rife as it is with military conflict, to set the timeframe for their work. The influential historian Eric Hobsbawn, for example, defined the twentieth century as beginning with the First World War and ending with the collapse of the Soviet Union; Hobsbawn named this period—1914 to 1991—The Short Twentieth Century. Indeed, the topic of war and rupture has dominated the discourse on Europe in the twentieth century—and understandably so.

We, the editors and authors of the *Making Europe* series, however, have taken an alternative approach to our subject. We offer a European history viewed through the lens of technology rather than war. We believe that a European history with technology at its core can help to understand the continuities that have endured despite the rupture of wars. *Making Europe* places continuities—from the rise of institutions like CERN to the evolution of hacker networks—in a longer-term perspective. The *Making Europe* narrative suggests that recent European history is as much about building connections across national borders as it is about playing out conflicts between nation-states. This view of technology from a transnational perspective has proven to be felicitous. As a phenomenon, technology has always been particularly mobile; this mobility has allowed new technologies to help shape international relations between countries, companies, organizations, and people.

To understand the role of technology in this history, we required ourselves to rethink the very meaning of technology: referencing far more than machines alone, technology also embraces people and values; ideas, skills, and knowledge. Technological change, in our view, is a deeply human process. Technology was—and still is—central to the creation of Europe. And given its centrality, technology has been hotly contested—politically, economically, and culturally—in the making of Europe.

Technology's role in shaping Europe coalesced around 1850, when a new era began, an era from 1850 to 2000 that we refer to as The Long Twentieth Century. It was during the mid-nineteenth century that a newly globalizing world began to emerge. This was a world in which the many new transportation and communication technologies played a decisive part. At this time, technology became a reference point for European superiority—both within and beyond Europe. Cross-border connections and institutions thrived; the knowledge-sharing practices that fostered these connections were widely circulated and adopted. This circulation of knowledge led to a worldwide imagining, negotiating, and experiencing of Europe that exists today. This was also the foundation for the formal process of European integration that gained traction in the 1950s. Our perspective simultaneously decenters the European Union and its direct predecessors—which, after all, comprised only one force of Europeanization—and places the process of European integration in long-term historical context. Acknowledging that this dynamic of integration continues today, *Making Europe* presents and interprets a history that is still in the making.

That said, it is clear to us as historians that the decade 1990–2000 marks another watershed: it was in this period that the digital revolution gained new momentum, as did shifting power relationships at the global level. This spurred the European Union to become a hegemonic force of Europeanization, and it helped globalization to enter a new phase. Simultaneously, however, the processes of integration and globalization in this apparent new phase have proven to be fragile: in light of the global economic crisis, Europe's future, called into doubt, has become a pressing issue, and one with a sharp political edge. Accordingly, Europe's past has also come under fresh scrutiny. We contend that technology will continue to play a central role in defining Europe; that the politics of Europe is the politics of technology as much as anything else; and that now is the opportune time to explore technology's historical role in the creation of Europe.

Making Europe provides a perspective on European history that transcends borders. The volumes in the series examine the linking—and, in some cases, the disruption—of infrastructures and knowledge networks that operate beyond nations and states. Also mapped here is the transnational circulation—and appropriation—of people, products, and ideas. The people and organizations featured in this series employed particular notions of Europe in building their cross-border connections. Indeed, they imagined and invented new Europes, often making clear distinctions between which people and places belonged and which were alien to the concept and the reality of Europe. *Making Europe* asks: Who projected their ideas of Europe? When did these projections take place—how, and why? The series looks at the people and the organizations that perceived themselves as central—and peripheral—to Europe, its colonies, and the transatlantic crossings that were part of the European imagination. Examined here are migrants and experts, foods and inventions, markets and regulations—virtually everything that was identified, experienced, and communicated as "European." This Europeanization, we find, had significant—and sometimes unintended—consequences: some connections between people and institutions were lasting, others broken, these continuities and ruptures shaping Europe as both an imagined place and a living community. *Making Europe* explores the stability and fragility of these European connections, communities, and institutions.

The majority of existing studies of Europe have been based on either of two approaches. First is the, often massive, single-author narrative. Second is the essay collection, which presents many voices, in some cases edited to align the authors' themes. In the field of European history, single-author volumes have tended to be broad-ranging and to address different timeframes and regions.

Often, single-author volumes are a compilation of national stories; at their best, compilations transcend their individual stories to posit a complete European picture. Essay collections, for their part, have generally assumed a sharper focus—on particular communities, ethnicities, and empires, for example. These usual approaches point to a distinctive feature of *Making Europe*: in this series, five of the six volumes have two authors; one book has three writers. These voices, thirteen in all, create multiple narratives. The six sets of *Making Europe*'s co-authors have worked as a team to draft a series of volumes with coordinated yet individual themes (see www.makingeurope.eu). These six volumes contain six distinct points of view; as editors, we have imposed neither uniformity nor the pressure to harmonize narratives. In our opinion, the most informative new contributions to European history embrace diverse actors and diverse meanings, a range of purposes and understandings. *Making Europe* captures this diversity, reflecting a dynamic European history that continues to unfold.

All of the authors in the series have drawn on the European Science Foundation's "Inventing Europe" collaborative research initiatives as well as the Foundation for the History of Technology's "Tensions of Europe" project, begun in 1998 (see www.tensionsofeurope.eu). They have profited from an intensive period of discussion and joint research and writing at the Netherlands Institute for Advanced Study in the Wassenaar dunes in 2010–11. The fruits of these initiatives include the *Making Europe* book series as well as a web-based exhibit "Inventing Europe, European Digital Museum for Science and Technology" that encompasses a dozen of Europe's technology and science museums (see www.inventingeurope.eu) and scores of scholarly publications. All aim to promote creativity in fostering a more inclusive understanding of technology's role in refashioning Europe—an ongoing process that is as fascinating as it is contentious. The authors of *Making Europe* have asked themselves what shape an open-ended European history of technology would take. They provide their answers in the form of this book series.

The first volume in the *Making Europe* series, entitled *Consumers, Tinkerers, Rebels: The People who Shaped Europe*, is written by Ruth Oldenziel and Mikael Hård. This volume spotlights the people who "made" Europe by appropriating and consuming a wide range of technologies—from the sewing machine to the bicycle, the Barbie doll to the personal computer. What emerges is a fascinating portrait of how Europeans lived during The Long Twentieth Century. Explored here are the questions of who, exactly, decided how Europeans dressed and dwelled? Traveled and dined? Worked and played? Who, in fact, can be credited with shaping the daily lives of Europeans? The authors argue that, while inventors, engineers, and politicians played their parts, it was consumers, tinkerers, and rebels who have been the unrecognized force in the making of Europe.

The second volume in the series, entitled *Building Europe on Expertise: Innovators, Organizers, Networkers*, is written by Martin Kohlrausch and Helmuth Trischler. Here the focus shifts from consumers of technology to a new breed of professionals: the technical and scientific experts whose influence soared from around 1850 onward. The authors show how these experts created, organized, and spread knowledge—enabling them to shape societies, create cross-border connections, and set political agendas. During Europe's Long Twentieth Century, technoscientific experts became a strategic resource for serving national, international, and transnational interests, the authors argue. They revisit experts' visions of Europe, showing how these visions manifested in the dictatorships of Nazi Germany and Stalinist Russia—as well as helping to build Europe's vast research networks during the Cold War. *Building Europe on Expertise* ends with today's efforts to reinvent the European Union—as a knowledge-based society defined by experts.

The third volume in the series, *Europe's Infrastructure Transition: Economy, War, Nature*, is written by Per Högselius, Arne Kaijser, and Erik van der Vleuten. This book elaborates on the first two volumes by introducing a new cast of historical actors: system-builders. These individuals and organizations helped to transform Europe by envisioning, constructing, and manipulating large-scale transport, communications, and energy systems. Their efforts reshaped Europe as a geographical entity by forming massive new material interconnections—and divisions—between places. This had far-reaching implications for European integration; for peaceful economic exchange; for military planning and logistics. System-builders challenged Europe's natural barriers, from the Alps to northern Europe's forests and the vast marshlands to the east. But

Europe's water, air, and land were not only connected, they were transformed radically, sometimes destroyed. In response, system-builders eventually turned much of Europe's environment itself into infrastructure, interlinking isolated ecosystems via human-made corridors and networks.

The fourth volume, *Making the Rules for Europe: Experts, Cartels, International Organizations* is written by Wolfram Kaiser and Johan Schot. Here, the focus becomes the norms and standards of technological innovation—discussed in depth for transport and heavy industry. Featured are the people and organizations that debated, negotiated, and regulated the cross-border issues raised by innovation. Presented here are individuals with special—and often interdisciplinary—expertise in technology, business, and law. Often, these experts sought to de-politicize issues by deeming them technical; this yielded workable solutions to shared problems. It also paved experts' way in rule-making for multiple, distinct yet overlapping, and frequently competing "Europes." In the pursuit of finding technological solutions, many institutions' transnational practices survived ruptures, including the two World Wars. After the Second World War, the European Union was obliged to accommodate—and to compete with—other institutions' established practices in order for the EU to gain greater influence in shaping Europe.

The fifth volume, *Communicating Europe: Technologies, Information, Events,* analyzes Europe's information and communication systems from roughly 1850 onward. Authors Andreas Fickers and Pascal Griset place these technologies at the very heart of European society. Presented here is a global vision of media, telecommunications, and computers that reveals the tensions inherent in designing and appropriating electrical and electronic devices. The authors argue that the control in the material realm by research and entrepreneurship and the emergence of new forms of creativity and new ways of life are two sides of the same coin, mostly driven by political and cultural forces. Examined in this volume are the political, economic, and cultural realities and meanings of information and communication technologies on a European level. This perspective, which extends over the long term, provides the tools for a new critical understanding of the digital revolution.

How did today's globalized, thoroughly mapped-out world emerge? What part did technology play in Europe's international encounters, colonial and otherwise? *Europeans Globalizing*, written by Maria Paula Diogo and Dirk van Laak, concludes the *Making Europe* series with a study of how Europe interacted with the rest of the world from 1850 until the close of the twentieth century. The volume details how technologies were applied and creatively adopted–from India to Argentina, South Africa to the Arctic. From the turn of the twentieth century onwards, we witness assumptions about Europe's technologically-based superiority being continuously challenged. And we discover that globalized Europe in its present form looks quite different from what Europeans once imagined.

Consumers and tinkerers; engineers and scientists; system-builders and inventors. Experts in technology, law, and business; communicators and entrepreneurs; politicians and ambassadors. This is a cross-section of the actors represented on *Making Europe*'s pages. These actors, through the institutions and organizations they cultivated, the connections they created, the rules and practices they fostered, co-created Europe. Narrated from contrasting as well as complementary viewpoints, the six volumes in the series create a collage of co-existent portraits that depict Europe's Long Twentieth Century; its technologies; and its meanings. Together, these histories form the view of modern Europe that we and the authors wish to contribute to the historical record at this time.

Johan Schot & Philip Scranton
Making Europe Series Editors
Amsterdam, the Netherlands & Camden, New Jersey, USA
July 2013

Acknowledgements

This book covers much ground—chronologically, thematically, and geographically. It is based on research and scholarship on topics ranging from nineteenth-century French fashion to late-twentieth-century Greek computer hackers. To be able to paint a broad picture of how Europe was crafted from below, we have relied on the expertise of a large number of scholars. Colleagues from an array of countries and language areas have kindly shared their knowledge with us. Some have read and commented on earlier versions or the ensuing chapters; others have provided the authors firsthand with papers and articles, presenting empirical material previously unavailable in the languages in which the authors themselves are fluent. Although the bibliography at the end of the volume might imply an English bias, this does not mean that the underlying scholarship focuses on the Anglo-Saxon world only. We have sought out early career scholars from various parts of Europe who have kindly provided translations and summaries of their original research. In fact, we have brought together many scholars from non-English based research communities through workshops, special issues, and collections of essays that form the backbone of *Consumers, Tinkerers, Rebels*. Furthermore, archivists, librarians, and student assistants have provided material in various ways.

The collegial knowledge exchange has to a large degree taken place at conferences and workshops. Several foundations and institutions have generously contributed to financing the organization of and the trips to such events. In addition, research councils have supported some of the research on which the book is based. Below, the authors list the collective bodies and individuals without whom this book could never have materialized. Hopefully, the list is all-encompassing; in case the authors have forgotten anyone, please accept their sincere apologies.

In the first instance we would like to thank eight colleagues who supplied original papers of direct relevance to the book: Adri A. de la Bruhèze, Marija Drėmaitė, Theodore Lekkas, Faidra Papanelopoulou, Emanuela Scarpellini, Karin Taylor, Milena Veenis, and Patryk Wasiak. They and others provided the authors with extensive advice and invaluable references on various topics, and in some cases generally shared unpublished material: Gerard Alberts, Irene Anastasiadou, Esther Anaya, Adri A. de la Bruhèze, Alexander Badenoch, Dagmar Bellmann, Simon Bihr, Robert Bud, Marie-Emanuelle Chessel, Peter Cox, Katarzyna Cwiertka, Pascal Delheye, Martin Emanuel, Els De Vos, Valentina Fava, Terje Finstad, Martin Franc, Santiago Gorostiza, Ferenc Hammer, Larissa Iakovleva-Zakharova, Dagmara Jaješniak-Quast, Jíra Janáč, Thomas Kaiserfeld, Stijn Knuts, Dick van Lente, Aleksandra Lipińska, Elena Kochetkova, Martin Kohlraush, Slawomir Lotysz, Gerrit van Maanen, Malgorzata Mazurek, Massimo Maraglio, Nina Möllers, Dobrinka Parusheva, Faidra Papanelopoulou, Enrique Perdiguero-Gil, Marcus Popplow, Martin Schiefelbusch, Frank Schipper, Emanuela Scarpellini, Luisa Sousa, Elitsa Stoilova, Laurant Tissot, Aristotle Tympas, Richard Vahrenkamp, Frank Veraart, Erik van der Vleuten, Margaret Walsh, Heike Weber, Andrea Westermann, and Len Winkelman.

Whereas Johan Schot, Philip Scranton, Andreas Ludwig, Annemarie de Wildt, and two anonymous reviewers commented on the entire manuscript, the following persons have read and criticized individual chapters: Gerard Alberts, Alexander Badenoch, Liesbeth Bervoets, Eileen Boris, Carolien Bouw, Marija Drėmaitė, Caroline van Dullemen, Martin Emanuel, Nicoletta Emmens, Rosa Knorringa, Hester Lenstra, Jan-Hendrik Meyer, Thomas J. Misa, Emanuela Scarpellini, Nathalie Scholz, Sonja Petersen, Milena Veenis, and Frederike de Vlaming. The authors are extremely grateful for their comments, as well for the invaluable and unflagging assistance that Lisa Friedman provided first and Phil Scranton toward the end of the editing process.

At the early, conceptual stage, Heinz-Gerhard Haupt, Thomas J. Misa, Kiran Patel, Pierre-Yves Saunier, and Frank Trentmann gave us important hints in which direction to go and how to design the manuscript. At later stages the authors of the other volumes in the *Making Europe* series, as well as the participants at several series workshops, contributed with immensely valuable comments and criticism. Throughout the process the members of the ESF-financed project "European Ways of Life in the American Century" (EUWOL) inspired the authors in various ways, especially when conceiving chapters 2, 4, and 5.

Of the archivists and librarians who assisted the authors, special thanks go to Cristina Cuevas-Wolf (Wende Museum, California), Monique de Hair (ANWB archives, The Hague) and Eva Hammerschmiedová (Organisation intergouvernementale pour les transports internationaux ferroviaires, Berne), Madelief Hohé (Gemeentemuseum Den Haag), Berthe Receveur (Spoorwegmuseum Utrecht), Roelof Jansma (Universiteit van Amsterdam), Erwin Nolet and Dindy van Maanen (both at the Netherlands Institute of Advanced Studies), and Arlette Strijland (Aletta E-Quality, Amsterdam). Equally important was the research assistance of Ana Bara, Giel van Hooff, Slawomir Lotysz, João Machado, Faidra Papanelopoulou, Marcela Rilovic, and Marie-Christin Wedel. And of immense importance to the design of the book were the image assistants Giel van Hooff, Slawomir Lotysz, Katherine Kay-Mouat, and Jan Korsten as well as the research team of the Inventing Europe, European Digital Museum for Science & Technology http://www.inventingeurope.eu/.

We express special thanks to the co-editors of special journal issues and edited books of direct importance to the volume:

- Liesbeth Bervoets and Mikael Hård, eds., "Coping with Modernity: European Ways of Housing in the American Century," Special Issue of *Home Cultures* 7, no. 2 (July 2010) [incl. case studies from Belgium, Bulgaria, Czechoslovakia, the Netherlands, and Sweden]
- Adri A. de la Bruhèze and Ruth Oldenziel, eds., Special Section on Global Cycling, *Transfers* 2, no. 2 (Summer 2012) [incl. case studies from Finland, West Africa, China, and Japan]

- Adri A. de la Bruhèze and Martin Emanuel, eds., "European Bicycling: The Politics of Low and High Culture: Taming and Framing Cycling in Twentieth-Century Europe," *Journal of Transport History* 33, no. 1 (June 2012) [incl. case studies from the Netherlands, Germany, Sweden, and Finland]
- Heike Weber and Ruth Oldenziel, eds., "Recycling and Reuse in the Twentieth Century," Special Issue *Contemporary European History* 22, no. 3 (2013) [incl. case studies from Germany, France, the Netherlands, and Sweden]
- Gerard Alberts and Ruth Oldenziel, eds., *Hacking Europe. From Computer Culture to Demoscenes* (New York: Springer Verlag, 2013) [incl. case studies from Great Britain, Czech Republic, Finland, Germany, Greece, the Netherlands, Poland, and Yugoslavia]
- Ruth Oldenziel and Helmuth Trischler, eds., *Re/cycling Histories: Paths towards Sustainability* [Oxford: Berghahn, forthcoming] [incl. case studies from Great Britain, France, Germany, Hungary, Japan, the Netherlands, Spain, and the United States]

In addition to these named individuals we would like to thank the participants of innumerable workshops and conferences, especially those that have been part of "Tensions of Europe," an international network of historians of technology. The most important conferences were organized in Sofia, Amsterdam, Lisbon, and Copenhagen. In addition to several meetings and workshops held within the framework of the EUWOL project, the following events deserve a mention: "Re/Cycling History: Paths towards Sustainability," Rachel Carson Center for Environment and Society (RCC), Munich, and "Towards a European Society? Transgressing Disciplinary Boundaries in European Studies Research," Centre for European and International Studies Research (CEISR), Portsmouth; "The European: An Invention at the Interface of Technology and Consumption," Munich Center for the History of Science and Technology, Munich, Germany.

We also wish to express our gratitude to the organizations that contributed financially to making this book possible: the European Science Foundation (ESF), the Foundation for the History of Technology (SHT), the Dutch National Science Foundation (NWO), the Netherlands Institute of Advanced Studies (NIAS), Darmstadt University of Technology, and the Technical University Eindhoven.

Finally, two persons deserve a special mention: Johan Schot—the visionary without whom this book and indeed the whole *Making Europe* series would never have come about—and Phil Scranton, who never gave up encouraging the authors to continue with their work. You seldom find more enthusiastic and inspiring series editors.

Ruth Oldenziel and Mikael Hård
Amsterdam and Darmstadt, March 2013

Introduction

Who could have predicted the shape of twentieth-century Europe—and which people were to shape it? In fact, one thinker came uncannily close to forecasting Europe's future government: in 1814, Claude Henri de Rouvroy, Comte de Saint-Simon, foreshadowed the European Union. In his treatise, *De la réorganisation de la société européenne* (*The Reorganization of European Society*), the French utopian socialist foresaw integration in the form of a single parliament governing the Continent's national governments. This integration, Saint-Simon suggested, would come about through the unifying forces of science and technology. Experts schooled in the scientific principles of the Enlightenment would help vanquish the political resentments and ideological struggles that were vestiges of the French Revolution and the Napoleonic Wars. According to Saint-Simon, these experts—philosophers, scientists, and engineers—would play the lead role in building the unified European utopia.[1]

While the philosophers, scientists, and engineers in Saint-Simon's vision proved to be important, they alone did not create Europe technologically and otherwise. As this book shows, consumers, tinkerers, or rebels were equally vital in bringing together the people and the nations of Europe and beyond. Further, the standard technologies of integration—from railways to computer networks—have not functioned exclusively in favor of Europe's unification. On the contrary, these so-called technologies of integration have often reinforced social segregation, cultural differences, and national divisions.

How Europeans have become who they are, through shaping—and being shaped by—the technologies of their time: that is the subject of this book. How have people in Europe adopted, re-purposed, molded, appropriated, even hijacked machines, gadgets, and infrastructures for their own purposes? Through the lens of such technologies, this volume projects a picture of how Europeans eat and drink; dress and dwell; travel and think; work and play—practically every behavior that comprises a "culture." Our timeframe for this book is the 1850s to 2000, that formative one hundred and fifty years that some historians refer to as "The Long Twentieth Century."[2]

As with the other volumes in the *Making Europe* series, *The People Who Shaped Europe* focuses on the European Continent. But what do we mean by "Europe"? The concept is usually understood as the sum of the nations located on the Euro-Asian Continent. Sometimes, the concept of Europe is limited to the people living in the Western part of the Continent. But the authors of this volume define "Europe" in a more encompassing way. The concept of "Europe" transcends geography to include other meanings, connotations, and identities. "European" is an ever-changing descriptor. And "European" can apply to *artifacts*, from objects like the touring bicycle and the train compartment; to *individuals* like Alva Myrdal and Elie Metchnikoff; and to *groups* like the Polish Union of Consumer Cooperatives and the German Computer Chaos Club.

The disciples of Claude Henri de Rouvroy—the so-called Saint-Simonians—championed canals, railroads, and telegraph lines as capable of breaking down national barriers between individuals and cultures. As this book shows, engineers and technologies themselves were seldom able to unite citizens of divergent nationalities, classes, ethnicities, and genders; their interests, intentions, and wishes were simply too different. This volume emphasizes the idea that technological innovations rarely produced uniform experiences of Europe. For example, while the advent of the train offered new freedom of mobility, emigrants on the lower rungs of the social ladder encountered an entirely different Europe from that of the *haute bourgeoisie* in their first-class cabins. Traveling across Europe in separate corridors, the classes rarely met. These separate but unequal transportation experiences were reinforced in Imperial Germany, where enclosed railway stations were constructed to prevent poor emigrants from mingling with other passengers—and to allow U.S. immigration officers to control and screen "undesirables" before they even left Europe. Switzerland perpetuated the system as well, building special sleeping quarters known as "Italian barracks" (*Italienerbarracken*) to manage the poor laborers who transited the country on their way from Southern Europe to the Ruhr region. To alleviate the situation, religious women's groups created shelters within train stations and at ports. These outposts, found all over Europe, were designed to help protect unsuspecting rural girls—entering the cities to search for jobs—from falling into the hands of pimps and resorting to prostitution.

Other volumes in the *Making Europe* series highlight the contributions of scientists and engineers in the process of unifying Europe. As a complement this book, instead, focuses on consumers and users of technology as the main actors. The authors examine how consumers in their role as users of technologies created different versions of "Europe." In the chapters that follow, we examine how sewing machines, Barbie dolls, and recycling bins, for example, became part and parcel of various "European ways of life." In the process of appropriating technology, networking, tinkering, and tweaking were necessary activities. The book spotlights scenarios in which consumers united to form transnational user movements; in which tinkerers modified machines to make them better fit users' needs; and in which rebels protested against the introduction of what they regarded as intrusive technological systems. The emerging user movements often opposed national laws, international agreements, and pan-European acts that tried to restrict user access to particular technologies.

Examples of powerful user movements abound in this book. For one: using liberal Britain as their model, bicyclists formed clubs in cities and towns across the Continent. Exceedingly well organized, they lobbied for better roads locally; set up networks of service stations nationally; and demanded easier cross-border mobility for cyclists internationally. The call for cross-border mobility transformed bicycle clubs from local and national initiatives into a trans-European—and even transatlantic—user movement. Another example: in the field of fashion technology, the concept of "Europe" coalesced in a less politically organized, yet similar way. Here, women around the world came to equate the city of Paris with Europe as a whole. In this case, "Paris" served as a *pars pro toto*—a part taken for the whole—of what Europe stood for; "Parisian fashion" became shorthand for "European fashion." And it was a phenomenon that, to a large extent, defied U.S.-style mechanization.

This book, then, delves into consumers' and users' experience of how Europe was understood and shaped. Why, for example, did leagues of nineteenth-century Europeans risk confrontation with customs officials by bringing bicycles across national borders? How did Eastern European Jews negotiate access to transportation technology in order to escape pogroms and establish new lives across the Atlantic? And how do we explain why 1980s hackers on both sides of the Iron Curtain felt more attracted to America's Silicon Valley counterculture than to the cultures of their home countries?

To some readers, the phrase "History of Technology" may suggest narratives about famous inventors like Gottlieb Daimler and Carl Benz; breakthrough inventions like the Concorde supersonic airplane; and ingenious information systems like the World Wide Web. The creators of these innovations certainly have their place in the history of technology, but they represent only one side of this history. In contrast to traditional historical writing, this volume tells stories about everyday technologies and the ordinary people who used them. These stories recount how user groups domesticated innovations—often in ways that the inventors had never envisioned. In terms of the authors' decision to pass over the well-known story of the Daimler-Benz car, the reasoning is this: only a relative few aristocrats had, in the early days, direct access to technologically exciting automobiles. Most European citizens used a personal mode of transport that was far cheaper, simpler, and more appropriate to their needs: the bicycle. Like the early automobile, the Concorde was similarly rarified: there were very few of these advanced-design planes that whisked businessmen and VIPs across the Atlantic; it was largely the Airbus A300 series—built by an international European consortium—that transported millions of tourists from Northern to Southern Europe. And in the case of the World Wide Web, its success cannot be explained by reference to Tim Burners-Lee and Robert Cailliau alone. The "real" inventors of the Internet are the users *en masse*.

Fig. 0.1 Technology as Europe's Unifying Force: *Historically, thinkers like Henri de Saint-Simon (born 1760) have believed that technology can unite peoples, nations, and, at its best, continents. This image—the cover of the Meccano toy company magazine from Christmas, 1923—charmingly reinforces this belief in technology: Flying high in a Meccano-made airplane, Santa Claus hands out construction kits to cheering, grateful boys. Included are the continents of Europe and North America. A LEGO precursor, Meccano metal parts enabled boys to build everything from miniature bridges to fierce-looking fighter tanks. Middle-class parents saw such building sets as helping to cultivate future engineers.*

The heroes of this narrative, then, are neither Justus Liebig nor Louis Pasteur; Alfred Nobel nor Guglielmo Marconi. Instead, among the heroes—and heroines—in this book are the intrepid housewives and home-economics teachers who appropriated the latest scientific findings and new-fangled cooking equipment. The protagonists of our narrative are the bold bicyclists who pioneered bicycle lanes and protection for their fellow riders on two wheels.

And the central characters are the young Dutch activists who reconfigured the burgeoning Internet for the purpose of reaching out to AIDS patients in the United States.

On the premise that engineers and designers do not alone create technology, this book emphasizes the user experience. Users do not merely influence technology: in fact, they co-produce it. The history of technology is no longer limited to the study of inventors, manufacturers, and factories. Consumers and activists, tinkerers and rebels—as well as the experts enlisted to speak and legislate on their behalf—all have challenged how engineers originally imagined the "proper" uses of technology. How users integrate technology into their daily lives—or, indeed, exclude it: this is the volume's main subject. In this new view of history, "consume" and "use" are not passive practices that follow pre-set rules; rather, consuming and using are vital activities encompassing deliberation and decision, modification and adjustment. After all, the act of consuming requires accessing economic resources, and the act of using depends on employing skill and knowledge. By focusing on consumers' and users' own innovations, this book shows that "users matter" in shaping technologies and cultures.[3]

The main argument of this book, then, is that the history of Europe, including the British Isles, Russia, and Turkey, can only be fully understood by including the study of consumers and users. That said, the authors treat machines and infrastructures only insofar as they play important roles in people's daily lives. This volume adds to the traditional narratives of modern, high-tech systems by exploring mundane, low-tech systems. And so, it is not the steam locomotive that is featured on these pages but the train compartment. When considering the Second World War, it is not the high technology of genocide and warfare that we write about, but the low-tech public-salvage campaigns to recycle old love letters and private pots and pans. Similarly, it is not Europe's famous architects and their visions that preoccupy us, but the way in which people heated and furnished their living rooms. We, the authors of this volume, are card-carrying historians of technology, responsible for placing machines and systems into a broader cultural context. To capture that context, we narrate the force that users exert in the making of technology.

Traditional historical narratives take a cause-and-effect approach to technology. For example, historians investigated how the television set as an artifact—originally a luxury item—trickled down to the poorer classes, and how it changed family routines. Again, *Consumers, Tinkerers, Rebels* challenges this approach to the past. While we do indeed trace machines and systems as they travel through the hands of various users, we do not interpret history in terms of cause and effect, invention and impact. Consider the sewing machine. With the introduction of this iconic invention, women garment workers who labored at hand-sewing ready-made clothes in sweatshops were pressed to develop new skills in order to retain their jobs. By contrast, wealthy, upper-middle-class housewives bought sewing machines not for sewing, but to decorate their parlors. Meanwhile, other users adopted sewing machines to make their own clothes—for reasons of frugality or fashion. The sewing machine was thus a much-needed tool for earning a living; a gadget for showing off wealth; and a handy device for creating a personal wardrobe. Indeed, there is no causal connection between innovation on the one hand and influence on the other. Furthermore, "inventions" may not be equated with "machines." An exclusive focus on the sewing machine as the defining invention would ignore the technologies of the fashion plate and the paper pattern; analysis of the sewing machine shows that it was a tool of diverse but limited utility. Arguably, scissors, pencils, and paper patterns—traditionally regarded as less spectacular and lower-tech than sewing machines—even more handily enabled women to creatively design, make, and reproduce clothing.

The practices of endless cutting and pasting, of shaping and tinkering are, therefore, on center stage. In this book, with the telling subtitle *The People Who Shaped Europe*, we focus on technology users in representative European locales. The narrative documents everyday activities, from the making of fashionable clothing to the designing of individual apartments; from traveling by train to touring by bicycle; from cooking evening dinners to sharing computer know-how. The history of how consumers copy patterns and styles, how rebels tweak novelty, and how countercultures define new areas of application, is the story of how material artifacts and technological systems found their place in society. This book, then, is about the history of technologies-in-use—and the people who used them.[4]

The making of Europe has been enacted not only by individuals operating independently, but also by groups. Users have organized themselves into political movements, like "hacktivists" or "pirate parties" of today. These are "user movements"—collective bodies, organizations with the power to challenge technocratic ideas of what is good for humanity. During The Long Twentieth Century, shared passions motivated enthusiasts of many different stripes to gather; form associations; and exchange information. Bicycle clubs, housewives' societies, and youth movements have tried to make sense of new technologies and to develop communal practices to turn unfamiliar technologies into familiar tools and toys. Analyzing these practices from a pan-European perspective offers insights into the plurality of the European experience. On the following pages we see European users reshaping their worlds in novel ways—at critical times in their personal histories and at crucial moments in European history.

The case of the computer and its users illustrates these points. From the very beginning, personal computing was as a subculture unto itself. This culture, although it germinated in the United States, traveled to Western and Southern Europe, and made its way behind the Iron Curtain. Members of the Greek and Polish comput-er-user movements created experiences distinct from those of their fellow hackers in West Germany and Finland. It was the decision to purchase computer hardware from abroad—and the move to develop local software in local languages—that distinguished Greek and Polish computer users. Similarly, personal computer aficionados in 1980s Yugoslavia wrote their own programs for their imported or do-it-yourself computers. In all of these activities, the investment of time, skills, and resources devoted to playing and tinkering was serious indeed.

This book, then, offers a rich array of cases that portray important but often forgotten scenes from European history. Whereas some stories are meant to bring out the transnational character of tech-nology user movements, other stories emphasize the local character of appropriation processes. Some cases illustrate the regulatory power of political and legal decision makers; other cases feature the extraordinary creative freedom that skilled users acquired.

The book's three parts follow three historical phases; like tiles on a rooftop, the examples from these periods overlap to produce a sturdy structure. Each part analyzes the possibilities that users identified and exploited in order to co-produce Europe's technolog-ical landscape. Beginning in the middle of the nineteenth century,

Part I, The Nineteenth Century: Shaping New Technologies, shows how companies, experts, and users co-produced modern machinery without the omnipresence of the state. This section, which covers the period before the First World War, shows the importance of class and estate in reconfiguring new technologies: in fashion and housing, the emerging middle and working classes challenged and emulated an aristocratic lifestyle. In the case of transportation technologies, class structures were reproduced by dividing passenger trains into first-, second-, third-, and fourth-class spaces. With imperial ambitions and aristocratic lifestyles, Paris, Vienna, and Berlin functioned as Europe's reference points—within Europe as well as in colonial settings.

Part II of this book, After The Great War: Who Directs Technology?, opens with narrating the devastating effects of the First World War, and moves on to the post-Second-World-War period. With the collapse of empires and aristocratic values emerged the competing model of classless European nations under state leadership. Classlessness became the foundation for Soviet communism, Nazi Germany, Fascist Italy, and middle-class America. In this new era, the state, in alliance with new professionals and commercial interests, intervened in the building of elaborate technical systems: large-scale housing projects, national and pan-European highway networks, and industrially-processed food infrastructures, for example. As best they could, user movements voiced their needs in the political arena, while trying to domesticate commodities and provisions to align with their own needs.

Finally, Part III, Beyond The 1960s: Users Empowered?, revisits the Second World War and the postwar years, ushering the narrative to the turn of the millennium. These were the very decades during which the colonies revolted and most European countries lost their empires. It was also the time when the Berlin Wall collapsed and the European Union expanded—events that offered users wider world perspectives. During these decades, users were by turns invigorated and overwhelmed by the power exercised by global corporations and by the emerging transnational government of the European Union. The European Union itself gained increased regulatory power to set technology standards. At the same time, users across Europe coalesced into consumer and user networks; as activists, European citizens also leveraged the Union's newfound power to suit their own purposes.

Together, these examples and scenarios create an intricate historical drama. The protagonists of this history are neither Europe's
kings, presidents, and governments, nor its Nobel Prize laure-
ates, technocrats, and civil servants; the spotlight is on Europe's
consumers, tinkerers, and rebels. The history of European technology as narrated in this book is one in which consumers and users
have successively retained their power to affect designs and innovations. Despite what the art historian Siegfried Giedion claimed in
the mid-twentieth century, mechanization has not taken command
of our lives. User movements have mobilized on local and regional,
as well as national and transnational levels. Throughout The Long
Twentieth Century and beyond, users have risked raising the ire
of governments and corporations—and actively shaped Europe in
their own images.[5]

I

THE NINETEENTH CENTURY: SHAPING NEW TECHNOLOGIES

1
Poaching from Paris

The steam engine may well be the icon of the industrial revolution.[1] Yet, for many people of the age, it was the power loom and later the sewing machine that represented the true markers of the new industrial era, if not modern technology. This was a time fraught with violent labor conflicts in the mills, where low-paid women and children stitched cheap clothing. Beyond the labor conflicts, these innovations also generated a world of fashion rife with conflicts played out in the streets and in churches, in the home and in the media. As our opening story shows, the nineteenth-century French aristocracy and British middle class served as role models for individuals the world over. In their attempts to copy Parisian fashion, creative women appropriated various techniques. Together with new fabrics and paper patterns, the sewing machine enabled a user movement to tinker with pencils, needles, and scissors. Salesmen from the famous Singer Company contributed to the international rise of the sewing machine. But it was the transnational circulation of magazines, pictures, patterns, and clothing that enabled fashion to develop into a truly global affair. European actors shaped fashion technology—for themselves and for their colonial subjects. But the outcomes of their efforts were often unintended.

In October of 1875, in Pretoria, South Africa, Friedrich Grünberger of the Berlin Mission Society railed against the Paris-inspired dresses of his African women parishioners. On the spot, the Lutheran minister forbade his female congregation to ever again wear fashionable crinoline. The stiffened petticoats that made the waist look thinner represented European aristocratic leisure. Some of Grünberger's church members heeded their minister's stern warning, but not all of the African converts were willing to give up their pretty dresses so easily. One church member even threatened to convert to the more liberal English Church if she could no longer wear her fashionable clothes to Sunday service. In response, Grünberger took radical action to discipline the resistant woman and others like her. "With two elders I went from house to house, starting off with the house of this very reluctant woman," he later wrote, reporting with glee that, "We confiscated all the crinolines to burn them to death before their eyes, with the others joyfully laughing."[2]

Seeing the female congregation wearing the likes of crinolines had not always incensed the German missionaries. Originally, they had believed it to be an outward sign of religious conversion when their African parishioners adopted European-style clothing. Ministers' wives were often mobilized in the sartorial effort, establishing classes to teach their African parishioners how to sew Western clothing for themselves and their families. In some cases, the most faithful converts were lent the mission's sewing machine. In transferring sewing skills, missionaries sought to maintain the colonial hierarchy and reserve for themselves the more prestigious skill of cutting cloth—a skill eagerly defended by male professional tailors and female dressmakers back in Europe. As another German Lutheran missionary put it bluntly in the mid-1860s, the main goal of transmitting European styles to the African converts had been "to make my congregation look nice and cover their nakedness" in exchange for "their old stupid hides and whatever they might drape around their hips."[3] Ten years later, though, the mood among the missionaries had changed radically. In their eyes, dressing according to the latest fashion of Paris and London was nothing less than trespassing the proper racial boundaries. Fashionable clothes represented a fake religious conversion that had to be suppressed at all costs. As a result, the missionaries' rage extended to the younger male African generation. Proud of earning high wages in the Kimberley diamond mines or on the sugar plantations, migrant

Fig. 1.1 On Strike:
In the 1800s, textile production was divided. On the home front were women sewing their own clothing. On the industrial front were factories and sweatshops producing clothing, most of it for men. For women in many of Europe's industrializing regions—from Barcelona to St. Petersburg—the textile industry provided the prospect of paid jobs. But given the dangerous working conditions and very low wages, textile work was a double-edged sword. This photo, taken in Lawrence, MA in 1912, shows immigrant textile workers from Europe striking to demand improved conditions. Labor strife was widespread throughout the century.

workers from Sotho and Tswana showed off their good fortune by dressing according to British fashion. It was a practice that did not amuse the white minority.

In the nineteenth century, fashion was an issue of race, gender, and class. In 1876, the appropriation of high fashion by the lower classes vexed a well-to-do British woman in London. "My income is small, but I have to keep up a good appearance, and am therefore obliged to keep two servants, a cook and a housemaid," she wrote to *The English Woman's Domestic Magazine*. To her consternation, her servants acquired information about the latest Parisian fashion through easy access to fashion plates and paper patterns, as well as to shops and department stores. Her cook had the audacity to come to church in "a black silk made exactly like a new one I had sent home in the beginning of winter, and a new bonnet which I am certain I saw in Madame Louise's window in Regent Street marked 25 s. She looked as if she had stepped out of a fashion plate." The middle-class woman, at a loss concerning how to re-establish proper boundaries with her cook, could only think of the most radical way out of the dilemma: "I feel certain if I remonstrate with her she will leave, do you not think I may as well discharge her at once?"[4]

The German missionaries' bonfire of the crinolines and the British lady's sacking of her cook were informed by the same, conservative ideology: the elite's disapproval of colonial subjects appropriating European styles and of the lower classes adopting upper-class ways. Both incidents shed light on the enormous social power aristocratic clothing commanded and the appeal of European fashion in other corners of the world. The episodes indicate the kind of access that African converts and British servants had to Parisian fashion models, knowledge, and techniques. It was the manufacturers of paper patterns, sewing machines, and other tailoring equipment who made it possible for women everywhere to transform their passion for Paris into sartorial splendor. Cheaper textiles and increasing media coverage made fashion more accessible to different classes. The circulation of fashion technology, knowledge, and skills allowed women to participate in a European aristocratic world of fashion they would never be able to visit in person. By developing their own cutting, sewing, and machine-operating skills, these women joined a transnational user movement in self-made fashion. They contributed to making European fashion—primarily French aristocratic fashion—the measure of civilization worldwide. Long after its political and economic influence peaked, the European aristocracy shrewdly wielded cultural power by defining good taste in women's fashion. Women all over the world and of all classes participated in reproducing the European aristocracy's taste and material symbols. Many middle- and working-class—and even colonial and rural—women became part of a fashionable transnational world through imaginary connections, although they had neither the means to travel nor the aristocratic titles.[5]

In this world of textiles and fashion, the user movement creatively tinkered with the global production of clothing. The demands of ever-changing styles in the women's fashion system and the clothing industry gave birth to one of our least understood informal cultures of innovation and modification, remaking and re-designing. Working-class and middle-class women reproduced what they believed to be aristocratic style elements. One prevailing system was the expensive, Parisian-based *haute couture*; the second, parallel system was the mass-scale production of ready-made clothing. Operating in between were women of all classes, engaged in a thriving, do-it-yourself movement of home dressmakers. The sewing machine and paper patterns became key ingredients in this culture—along with the tacit knowledge, techniques, and skills these technologies imparted.

Fig. 1.2 Textile Technology's Far-Flung Influence: *In the twentieth century, white European missionaries sought to convert black Africans to Christianity. But religious ideas were only part of the cultural experience. Through missionaries, parishioners appropriated textile technology. Learning the specialized skills of cutting cloth, using paper patterns, and operating sewing machines, Africans made their own European-style clothing, including crinolines and suits. This photo of parishioners was taken around 1900 by the Basel missionary in Cameroon. The photographer's original caption reads, "At the sewing machine. Heathen with 2 wives and children." Note the missionary's ambivalence about Africans adopting European dress codes.*

This chapter describes how Grünberger's African converts came to subscribe to the Parisian standard of crinolines, and how workers came to imitate the dress code of their employers. Where did South African women and British domestic servants find their role models—and where did they acquire the skills and materials to participate in this transnational world of fashion? We trace the roots of this phenomenon to French aristocratic circles, a culture characterized by idleness, opulence, and expressiveness. We have elected not to map the history of famous designers and fashion houses, which sought to protect their work through copyright and patent laws. We have also bypassed the oft-narrated violent history of the textile mills, where low-paid women and children turned out cheap clothes. Instead, the chapter focuses on the largely forgotten but widespread culture driven by the wish to domesticate Parisian ways of life. The backbone of this culture was the very large group who could not afford to buy the latest fashionable clothing directly, and so tried to copy the Parisian exemplars. The skill of tinkering was at the core of this vibrant culture: as workers and home-makers, women sought to mediate between the parallel systems of *haute culture* and ready-made clothing, so as to participate in the period's aristocratic culture of fashion. Adopting various techniques in the manner of *bricolage*—using whatever materials were available—these tinkerers comprised a user movement that appropriated, contested, and changed fashion to suit their own tastes and circumstances. The movement was indeed socially heterogeneous, though it was held together by a shared ideal of good taste and good manners. Providing women all across Europe with a powerful point of reference, French fashion contributed to the creation of a transnational sense of communality that often masked national and class differences.

Made in Paris, Made in Europe

Since the late eighteenth century, textiles had been one of the industrializing world's leading sectors. Textile manufacturing provided more jobs than other industry until the Second World War, if not later. From then on, textiles became a global affair.[6] Replacing home-spun and home-made clothing were a number of innovations in agriculture, crops, and textiles. These innovations connected the wool-producing merino sheep in Australia and the cotton plantations in the American South with German dye producers and British and U.S. textile mills. Spinning and weaving became highly mechanized, and textile machines became symbols of progress. The next production phase in the transnational production chain, however, remained craft-based well into the twentieth century: the garment industry. The manufacturing of clothing—and its kissing cousin, the fashion industry—did *not* simply develop from craft to mass production, from home to factory, from being in the hands of men to residing in the hands of women, as traditional narratives of industrialization would have us believe. Organizationally, the garment industry had little in common with the mass-production methods of Frederick Winslow Taylor or Henry Ford. From Paris to New York, from Manchester to Lisbon, piecework and urban workshops dominated the dressmaking scene. The industry attracted waves of migrant men and women ready to work for low wages, following the ebb and flow of seasons and fashions.[7]

Garments became the first widely available consumer goods on which the working class spent its meager disposable income. In the nineteenth century, rising working-class wages and more efficient textile-manufacturing processes brought ready-made clothing within reach of an ever-growing number of families. The shift from wool to cheaper cotton furthered this process. Clothing and fashion became arenas where people encountered the industrial world and learned to tinker with the new goods to suit their own needs. The year the German missionary faced his assertive African parishioners, 1875, marked an important sartorial shift. The nineteenth-century French sociologist Frederic Le Play and his research team discovered that male artisans and skilled workers at this time began to appropriate middle-class accessories. The actors here included tailors, glassmakers, and glove makers. The accessories included three-piece suits, and overcoats, watch chains, canes, and gloves. After the mid-1870s, even unskilled workers in Paris and provincial towns had enough money for fashionable clothing and were willing to spend it. In the United States, fashion trickled down still further—all the way to farmers and their laborers.[8]

Paris provided role models for women around the globe; London served as the fashion capital for men. The sober, standardized, and ascetic British attire of dark suit, watch-chain, and umbrella referenced the power center of the business world. Unlike intricate female apparel, the London-based male fashion quickly became standardized, industrialized, and easily available in shops. Male attire stressed middle-class values of productivity, practicality, efficiency, and uniformity—generating the prototypical image of the middle-class, white-collar worker.[9]

The women's garment industry was more complex. Crowning the pyramid was the exclusive *haute couture*, custom-made clothing with unique designs that no one else could buy, each a work of art and ingenuity. At the bottom was the standardized clothing, produced in bulk for all. The creation of craft-based *haute-couture* flexibly accommodated the never-ending vagaries of the female fashion world. The demand for new styles posed incessant challenges for designers and engineers. With its urban workshops and many workers, the garment industry was positioned to respond quickly to each change in fashion. For factory workers and subcontractors, the rapidly shifting fashion fads meant high turnover, seasonal employment, and poor working conditions.[10]

Ready-made clothing bought off the rack catered primarily to men. To save on materials and money, women all across Europe bought some clothing in stores, but continued to produce their

Fig. 1.3
Manufacturing *Haute* *Couture*: *It began ON small scale, with the Parisian court's appetite for one-of-a-kind fashion. In the 1800s,* haute couture *evolved into a larger-scale industry. Sewing machines and paper patterns enabled tailors in London, for example, to offer wealthy customers a wide choice of fabrics for custom-made clothes. And from circa 1850 onwards,* haute couture *houses exhibited their portfolio of creations on mannequins and live models. Women clients could then see, select, and order the styles they desired. This 1910 image depicts the famous atelier of* Galerie de vente *in Paris.*

own underclothing, caps, bags, and aprons. They made everyday clothing for their children, as well as shirts, handkerchiefs, and underwear for their husbands. If women could afford it, they outsourced the more elaborate work to seamstresses, tailors, and professional dressmakers. A century after men's clothing had become standardized and mass-produced in the 1840s, women continued to design their own dream worlds.

Throughout this time, and practically all across the industrializing world, Paris represented the fashion capital for women. Even in a country as distinctly anti-aristocratic as the United States, Paris symbolized consumption. Interested in the airs of distinction, individuality, and exclusion, the *nouveaux riches* of New England looked to France as the ultimate arbiter of taste. Aristocratically oriented Parisian fashion produced a sculptured female body with a wasp-like waist and an exaggerated bust and hips. Its tightly-laced corsets, wide crinolines, and long trains were distinctly hostile to a culture of work; rural and working-class women typified by the African parishioners and the British cook could sport these silhouettes only on Sundays. Celebrating aristocratic idleness, female fashion created clothes that were restrictive, impractical, and orna-mental. Craft-based, the clothing breathed individuality.[11]

The global appeal of Paris and London referenced a community that knew few national boundaries. As part of the transnational fashion world, journalists in St. Petersburg, Bucharest, Sofia, The Hague, Madrid, New York, Kansas City, Buenos Aires, Batavia, and Pretoria reported in their society pages on the aristocratic life, including the fashion balls on the French Riviera and at the Paris Grand Opera. Writers laced their reports on women's fashion from Paris with Francophone terms such as *cachemire des Indes* and *crêpe de Chine*, thus indicating the colonial basis of textile produc-tion. Fashionable royals like Empress Eugenie, the younger Queen Victoria, and Queen Alexandra attracted attention with their exquisite gowns and boosted the sales for the *haute-couture* houses. During the *Belle Époque,* the fashion houses *Maison Worth* and *Jeanne Paquin* were the most influential designers in commercializing *haute couture.* Although the couturiers created craft-based clothing, they organized their ateliers around the principles of division of labor, using quasi assembly-line production methods to meet the growing demand. The largest of the houses each employed over 1,000 seam-stresses to satisfy the desires of European and global elites. An invitation to visit the Empress Eugenie for a week could mean a new wardrobe of twenty-one dresses. That meant three dresses for each of seven days: one dress for daytime, one for evening, and one "extra."[12]

Despite the French Revolution, the Parisian court culture's insa-tiable desire for fashion was restored during the Second Empire of Napoleon III. Charles Frederick Worth, a British immigrant draper, turned Parisian *haute couture* into an efficient manufacturing system. In doing so, he resolved the tension between aristocratic demands for exclusivity and democratic desires for wider access. Worth, the principal couturier for Napoleon III's wife, Empress Eugenie, survived the end of the Empire in 1871 because he commercial-ized his business for a clientele beyond the court, where only one-of-a-kind designs counted. He is best known for the seasonal portfolio—and for presenting his portfolio on live models, at that—rather than for creating unique pieces. The client selected designs from the portfolio, specified the colors and fabrics, and bought duplicate garments tailor-made in the couturier's workshop. Worth thus combined French individual tailoring with standardization methods more characteristic of the British ready-to-wear clothing industry for men.[13]

The American *nouveaux riches* and the wealthiest members of the Russian aristocratic elites were *Maison Worth's* best clients; the atelier also received orders from Mexico, South America, and Japan. For Americans like the Rockefellers, Morgans, and Vanderbilts, whose fortunes had been built on railroads and banking, Parisian fashion represented a status symbol confirming their newfound social standing. Integrated as it was in the European aristocratic culture, the nobility of Czarist Russia generated the most insatiable demand. Fashion had played a key role in Russian court culture ever since Peter the Great had articulated a policy to westernize his empire and join the European family of nations. In 1701 he decreed that his nobles should henceforth dress like their West European counterparts and abandon traditional clothing like caftans. By 1900, the Frenchified Russian court probably displayed the world's most spectacular form of aristocratic consumption. Members of European courts had their couture ordered by telegraph, made in Paris, and delivered by train or ship. Local elites from around the world who had the time and leisure to do so, traveled in person to Paris using new transport technologies – ocean liners and trains. Having ordered their fashions directly from the couture houses, elite women worked with professional dressmakers to achieve the perfect personal fit, the dressmaker and the lady collaborating closely and intimately on the creations.[14]

Protected by the *Syndicat de la Haute Couture* and supported by the Ministry of Economics after 1868, fashionable clothing became one of France's most important and successful government-sponsored export items. French copyright laws defined fashion designers as artists and safeguarded their intellectual property. The 1896 Berne Convention of the International Union for the Protection of Literary and Artistic Work, an early predecessor of the ACTA agreement, took the essence of the French initiative to a transnational level. Despite the *Syndicat's* tight control over copying, the Paris fashion system depended on standardization and duplication to attain its spectacular and enduring success.[15]

Parisian fashion found its way into the wider world in many ways. Local elites who could not travel to Paris bought copies from licensed fashion houses in their own countries. A Czarist decree created a home-grown fashion industry that fueled the empire's economy and motivated West European tailors to establish ateliers and fashion stores in St. Petersburg and Moscow to meet growing demand. In Belgium, a country with a long history of textile manufacturing, the couture company *Hirsch & Cie.* (1869–1962) imported French prototypes, manufactured reproductions in its own workshops, sold them in their Brussels stores, and established branches in Germany, Austria, and the Netherlands. From 1882 onwards, *Maison Hirsch* in Amsterdam sold custom-made original Parisian models produced under its own labels in the firm's local ateliers. Clients were addressed in French by highly-educated saleswomen and treated with special regard to convey the idea that they were participating in a transnational culture radiating from Paris. Members of the firm's extended Jewish family and its network of former employees opened similar houses at other locations (Brussels, Hamburg, Dresden, Cologne, and The Hague) under a licensing system, thereby creating cosmopolitan enclaves of Parisian culture and manners outside France. Until the Second World War, the rules of the French *syndicat* made style piracy a criminal offense and guaranteed that the established Parisian couturiers could maintain their monopoly—at least in France.[16]

Copying & Pasting Paris

Copying Paris fashion was strictly forbidden by the *syndicat*, yet there were many informal and illegal routes for reproducing Parisian fashion. Those women who could neither travel to Paris for personal fittings with couturiers, nor access licensed houses like *Hirsch*, collaborated with skilled local tailors and dressmakers to rework the designs according to their own taste or class standing. When it came to acquiring the new products and innovations, consumers in the female fashion world showed tremendous ingenuity, skill, and resourcefulness. The impact of Paris was impressive. Everywhere in the modernizing and even the colonized world, working-class and middle-class women—except the very poor, the landless, or those purposefully resisting modernity—sought to adapt Parisian *haute couture*. As we have seen, Parisian fashion spread even to South Africa, where it came to symbolize European civilization. In the 1840s, U.S. women were already discussing and trying to adapt last year's Parisian styles. So did Russian wives and daughters of the business, civil service, and artistic elites. On the eve of the revolution, they could shop in hundreds of fashionable houses run by French and other West European entrepreneurs in Moscow's Popov Arcade or on St. Petersburg's Nevskii Prospect, close to the Winter Palace. For everyone from the humblest New York shop girl and the rural-dwelling daughters of Russian notables to the wealthiest lady in Istanbul, being modern and cosmopolitan meant looking French.[17]

The *Belle Époque* was a golden age of intricate style: 'fluffy' hats, tight bodices, each dress an original creation unto itself. For this reason, fashion of the "Beautiful Era" resisted mass production—and a great deal of skill and ingenuity was involved in the art of copying. Inspired by fashion plates and paper patterns available to the members of this emerging user movement, women measured up and sewed their clothing with whatever material they had. Such poaching was never done "wholesale," but tweaked to circumstances, personal tastes, individual sizes, and local variations. The British cook could gain access to the latest fashions through a complicated web of transnational circulation, appropriation, adaptation, and refashioning. Department stores, fashion magazines, mail-order companies, family sewing machines, and paper patterns made the copying of Parisian fashion possible.

Stores and catalogs offered two indirect avenues to Paris. Those living too far away from bazaars or provincial stores had the option of selecting clothes through mail-order houses that could offer lower prices by cutting out the middlemen. In France, the department store *Bon Marché* started a successful mail-order service in 1871. For the 1894 winter season, it sent more than 1.5 million catalogs to potential customers in the French provinces. Translated versions targeted foreign markets. In 1902, *Bon Marché* sent packages all over the world, including South America, Russia, and the colonies. By that time, mail orders made up about 17 percent of its total sales. As of the 1890s, American home-delivered mail-order services covered rural areas and reached provincial elites and countryside women who wanted to keep up with Paris. *Muir* and *Merilees* shipped goods to all corners of European Russia, the Caucasus, and Siberia. Its return policies for ill-fitting or flawed clothing were meant to make consuming fashion as easy as possible for clients. Milan's trendy department store *Alle Città d'Italia* had been shipping its merchandise throughout Italy since the 1880s. For ordinary shoppers, many accessories were too expensive, but *Alle Città* sold some low-quality basic models consumers could personalize and make fashionable. If these avenues to fashion were still too costly, women could buy old or outmoded clothing secondhand and refashion it at home, based on images in catalogs and magazines.[18]

Our British lady had found it disconcerting that her cook looked as if she had stepped out of a fashion plate. Between 1850 and 1870, the circulation of fashion magazines exploded, some publishers selling 100,000 copies. The female readership expanded with each year: by 1894, the English women's magazine *Forget-Me-Not* had 140,000 readers; by 1910, the American *Ladies' Home Journal* claimed over one million subscribers. Publishing houses collaborated across national borders. For example, the American *Harper's Bazaar* and the Russian *Novyi russkii bazaar* were adaptations of the German *Der Bazaar*, which in turn was a collaboration between the trendsetting French *Mode Illustrée* and the Dutch magazine *La Gracieuse*. In Germany, Franz Lipperheide revolutionized the fashion magazine industry by catering to the middling class with *Die Modenwelt: Illustrirte Zeitung für Toilette und Handarbeiten*. From 1865 onward, he adapted French publications to offer practical,

Fig. 1.4 The Technique to Copy Paris: *For most people, the term "paper pattern" does not necessarily bring to mind technology. But paper patterns were a technical innovation— and an intricate one, at that. Using a paper pattern is as complex as following an engineering drawing. Required is the dexterity to transform two-dimensional designs into three-dimensional garments. This image, part of a paper pattern, issued in French, German, and English, comes from* La mode Parisienne *(ca. 1895). Note the painstakingly detailed drawings and measurements. The pattern required marking, pinning, cutting, and tacking fabric, among other specialized skills.*

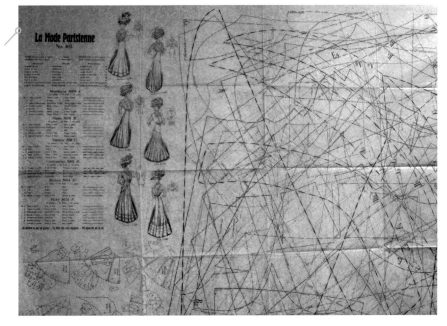

do-it-yourself instructions on how to make Parisian clothes at home. Within twenty years, the magazine had a circulation of more than 300,000. *Modenwelt* established contacts with fourteen foreign publishers. For example, the Russian *Modnyi svet* copied all or part of the German magazine, always adapting Parisian examples to suit Russian tastes and circumstances. Some magazines in Europe translated all the content and copied the plates, others were bilingual. In Italy, for example, the magazines were published partly in French, partly in Italian. By copying and adapting images and rewriting texts, the magazines also represented a form of *bricolage*—just like their readers' designing and sewing activities.[19]

Fashion magazines mediated between Paris salons and the transnational world of clothing production. These publications also contributed to creating a common European fashion experience. Local tailors and seamstresses copied the dresses based on paper patterns or images in magazines. Readers used the magazines to exchange information and experience. Given that cutting cloth was economically risky and the most difficult of all skills, women of lesser means bought fabric and had a professional dressmaker cut it to measure. The customer then sewed her pieces together at home. There were several creative ways to acquire, design, assemble, make, or remake garments. Those working-class women closest to the Parisian fashion system of production and selling—garment workers, department store clerks, and servants—were at the forefront of developing strategies to keep up with the latest fashion on their meager salaries. Through their jobs, women workers interacted with the upper and middle classes. They appropriated upper-class styles, acquired upper-class taste, and began to copy examples they saw at the workplace or in department-store windows. They tinkered expertly with resources, skills, and materials to match or elaborate on the upper-class clothing they encountered in their jobs. Some apparel workers designed their own versions of *haute couture* by using scraps of fabric left on the factory floor. Others received help from family members and friends in the garment trades and with whom they could trade skills—an aunt who worked with bonnets or a cousin employed in a workshop sewing sleeves. Others still relied on reconstructing old clothes to match the latest fashion.[20]

Department-store saleswomen were the second important group who appropriated Parisian fashion. Sales clerks at *Bon*

Marché acquired intimate knowledge about the latest style, even though they could not afford it. In U.S. stores, management often demanded that their shop girls dress fashionably—something they could barely achieve on their small salaries. Russian sales women who served the upper classes faced the same challenges, and were aided by local bazaars offering the necessary material and paraphernalia. Intimidated by the fashionable store near Moscow's Kuznetskii Bridge, one Russian seamstress in the 1840s instead went with her sister to a market that had begun selling items from Western Europe. Here she was able to find the scarf she had been looking for, as well as tulle, hairpins, and whalebone for a corset. The kind of servants who we have already encountered in South Africa and Britain made up the third group of workers who showed their ingenuity in copying fashion from their lady employers.[21]

The transformation of sewing into a typically female skill was not a foregone conclusion. For centuries, men had dominated the tailor's craft; they also made up the majority of employees in the early days of manufacturing. When women entered the industrial world of garment making, they did so either as professional dressmakers or as low-paid finishers. Traditionally, sewing skills had been passed on either formally through apprenticeships in the dressmaking trade or informally from mother to daughter. Sewing became a valuable skill women brought with them when migrating from the countryside to the city or from Europe to the United States. In the emerging industrial world, a young immigrant woman who had learned sewing at home and/or later at school could find a job in the booming clothing industry as a seamstress, dressmaker, or needlewoman, stitching together fabrics that had been cut by male tailors in garment districts in New York, Paris, London, Leeds, or in the Moscow and Ryazan provinces. Of course, getting a job in the factory or a sweatshop did not mean that one became an expert dressmaker. The art of tailoring—designing, drafting, patternmaking, and sewing—was a highly skilled and technical trade catering to the upper classes.[22]

During the nineteenth century, sewing education became increasingly formalized and feminized through private schools and special government programs. In Russia, sewing was made mandatory by 1870 for all girls attending primary and secondary school, while charitable institutions taught poor women sewing skills so they could support themselves and their families. In the United States, sewing education became institutionalized through the land-grant colleges between 1890 and 1917 to help improve rural life. The curriculum included plain sewing like seaming, hemming, making buttonholes, mending, or knitting, as well as ornamental work like embroidery and muslin work. Philadelphians developed a model system of teaching, creating samples for seams and seam finishes, darning and mending techniques, methods of applying lace, cuffs, and the insertion of plackets. Students pasted miniature garments into sampler books for future reference. These institutions spread to several European countries. In Greece, domestic-science instruction was initially intended for daughters of the working class and the poor, to train them as servants for bourgeois households. Based on Swiss and German examples, the businessman Panagis Charokopos in 1906 started plans for a "Professional School of Housekeeping." His vision was to turn Greek women into what he considered European middle-class housewives, but he also planned to offer unskilled women vocational training to prepare them for employment in the service, textile, and garment industries. Although the school was based on West European models, the Greek staff was trained in Home Economics in the United States (Boston) before the school opened in 1920. By 1931 it became public, offering programs from the Ministry of Education. In Italy, the international Home Economics movement arrived by way of simple handbooks for household management, guides that would become particularly successful during the Fascist period of the 1920s and 1930s.[23]

Paris remained the fashion capital for craft-based *haute couture* and the luxury trade, but it faced increasing competition from the United States, where the expanding garment industry provided jobs to hundreds of thousands of Jewish and Italian immigrant women. Because American copyright laws benefited the commercial interests of businessmen and users rather than artists and designers, the U.S. industry served as a basis for a commodity culture of fashion: mass production, paper patterns, and sewing machines helped turn an aristocratic culture into a modern middle-class consumer culture. Like so many consumer technologies before and after, these innovations were originally European, but it was American entrepreneurs who turned them into mass-produced items and marketed them globally. The sewing machine made its way into homes and helped, together with paper patterns, helped to establish a new culture of home sewing that bridged the dream worlds of the magazines and the humble circumstances in which most women and their families found themselves. By domesticating these two techniques, members of this user movement forged a path between the worlds of *haute culture* and ready-made clothing.[24]

The Sewing Machine at Work

In *Capital* Karl Marx described the factory sewing machine as "decisively revolutionary." He erroneously believed the garment industry would soon be fully mechanized. While the machine helped speed up running the long straight seams of a skirt or coat, it was useless when applied to more intricate tasks. Making up a man's jacket demanded up to 25,000 stitches—a process that New York garment industrialists by 1905 had standardized by dividing it into thirty-nine different operations. Berlin, Paris, and London became the main European production centers, followed by Milan and Turin. Industrial sewing had been a man's job and the first sewing machines became their exclusive tools. When the Singer Company framed the sewing machine as a consumer good, however, it aggressively marketed the machine to women.[25]

The family sewing machine turned into the first standardized consumer durable serving a global mass market. Prior to the First World War, the American company Singer was largely responsible for its remarkable world-wide success. The first sewing machines—patented and manufactured in 1830 by Barthélemy Thimonnier to produce French army uniforms—met strong Luddite resistance from hand sewers. Convinced the machines would put them out of work, artisans in several countries threatened to destroy them. After two decades of unsuccessful attempts to sell sewing machines in the United States, France, and Britain, American manufacturers finally began to understand that the machine would not find its niche in garment factories or tailors' workshops, but with the many seamstresses doing piecework at home. The family sewing machine developed into the workhorse of an infinite number of working-class and rural homes.[26]

Singer succeeded commercially through its overseas operations in Britain, Continental Europe, and beyond—regions that fell outside the ruling U.S. patent pools. In the United States, about 60 percent of all family sewing machines were Singers. In the early 1910s, the company controlled an estimated 90 percent of the market in in Western Europe, Czarist Russia, the Ottoman Empire, and the Balkans, and it dominated parts of Southeast Asia, the Middle East, and Southern Africa. Worldwide, the company name became synonymous with sewing machines. For its overseas operations, Singer opened two strategically positioned assembly plants in Europe: one in the transatlantic shipping port of Glasgow to serve global markets, and another in Floridsdorf near Vienna to circumvent the Austrian-Hungarian Empire's tariff barriers. Although it developed into the world's seventh largest multinational company, Singer made neither technologically superior machines nor a particularly competitive product. For that matter, the design was also not very appealing. In fact, Singer machines were technologically ordinary and relatively expensive. For French seamstresses, a Singer cost one fifth to one half of their annual earnings. The working class and rural families could only afford such a once-in-a-lifetime-investment because the company offered attractive credit plans and charged relatively modest annual interest rates. The firm's large market share came from its elaborate sales system. Importantly, however, Singer invented neither the sewing machine nor the art of domestic dressmaking.[27]

What the Singer Company did pioneer was the art of direct selling: knocking on people's doors all around the world. Ironically, this intrusive but successful retail system had been developed not in the United States, but at its global headquarters in London and Hamburg, where sales subsidized the inefficient American mother company and accounted for overall commercial success. Singer broke new marketing ground through a sophisticated cadre of salesmen-cum-debt collectors. In 1906, more than 62,000 salesmen, managed through 4,500 branch offices, visited people's homes worldwide. In rural areas, the canvasser traveled in a horse-drawn wagon with a Genuine Singer Sewing Machine in the back. In cities, Singer established a network of shops that managed a staff of salesmen, installment collectors, female teachers, repairmen, and transport workers. During the mornings, salesmen identified potential clients by scanning relevant residential areas; in the afternoons, they visited customers with the highest potential. Less documented, but equally important, was the follow-up visit of a woman teacher—often trained in home economics—who

Fig 1.5 Sewing Local, Sewing Global: *By most accounts, the Singer sewing machine was the first consumer product to push its way into practically all countries around the world—a feat the multinational company eagerly advertised. Proclaiming that "ALL NATIONS USE SINGER SEWING MACHINES," this international collection of business cards was put together for the 1893 World's Fair in Chicago, U.S.A. Each business card portrays a sewing-machine user—dressed in the traditional costume of the local region. Represented are countries like Japan and Italy as well as regions like the Iberian Catalonia and Castile. The United States itself is not represented.*

demonstrated how to use the machine. During the trial period, free sewing lessons were offered—a helpdesk *avant-la-lettre*. After the sale had been successfully closed, a different Singer employee would visit to collect installment payments weekly or monthly, depending on the wage system in the country. The salesmen and women instructors made home visits in towns and cities in industrial Britain and Germany, and they tracked down customers in the countryside and rural communities practically everywhere, from

Fig. 1.6 The Men Who Sold Sewing: *The Singer Company is acknowledged as having invented a new sales model: the credit system. Sewing-machine salesmen like this stalwart Finn sold the company's wares on credit. More than 20,000 such salesmen canvassed all corners of Europe. The sales pioneers braved icy winters, hot summers, and long distances between rural villages. Sewing machines took root in the countryside as well as cities. This enabled women to do piecework at home; to make their own clothing and draperies; and to keep up with fashion trends.*

Norway to the Ottoman Empire. In Czarist Russia, the company built a factory in Podolsk; its sales system covered all corners of the Empire. Throughout Europe, Singer employed more than 27,000 salesmen who demonstrated the machines in homes, signed credit

Fig. 1.7 The Satisfaction of Sewing: *Before the era of mass production, most people designed and made their own clothes. Necessity aside, creating and wearing "home-made" garments imparted self-confidence and personal identity to women, men, and children. In this 1922 photo of the Telšiai Vocational School in Lithuania, students take a sewing class. In Yiddish, the photographer meticulously recorded the names of the Jewish girls making their own dresses: "...Some of the children wear clothes that they themselves have sewn... Do you see? Hinde Kats and Taybe Rubin wear clothes they've sewn for themselves." Note the world map in the background.*

installment plans, collected payments, and were responsible for maintenance and repairs.[28]

No village was too small, no distance too far. Singer hired agents who were intimately familiar with local conditions and usually rented space from local peasants. The exacting sales system provoked Singer salesmen and women from Warsaw to Tokyo to go on strike, demanding better working conditions. From the company's point of view, though, this marketing system enabled a sustained, long-term relationship with the women users and their machines. Through regular feedback, Singer kept a close eye on the users of the sewing machine, creating a customer relationship that no competitor was able to match. The sales system became the model of many other technologically complex, durable consumer goods, from vacuum cleaners to cars.[29]

The success of the family sewing machine cannot be explained by Singer's business strategy alone, nor does the sewing machine's role as a status symbol explain its universality. Its status in the home was rather ambiguous: it was simultaneously a symbol of consumption and an industrial tool. As such, it was both a means of keeping pace with the latest fashion and a way of making ends meet. To reduce the ambiguity and make the machine fit into middle-class parlors, manufacturers ensconced their products in decorative designs. Sewing machines were sold as consumer durables embedded in the middle-class value of domesticity. Many lightweight, highly-adorned sewing machines graced the drawing rooms of well-to-do women. Entrepreneurs offered sewing machines as pieces of furniture with elaborate decorations to cater to the Victorian ideology of domesticity—a counterpoint to the evils of industrialization. Sewing machine cases came in lush mahogany, with cupids and bows, curves and floral decorations. Such whimsical machines were sold to aristocratic and upper-middle-class women who could show them off as toys to visitors—the centerpiece of the parlor, boudoir, or drawing room. After all, these women did not perform their own sewing work but outsourced the job to tailors, seamstresses, and dressmakers invited to their homes. The decorations masked and mediated the machine's industrial origin and purpose.[30]

The majority of sewing machines served as manufacturing tools, providing working-class and rural women the opportunity to generate income by working at home for the flourishing garment industry. The machine fitted into the household economy in an industrializing world. It owed its commercial success to working-class and rural subcontractors, who did piecework at home by sewing parts for the ready-made industry near urban industrial centers. One of Singer's British auditors indicated as much: when in 1880 he examined the firm's high sales for South London, the auditor credited the figures to the district's "vast population of the working classes, who are the back-bone of our business."[31] Women transferred the skills and tools required for their industrial assignments to the design and production of garments for private use.[32]

Often, the boundary between the sewing machine's private and industrial applications was porous. In Northern Italy of the 1870s, for example, female dressmakers who worked in the ateliers also catered to private customers and actively disseminated fashion models and ideas among neighborhood women. In early twentieth-century Turin, in excess of 30,000 dressmakers worked in more than 1,000 ateliers, boutiques, workshops, and back shops, as well as in many private homes. The dressmakers creatively combined traditional craftsmanship with technical skill. Many of the men and women—both in the cities and the countryside—who invested in a sewing machine on installment plans were hired by textile manufacturers and forced to meet output quotas. Government reports in Sweden, France, and the United States from the turn of the century indicate that earnings were low and working hours long. The work was often done in the kitchen, a multi-purpose room that was used for cooking, sewing, and sleeping.[33]

Competing manufacturers, primarily in Germany, emerged throughout Europe. In the 1870s, the Swedish arms manufacturer Husqvarna Company entered the sewing-machine market as part of a diversification strategy to reduce its dependency on weapons-production revenue. Some of Husqvarna's products were based on Singer licenses, others were new models. Reflecting the situation elsewhere in Europe, working-class women were the company's main customers. For example Betty, of Sweden, was a prototypical example of a Husqvarna sewing-machine user. As a young woman, Betty worked as a maid for a baker's family, where she learned to operate the family's sewing machine. In 1880, after saving enough money, she managed to buy her own Husqvarna machine for a price equivalent to one year's wage. Pulling her machine on a wagon or sled (in the winter), she went on to offer her services to families in the countryside on a regular basis. When she married, she gave up her occupation as a traveling seamstress and continued to serve her customers from home.[34]

By 1914, sewing machines had taken up residence in many homes: one-quarter to one-third of households across Europe had bought one. At the European periphery, diffusion began later but ultimately moved much faster. A man born in 1937 remembered that during the Soviet period, "every family probably had a Singer sewing machine and used it to sew everything."[35] In South Africa, an estimated 25 percent of households had a Singer machine in 1920. In the Philippines, it was about 15 percent; and in the Ottoman regions, it was 12 percent. By contrast, only 4 percent of all Japanese households bought a Singer; many potential consumers resisted the sewing machine because the appliance fell short when sewing traditional clothes like kimonos. No matter how good sewing machines were for straight lines, hand sewing remained essential for delicate and intricate tasks. The cutting of cloth, the handling of fragile fabrics, the stitching of curved seams of the bodice, and the fitting and finishing of a dress in aristocratic fashion defied easy mechanization.[36]

Paper Patterns as Tools

After the 1850s, the commercialization of paper patterns helped transform the designing and making of fashion. In wide circulation, paper patterns found their way to homes all over the industrializing world through specialized pattern shops, mail-order services, magazines, and newspapers. The trade's development was linked tightly to the explosion of the print media and the manufacturers' promotion of family sewing machines. French women's periodicals first featured fashion plates describing the latest Parisian styles. These publications offered small-scale and fully-sized diagrams, sometimes with measurements and instructions on how to draft the pattern and cut the cloth. Journals added such do-it-yourself instructions as supplements. The patterns often resembled a maze that featured overlapping pieces of a garment: bodice, sleeves, collars, and cuffs. Distinguished by separate lines, the various parts first had to be traced onto another piece of paper to measure before they could be cut and laid out onto a piece of fabric. Later the publications included easier-to-handle tissue-paper patterns.[37]

The early publications catered to the professional market of dressmakers, but in 1852 Samuel Beeton started *The Englishwoman's Domestic Magazine*. This "Illustrated Journal, Combining Practical Information, Instruction and Amusement" included do-it-yourself instructions for making dresses, complete with the promise that "any lady may readily furnish herself with a paper pattern from them."[38] It was Beeton's magazine to which the English lady turned when complaining about her servant's audacity in dressing "up" and crossing social boundaries. When Beeton and his wife Isabella offered via mail order tissue-paper patterns (based on patterns they bought in Paris from Adolphe Goubaud), including full-size pattern supplements, a new magazine format was born. Customers could even order cut-tissue-paper patterns. Many other London players followed Beeton's lead. Weldon became a prolific publisher of magazines and catalogs selling paper patterns. Between 1880 and 1950, his company would issue nearly fifty different kinds of publications. Entrepreneurs in Britain and overseas sold do-it-yourself manuals and fashion periodicals to middle-class women who could not afford the services of clothing professionals.[39]

Fig. 1.8 Women & the Sewing Machine: *Traditionally, machines have been associated with men and male attributes. And traditionally, household machinery is thought to be adopted first by the upper classes—and then by the lower classes, in a trickle-down effect. But the case of the sewing machine overturns both notions. In fact, the first to appropriate the sewing machine were working-class women, either as independent seamstresses or as employees in factories, sweatshops, and middle-class homes. Only after 1900 did members of the middle class, such as these Hungarian women (ca. 1911), no longer regard using the sewing machine as beneath their dignity.*

The American paper-pattern industry led the way globally—a prime illustration of the trans-Atlantic circulation of ideas and technologies in this trade. When the U.S. industry expanded, women at the frontier were eager to get information about methods for making fashionable garments. In fact, pattern-making companies were instrumental in establishing the idea of mail-order catalogs in the first place. On the basis of Parisian examples, the leading American entrepreneurs Ellen Demorest and Ebenezer Butterick mass-produced their patterns and exported them to Europe. The Butterick Company established offices in London, Paris, Vienna, and Berlin as early as the 1870s and was ready to claim a foothold in Dresden and St. Petersburg. The firm published full descriptions in English, Spanish, and German, thus stimulating the international circulation of patterns within Europe. What is more, Butterick also allowed users to tinker with the designs by introducing a modular system. Their business was the first to introduce patterns in sizes. The company and its competitors published specialized sewing manuals that included instructions on how to alter a pattern to fit one's own body. Soon, hundreds of French, British, German, and Russian paper-pattern designers promised women of all stripes and shapes that they could make their own clothes at home according to the latest fashion. By the 1890s, pattern companies produced 30–40 pattern styles each month.[40]

The dream of simple replication and quick reproduction was a marketing ploy. Sewing companies like Singer offered classes. Sewing-machine manufacturers and fashion magazines sought to boost their sales by pushing paper patterns. Despite the promise of ease, paper patterns were technically complex affairs. The blue-prints required of the user considerable spatial and visual under-standing; knowledge of fabrics; and dexterity in handling scissors, needle, thread, and machine. Advocates of "scientific" methods sought to capture the dressmakers' tacit craft knowledge, prom-ising that uninitiated women would be able to replicate upper-class fashion easily. And indeed, the innovation of paper patterns greatly facilitated semi- and non-professional women's ability to copy and tinker with Parisian fashion. Paper-pattern making challenged the dressmaker's monopoly on the highly skilled job of cutting cloth to measure.[41]

Paper patterns promised women of all classes that they could participate in the transnational community of fashion. This contributed to the creation of a pan-European and trans-Atlantic user movement in the area of garment manufacturing, in which Paris was the center. Like later-day computer hacking, home-sewing called for skill and experience. The numerous drafting systems deceived sewing women, asserting that they could become professional dressmakers by simply following a set of instruc-tions without having to acquire any formal skills and without developing the mental ability required to envision the shape of the finished garment. The firms offered thin tissue paper with complex and intricate systems of circles, squares, and notches based on the tailor's marking system. Small circles indicated tucks, pleats, darts, and hems. Double circles in a triangle signaled there was no seam or that the edge of the fabric had to be placed on the fold. Such systems did away with the laborious fitting sessions a dressmaker had with her client, helping to speed up the dressmaking process. The demand for paper patterns was staggering.[42]

It was the technical innovation of paper patterns rather than the popularization of the sewing machine that challenged the monopoly held by highly skilled dressmakers. By the same token, however, paper patterns also generated new business for dressmakers, helping them to keep up with Parisian fashion and easing the difficult task of cutting expensive fabric to measure. Supported by fashion plates, paper patterns allowed the widespread copying of and tinkering with Paris-inspired fashion. The outcome was a transnational fashion culture. In nineteenth-century Czarist Russia, over a hundred journals were published that dealt with fashion. The most prominent were the *Novyi russkii bazaar* (New Russian Bazaar), *Modnyi svet* (Fashion World), and *Vestnik mody* (Fashion Herald), as well as a specialized journal that catered to professional dressmakers and tailors. The most widely read Russian literary magazine, *Niva*, with a circulation of more than 200,000, had a special supplement appropriately called *Parizhskie mody* [Paris Fashion] that included dress patterns allowing women to create their own Parisian clothes at home. *Niva* reached not only the rural elite, but also prosperous peasants. Russia lacked its own specialized pattern companies, but the fashion and literary magazines filled the gap with pattern supplements that helped enterprising urban and provincial women to create their clothing or outsource the work to local seamstresses. In the Netherlands, the weekly *La Gracieuse* offered the latest patterns from Paris. In

Fig. 1.9 Customized Copying: *Technical though they were, paper patterns allowed consumers of many kinds to control the manufacture of their own clothing. Middle-class women, for example, hired seamstresses to create custom-made garments at home. This 1907 German photograph may represent this scenario. The woman at left holds a magazine,* Die Praktische Berlinerin *("The Practical Berlin Woman"), which contained paper patterns. To her right is a figure with a measuring tape around her neck—most likely the seamstress. Members of the upper class bought* haute couture *at ateliers, while working-class women sewed their own and their families' clothing.*

Britain, the journal *Fashions for All* dedicated its publication entirely to home-sewn fashion for budget-conscious women. The designing and making of fashion by means of paper patterns enabled women of all classes to create their own versions of high fashion. In a truly transnational manner, the American pattern industry allowed Parisian fashion to pass through New York and London before returning to Europe and spreading to the rest of the world. And it was not the democratic but the aristocratic values of distinction that were disseminated in the process, namely the values of exclusion at the heart of what was purported to be the Paris fashion system.[43]

Tweaking Paris

Intellectuals criticized the breathless admiration for Paris and the European aristocratic life. French novelist Emile Zola's *Au Bonheur des Dames* (1883) famously ridiculed the female fashion world, epitomized by the department store *Bon Marché*. American author Theodore Dreiser followed suit with a similar critique in *Sister Carrie* (1900). In *The Kreutzer Sonata* (1891), Russian writer Leo Tolstoy mercilessly portrayed how high-society women married to acquire money in order to shop and dress. Iulii L. Elets, a Russian journalist and social commentator, went further in his book, *Epidemic Insanity: Toward the Overthrow of the Yoke of Fashion* (1914). He believed shopping and clothing ruined many husbands financially because of their wives' and daughters' reckless spending.[44] After seeing the Exposition at the Chicago World's Fair, the Norwegian-American social commentator and anthropologist Thorstein Veblen, in his 1899 *Theory of the Leisure Class*, built an economic theory around the notion that bourgeois women consumed and dressed in expensive clothing to enhance their husbands' status. The critique these men articulated had much earlier origins. For one, the famous luxury debates in the eighteenth century—from Rousseau's condemnation of opulence in the name of a more "natural" way of living, to Mandeville's defense of luxury in the *Fable of the Bees*—questioned the celebration of a culture of consumption.[45]

Veblen had his predecessors in the United States. As early as the 1850s, dress reformers, nationalists, and anti-aristocratic thinkers deplored the culture of pomp, exclusiveness, and excessive leisure, while at the same time celebrating America's republican values. The reformers focused on the impracticality of tightly laced corsets, wide crinolines, and long trains. In 1856, American feminist Amelia Bloomer established the *National Dress Reform* to liberate women from restrictive clothing and offered alternative clothing designs. These early attempts met with considerable resistance and ridicule. Protests failed to gain traction until the 1880s, when a coalition of women activists, artists, and medical reformers launched an international effort to found an anti-fashion user movement. Through personal contacts, the American initiative fostered the British Rational Dress Society (1881), the German *Verein zur Verbesserung der Frauenkleidung* (1896), and the Dutch *Vereeniging voor Verbetering van Vrouwenkleeding* (1899). World's Fairs offered platforms of exchange for these transnational leaders. Directly contesting the Parisian aristocratic fashion ideals, these social clubs came to function as proto-consumer organizations.[46]

During the 1880s, a third source of political resistance came from middle- and upper-class consumer leagues. Well-to-do members criticized the labor conditions under which their working-class sisters slaved. In a fair-trade manner *avant la lettre*, they challenged the working conditions of the garment industry that sustained the fashion system. With their so-called white-label campaigns, the women activists sought to identify goods produced under the right labor conditions for fair wages. Such leagues were the first associations that organized middle-class women as politically active consumers. Borrowing the idea from England, the American National Consumers League (1891) was the first of its kind. The League made middle-class consumers aware of the economic realities behind the products they bought; educated them about the circumstances under which workers labored; and pressed them to buy or boycott products based on whether they were made under decent conditions. The international press reported widely on the League's efforts. Through personal exchanges at—again—World's Fairs, in Chicago and Paris, the pioneers proselytized by means of displays and lectures. This is where Veblen found his inspiration. The American example encouraged French women to establish their own *Ligue sociale d'acheteurs* in France in 1902. Sister organizations followed suit in Switzerland (1906), Germany (1907), and Belgium (1911), as well as in Milan and Barcelona. The women activists, echoing union and cooperative movements, did not just seek political leverage. They defined and

professionalized what it meant to be buyers and users of manufactured goods in an emerging industrial world. Their goal was
to educate knowledgeable consumers who used their purchasing
power both as an economic and a political tool.[47]

A subtler and ostensibly less politically rooted approach to
Parisian fashion emerged below the radar. Women began to develop
alternative forms of clothing by taking cues from bespoke British
men's tailoring. These British sartorial signals telegraphed a subtle
form of resistance to the French aristocratic style—a style deliberately designed to maintain existing social class boundaries—was
explicitly meant to be inaccessible to working women. As the
examples of the African parishioners and the British cook suggest,
working class women defied those aristocratic strictures, particularly in church on Sunday—the day of rest. One element of the
British style was the modification of men's wear to suit women.
Beginning in the 1870s, women of all classes began to wear black
velvet neck ribbons. Middle-class women wore them with school
and nurse uniforms and business dresses; photographs of servants
also show them sporting ribbons. Women also appropriated hats,
perhaps the strongest symbol of masculine power. They wore
men's jockey caps, hunting caps, and peaked yachting caps during
sporting activities. After the 1880s, the straw hat, or boater, also
became a fashionable, but subtle, statement of independence among
young women in such new occupations as office work. Tailored
suits fashioned from fabric used for men's clothes gained in popularity toward the end of the nineteenth century. A French woman
who wore such an outfit was appropriately called *l'anglaise*, a term
that on the European Continent not only connoted Englishness but
also signaled independence. And in the United States and Britain,
the shirtwaist—a man's shirt adapted for women with a stand-up
or turn-down collar—sold widely. A woman who wore them all
together (tie, straw hat, tailored suit, and waist shirt) made a strong
statement of personal autonomy.[48]

Even more influential, modern sports questioned the impracticality of aristocratically inspired fashion. Women's sports questioned traditional dress codes. In long skirts, tight corsets, and
large hats, middle- and upper-class women engaged in tennis,
croquet, golf, and ice-skating as social rather than sports activities. Horseback riding and swimming became the first arenas for
clothing experimentation; toward the end of the century, cycling

generated the most intense discussions and calls for more "rational clothing." Cycling was a new sport with technology that had not yet been fully socially coded. Choosing sports clothing became the moment when women started to contest and rethink their femininity, challenging traditional etiquette and morals. In the 1870s,

Fig. 1.10 Copying Britain: *In the case of sports clothing, the model to copy was liberal Great Britain. Through appropriation, European women of the* haute bourgeoisie *rebelled against the elaborate, cumbersome dresses that Paris prescribed. Via paper patterns, they adopted casual, practical, British-style sportswear for playing, tennis, bicycling, and other sports. This turn-of-the-century page from the French magazine—seen here in the Dutch translation and adaptation as* De Gracieuse—*provides patterns for streamlined, sew-it-yourself bicycling and tennis outfits.*

schools and women's colleges started to teach women physical education. Twenty years later, when biking became wildly popular among women, activist magazines offered their readers paper patterns to make their own easy-to-wear sport clothing.[49]

Commercial pattern makers and advertisers played their role in responding to middle-class women's new practices and novel demands for fashion. American women's magazines like *Godey's Lady Book* and pattern makers like the Butterick Company (still in business today) sold sports-clothing patterns for walking and biking: skirts and knickers that women could make at home with instructions allowing for individual adaptations. Giving ample leeway to women's own creativity and sense of style, *Godey's* in the early 1870s featured a bathing-suit pattern that could be adapted according to one's age, size, and taste. Its 1880s patterns for calisthenics suits included more details to define specific sizes. According to Butterick's correspondence, aristocratic European women ordered patterns for a bathing suit that could be used for calisthenics. Such suits would probably be made by a seamstress at home. The paper-tissue pattern offered eleven options allowing for individual interpretation and tinkering with design. One could cut the top and then have a choice of attaching bloomers; short or long sleeves; a scoop-neck, V-neck, or high collar; a skirt or a peplum; and the option of adding a ruffled, cape-like collar for around the shoulders.[50]

In the 1890s, as a response to the popularity of biking, entrepreneurs advertised cycling corsets, skirts, and knickerbocker suits. The discussion about proper and comfortable clothes was heated. Cycling's popularity generated a host of fashionable choices like skirts with extra pleats to make the fabric act and look divided; full skirts that were indeed divided but masqueraded as an undivided garment; and even unapologetically divided skirts. One linen item of the era, preserved in New York, documents how the maker had sewn a costume which, from the back, worked like trousers, allowing the rider to sit on the bicycle saddle—but from the front appeared to be a modest, feminine skirt. In France, where reactionary authorities had issued a law in 1800 forbidding women to wear trousers, the ban was lifted for sports, first in 1892 for female horseback riders, then in 1909 for women cyclists. In 1893, a French department store offered bicycling costumes with divided skirts or with trousers concealed as skirts. Fashion magazines responded to the changing practices by offering patterns for sports clothing. In the Netherlands, *Gracieuse* featured paper patterns for women's cycling clothes between 1893 and 1906. These trends indicate that women in many corners of Europe were tinkering with fashion to suit new practices and new liberated ways of life. The exploding popularity of sports produced new lines of paper patterns creating a more relaxed fashion that allowed more freedom of movement than long skirts and tight-fitting blouses. Some practical features of the new sports clothing spilled over to women's fashion in general. The tinkering with the alternative Anglo-Saxon style marked the resistance to the elite Parisian culture of idleness. By 1910, mainstream women's clothing had become streamlined and simpler.[51]

Artistic and business communities also attempted to dislodge or at least compete with the Paris fashion system by launching alternative designs. In the capital of the Austrian Empire, the *Wiener Werkstätte* (1903–32) developed a fashion challenge characterized by the vibrant colors of its fabrics. The new style attracted support from the famous Parisian designer Paul Poiret; the fashion establishment in London and Berlin; and was further disseminated through American magazines like *Vogue*. In 1909, the influential Dutch-American editor of the *Ladies' Home Journal*, Edward Bok, together with *The New York Times* began a campaign for American fashion to encourage fashion nationalism and domestic industries. Bok asked his female readers to submit their creative, American designs. Although hundreds of U.S. newspapers, magazines, and trade journals participated in the design campaigns "made in America," by the beginning of the First World War Bok had to admit that the effort had failed because women simply continued to follow Paris. Once war broke out, however, the situation changed. Nationalists in several countries tried to mobilize the Vienna style, *Los von Paris!* "The Austrian nation, its audience and producers, frees itself from the bands of centuries of dependence on France's culture."[52]

When the war ended, Paris designers like Poiret and Jacques Worth frantically tried to recapture their position. They had discovered that U.S. copyright laws favored consumers and failed to protect them as designers. Counterfeit labels and pirated copies of couture dresses circulated widely. To make the situation worse, when the Russian aristocrats scattered and the war brought international trade in Europe to a standstill, the couturiers realized the importance of the transatlantic market to the survival of French *haute couture*. Securing U.S. copyright protection was one strategy; offering lower-priced copies with more emphasis on comfort and ease was another. The war had a major impact on fashion in many other ways. Short supply encouraged (or forced) women to use and re-use old clothes or adapt uncommon materials and poor substitutes to achieve stylish effects. Women working in factories and offices, often as substitutes for soldiers, wore simple dresses or uniforms. The war experience altered fashion's notion of the ideal female body, transforming women's clothing from decorative to more functional wear.[53]

In the interwar period, then, the prototype of the modern girl began to fill the streets in cosmopolitan port cities from Shanghai to Cape Town. She belonged to a transnational global modernity in which European aristocrats, immigrants, migrant workers, and members of the middle class participated. At this time, the modern woman continued to feel like a part of a larger cosmopolitan world, even after the transnational culture of the aristocracy had largely collapsed. In many ways, the flapper style represented the antithesis of the *Belle Époque*. The 1920s silhouette of the flapper girl—made famous by the artist Charles Dana Gibson—de-emphasized womanly curves, referencing pre-teen and teenage girls, an ideal adopted by the J. Walter Thompson advertising agency in the 1920s for their younger adult clientele. The androgynous, prepubescent figure downplayed women's natural curves through the use of a bandeau brassiere. Its simplified lines liberated younger customers from the tight lacing and corsets of the *Belle Époque*. The new fashion turned out to be a boon for the garment industry in that it encouraged easier production through simplification and standardization. Entrepreneurial fashion designers responded to women's desire for more mobility with flowing dresses and tunic-shaped costumes that defied the narrow waists and corsets. The elaborate, aristocratic European style of women's *Belle Époque* dresses—and traditional Japanese clothing, for that matter—could not easily be copied using the sewing machine. In contrast, the new 1920s style of designers like Coco Chanel recognized U.S. women's demands and introduced straight lines: this allowed the sewing machine to break through in manufacturing. The demand for skilled dressmakers decreased accordingly.[54]

By the end of The Long Twentieth Century, it looked as though mass production had won the game and that women had

been reduced to being consumers only. The sewing machine—
itself a mass-produced object—had found its way into the garment
industry as well as into homes around the globe. Certainly,
customers profited from the mechanization of production. Still,
these developments did not herald the demise of the Parisian tradition or the do-it-yourself culture. Paris gradually lost its indisput-
able place, despite the French *syndicat's* success in protecting French
industry against illegal copying through an elaborate system of
rules and regulations. Users now had a multitude of role models
to choose from: Hollywood set one standard for women around
the world, as did Italian fashion houses after the Second World
War. Yet, ultimately, the very success of Paris, and Hollywood later
on, depended on resourceful women to bridge the world of *haute
couture* and ready-made clothes. Women of all classes around the
globe continued to read fashion magazines, experiment with their
sewing machines, and, aided by paper patterns, adapt or tinker
with new designs—just like African converts and British domestic
servants before them. Thanks to textile technologies, individual
middle-class women were able to innovate creatively. The politically
organized user movement also included women's dress reformers
who, decades earlier, had questioned and tweaked the restrictive
clothing ideals and forced a major re-think in clothing. Women—as
individuals and as a politically organized user movement—challenged Parisian hegemony. By tinkering with its designs, women
contributed to the emergence of a European fashion culture that
was distinctly middle class.[55]

2

Creating European Comfort

Throughout the nineteenth century, an aristocratic way of life remained a distant dream for most Europeans. This was the case not only for fashion but also for housing. Although domestic servants came into direct contact with the manners and lifestyles of the aristocracy and the upper middle class, servants understood that they would never be able to become accepted members of the rich man's world. Whatever farmers and day laborers knew about the transnational world of conspicuous consumption, they never considered such a life to be a real option for themselves. Still, the working poor tried their best to emulate the upper classes' lifestyles. By adopting various technologies, the working class tried to make their living quarters attractive and pleasant—or clean and comfortable, at least. In this effort, workers received vocal support from middle-class experts, such as medical doctors, social scientists, and progressive politicians. These experts sought to speak on behalf of residents on the housing question. In the second half of the nineteenth century, these experts developed a Europe-wide discourse around improving housing quality. Although living conditions of the lower classes differed from country to country and from region to region, experts across Europe defined problems and solutions in a surprisingly similar manner. Professionals and politicians sought to create social and public housing for the working classes. These plans, based on ideologies of hygiene and modernism, demanded that the working class conform to a specific lifestyle. For their part, the working-class people had their own ideas about how to create comfort and improve daily life. The conflict over housing and comfort was in the making.

A famous Russian short story illustrates how the comforts of home differed dramatically across Europe's classes. This literary artifact also reveals how people experienced those stark differences. The protagonist, Nikolay Chikildeyev, had experienced the worlds of both the rich and the poor. After growing up in the Russian country-side, he decided to seek his fortune in the empire's capital, Moscow. While working as a servant for the Slavyansky Bazaar, an elegant hotel and restaurant, Chikildeyev's health failed and he was forced to quit. Without an income, he and his wife had no choice but to return to his home village. On crossing the threshold of his parents' house, Chikildeyev was overwhelmed by the dramatic differences between the city and the country, the rich and the poor.

> [Chikildeyev] had always remembered his old home from child-hood as a cheerful, bright, cozy, comfortable place, but now, as he entered the hut, he was actually scared when he saw how dark, crowded and filthy it was in there...the huge neglected stove (it took up nearly half the hut), black with soot and flies—so many flies! It was tilting to one side, the wall-beams were all askew, and the hut seemed about to collapse any minute...labels from bottles and newspaper cuttings had been pasted over the wall next to the icons. This was *real* poverty![1]

Despite the darkness, the Chikildeyev couple could make out a table, a few benches, a lamp, some pots, and a *samovar*, the typical Russian metal urn used for boiling tea water.

All aspects of daily life took place in this room, which was, in fact, the only room in the house. At the centrally placed wood-burning stove, grandmother prepared meals and made tea. The enormous, multifunctional heat source that was the stove practically filled the multifunctional room. At night, the older generation slept on top of the oven. In this "cramped, stuffy and evil-smelling" space, people ate and quarreled, children screamed, and Chikildeyev's brother regularly beat his wife.[2] This was certainly not an aspirational

Fig. 2.1   The Woman & her Stove: *The classical Russian wood-burning stove, here depicted in a 1869 oil painting by Vasili Maximovich Maximov, served several purposes at once. It was used both for cooking and space heating, and it also provided the often small and dark huts with some dim light. In the daytime, the stove was women's precinct, and at night the older generation slept on its flat top. During the long, cold Russian winter, this kind of stove was a highly appropriate form of technology.*

household: here, no one expected that life would ever improve. The conditions were so impoverished that grandpa found himself longing for his days of serfdom, when they ate cabbage soup and porridge twice a day. The situation grew still worse when the family could no longer pay its taxes and the authorities confiscated the *samovar*: "Life became completely and utterly depressing without a samovar in the Chikildeyevs' hut. There was something humiliating, degrading in this deprivation, as though the hut itself was in disgrace."[3] Despite the poverty and filth, the traditional Russian appliance guaranteed a certain degree of pride and reassurance. Its social and cultural role was similar to that of the open hearth in England and the tile stove in Continental Europe.

We owe this frank description of life in rural Russia to Anton Chekhov's 1897 short story, *Peasants*. As a rural doctor and the grandson of a serf, Chekhov was all too familiar with the misery of the impoverished countryside. Unlike his aristocratic contemporary, Leo Tolstoy, Chekhov never romanticized peasant life. Despite being a fiction writer—and neither an anthropologist nor a sociologist—Chekhov's dramatically described poverty in

ways resembling accounts produced by the social scientists and
modernist reformers who were his contemporaries.

The Chikildeyevs' hut was a typical *izba*, a small house with a
single, multi-purpose room and a cellar. Prominent in every *izba*
was the huge Russian stove, known as a *pech*. The stove was where
women cooked, baked, dried clothes, and heated water; it also
provided the *izba* with heat and light in the long, cold, and dark
period of the year. In the daytime, children played on top of the
stove; at night, as many family members as physically possible
slept there. Everyday life circulated around the *pech*.[4]

In the gendered order of rural life, the stove and its immediate
surroundings were female spaces. Like his social reformer contemporaries and medical colleagues, Chekhov associated such interiors with dirt and darkness. To the peasants, however, the woman
in command of the *pech* was so central to their existence that she
retained a position of worldly power. Evidence of the household's
relationship to heavenly powers was to be found on the opposite
side of the room. Here, icons, candles, and—as in Chekhov's story—
simple pictures graced the wall. Chairs and beds were uncommon.
Instead, people sat on benches along the walls and slept on
straw mattresses. Carpentry tools, and perhaps even a loom or a
spinning-wheel, were kept in one corner. The Chikildeyevs earned
a few extra kopecks by spinning silk for a local mill.[5]

In Russia's harsh climatic conditions, the easy-to-heat *izba*
made perfect sense. Nevertheless, in Russia and elsewhere in
Europe, medical specialists, hygiene activists, and social reformers
condemned the sanitary and moral problems that accompanied
this mixing of activities—the key characteristic of single-purpose
living quarters. Architects, in particular, sought to solve the
problems of poverty by means of building reforms. A historical
anthropologist summarizes this view: "Winter in the *izba* was
dark, dirty, poorly ventilated, damp, and odoriferous. Livestock
brought their special aromas to the *izba*, and not all peasants were
able to perform their excretory functions outside the house."[6]
Summer brought relief, as many activities could then be carried
out elsewhere. Better-off peasants even moved into a separate,
brighter, and cleaner *izba* in the warm season. Towards the late
nineteenth century, the massive piece of masonry referred to as
Dutch stove began to enter the homes of rich Russian farmers. It
not only improved the lives of many families in the wintertime,

but symbolized the spatial separation of the house according to function. Specifically, with this *gollandki* stove, families could heat an adjacent room, thus allowing the construction of houses with separate spaces for different activities. In most parts of Europe and the United States, the separation of functions signaled a separation of class embodied by two appliances: the heater to warm the room and the stove to cook meals.[7]

Chekhov dramatized the hardships of rural life at the end of the nineteenth century, showing how Russian peasants struggled to maintain their dignity and autonomy despite poverty and suppression. His short story can prompt the exploration of how people of different classes—and in various European countries— tried to turn houses and apartments into homes, and how they attempted to create spaces of their own in a rapidly changing world. In this process of domestication, users often appropriated items and ideas from the upper classes—and even from foreign countries. As in the world of fashion, Paris—with its flamboyance and wastefulness—initially also played a predominant role in the realms of furniture and homemaking. By the time the century drew to a close, however, other exemplars replaced Paris as the mecca for homemaking and home fashion. In fact, citizens in other continental cities found inspiration in the décor of the German upper middle classes. To many, the straight lines and clear purposefulness of the Biedermeier style were more attractive than the frou-frou style of Paris.

Compared to Chekhov's fictional Nikolay Chikildeyev, Leo Tolstoy's character, Alexis Alexandrovich Karenin, lived a completely different life. In the following excerpt, Karenin is alone in his lavishly furnished St. Petersburg apartment. He is agitated by the news that his wife, Anna Karenina, has fallen in love with another man:

> Without undressing he paced back and forth with his even step across the resounding parquet floor of the dining room, lighted by only one lamp, across the carpet of the dark drawing room where some light was reflected only on the large portrait that had recently been done of himself and hung over the sofa, and on through her sitting room, where there were two candles burning, lighting up the portraits of her relatives and friends and the pretty knickknacks on her writing table he had known so well so long. He walked through her room to the door of the bedroom and turned back again.[8]

In the noble and upper-bourgeois circles that Tolstoy described in *Anna Karenina*, each room served its particular function. The apartment was divided along class and gender lines. The kitchen comprised the servants' domain and was separated from the dining room, where only the master of the house and his family had their meals. After dinner, Karenin withdrew to the drawing room (sometimes referred to in English as the parlor or the salon). Alexis and Anna enjoyed the luxury of having their own realms, including separate bedrooms. Like many real-life members of the European aristocracy who drew their examples from Paris, Anna had a dressing room and Alexis a study. Anna's younger sister could even boast her own space within her sister's home: "[a] little boudoir, a pretty pink room decorated with old Dresden figures." And, after his infidelity, Karenin's brother-in-law preferred to sleep in his study instead of "in his wife's bedroom."[9]

Tolstoy's upper class never discussed heating technologies, presumably because their wealth insulated them from such practical concerns. The most communal room was the parlor, which served the functions we currently associate with a living room. At the table, draped with a "white table-cloth," children did their homework and grown-ups conversed; in the evening, father retired with a good book and a cup of tea to his comfortable armchair. In these contexts of the nobility, the *samovar* was made of silver and the teacups of "translucent china."[10] In his famous novel, Tolstoy captured the comforts of the aristocratic home.

Inventing Individuality

The upper-class lifestyle evident in pre-revolutionary St. Petersburg and Moscow was by no means limited to Russia. In Western Europe also, houses built with rooms for specific purposes signaled prosperity. Although often associated with the industrial revolution, the breadwinner-homemaker household, in which women manage the world of consumption and men run the world of production, emerged before the industrial revolution: this model already existed in the Netherlands of the seventeenth century. The housing and the lifestyle associated with the wealthy Dutch Republic has been a marvel ever since. The bourgeois housewife, her servants,

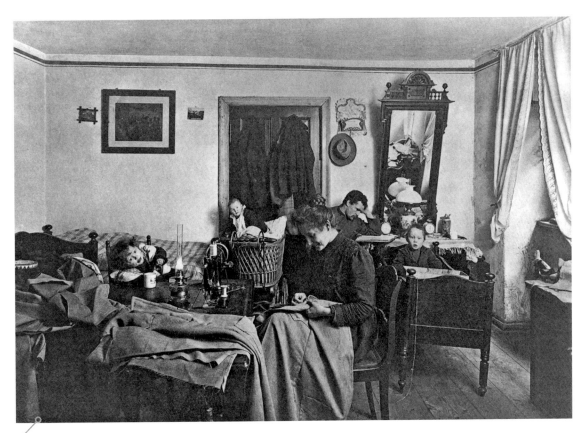

Fig. 2.2 Multi-Functional Living—in One Room: *For most of Europe's working-class and rural families, daily life revolved around a single, multi-functional room—either the kitchen or the living-room. The middle class, in contrast, could afford apartments and houses with a separate kitchen, parlor, study, and bedrooms. In this Berlin scene of a basement apartment, captured in 1905 by a reform-minded photographer, the mother does her income-producing sewing while sitting with the children; the children play; and the father reads.*

and their reputation for cleanliness, are the actors in the scenes that figure prominently in many Dutch works of art. Paintings by Johannes Vermeer and Samuel van Hoogstraaten feature the narrow hallways and domestic enclosures in minute detail, with separate rooms designated for eating, sleeping, and entertaining. The Dutch experiment with the Republic astonished foreigners. The British critic Owen Felltham believed the wealthy Dutch spent too much money on their houses, commenting that their homes were cleaner than their bodies, and their bodies cleaner than their souls.[11]

The trend towards a functional differentiation of rooms in middle-class homes thus started in the seventeenth century. The topology of the home reflected the divide between dwellers as well as the social relationship between residents and visitors. Separation between sleeping and eating; as well as work, rest, and play, characterized daily life. The separation of functions worked along class, gender, and generational lines. In many cases, individual family members and even servants expected (heated) rooms of their own.

Earlier the man of the house often had his working space in the hallway, in his wife's bedroom or in his children's room. This was to change. The wife of a German professor reported in 1796 that she had "three rooms beside each other, all of which I can use myself. One is the reception room, the middle one is a smaller room followed by my living-room [...]. My husband lives on the other side of the reception room."[12]

Individualization and gender differentiation went hand in hand with a distinct separation of private and public spheres, especially in apartments built for the upper and middle classes. Until the eighteenth century, residents of so-called *maisons à allée* in Paris often had to pass through rooms belonging to other tenants in order to reach their own apartments. The first six- to eight-story apartment buildings erected in the 1820s avoided such an unruly living situation. To make sure that unwanted visitors did not enter the building in the first place, a janitor controlled the comings and goings. Servants were assigned separate entrances to minimize their contact with the upper classes. And the apartments themselves were strictly divided. Reception rooms and the bedroom of *monsieur* and *madame* faced the street, whereas the servants' area—including the kitchen—faced the courtyard.[13]

Similar strategies to regularize social life through architectural structures and social rules developed in other parts of Northern and Western Europe. Domestic technologies and ideals circulated throughout Europe and across the Atlantic. In some cases, techniques traveled from the German lands towards the north, the east, and the south. In other cases notions and ideas spread from the British Isles to the Continent. In Sweden, the maid in most scenarios did not have her own room until the late nineteenth century; she slept either in the kitchen or in the same room as the children. In Germany, the set-up was different: the separation between servants and those they served remained in place. Berlin families from very different walks of life lived in the larger apartment buildings where the most expensive apartments were on the first or second floor facing the street. When the better-off middle classes began constructing elaborate villas on the outskirts of Berlin, the tradition of separate entrances was upheld. In his 1864 book, *The Gentleman's House*, architect and engineering professor Robert Kerr criticized the British habit of designing houses with only one entrance, emphasizing that rooms where servants worked and rested should be completely separate from the rest of the home.[14]

The history of the hall as a space within the home nicely illustrates the shift toward individualization and spatial separation. In earlier centuries, the entrance hall was relatively large and served various functions. Here, families worked, enjoyed their meals, and conducted important rituals of their social life. The room often included a desk used for working on the household accounts and for conducting correspondence. In Northern Germany, traditional houses were designed with one large, centrally placed hall. A late-eighteenth-century source describes a typical such room in the hanseatic city of Rostock: "The hall makes up about two-thirds of the whole house; this is where the family spends its time [...]. These halls are partly nicely decorated; I have even seen that people receive their guests here."[15]

Over time, however, the hall shifted from being a room of work and leisure into a transitory space, an area through which guests and residents merely passed in order to reach the main sections of the home. In Britain and France, as well as in some Dutch cities, the entrance hall developed into a corridor for access to the individual rooms for sleeping, working, and socializing. The hall became a place where visitors were assessed before being channeled in the appropriate direction or refused further access. In Victorian Britain and the United States, hallways had a gate-keeping function. Many middle-class dwellings in the United Kingdom and Northern United States—unlike in the German lands—had only one entrance. This meant that the visitor inevitably encountered the hallway before any other space inside the home. To make a good impression on distinguished guests, the space could not be too bare and simple. On the other hand, given that peddlers and servants were apt to wait in the hallway for substantial periods of time, the entrance hall could not be too lavishly furnished. In this sensitive situation, the hallstand, in combination with a chair or a bench, came in handy. These pieces of furniture were usually made of nicely carved wood, meant to appeal to the aesthetic sense of well-educated visitors. The hallstand—which was especially popular in Britain and the United States in the second half of the nineteenth century—also served practical functions. It had hooks for hats or capes; an umbrella stand; a mirror; and a small table or bench.[16]

If middle-class peers used the bench or the chair at all, they did so only briefly, in order to take off or put on their gaiters or overshoes. For servants, these chairs were meant to signal submission. Seats had no upholstery. Waiting half an hour for a reply, the telegraph boy had to put up with the hard and uncomfortable bench or chair. Clarence Cook, U.S. author of a contemporary handbook on interior decoration, emphasized that the bench or the hall chair was in the first instance meant for members of the lower classes—"messenger-boys, book-agents,...[the] census-man and...[the] bereaved lady who offers us soap"—and went on to explain that, "as visitors of this class are the only ones who will sit in the hall, considerations of comfort may be allowed to yield to picturesqueness."[17] On both sides of the Atlantic, the hallway was not a neutral space but a room filled with artifacts that bore signs and transported meanings. In this way, even an ostensibly innocent hallway bench contributed to cueing every individual—of every social class—related to the household as to the behaviors and the rituals that were expected.[18]

Heating the Home

The shift toward rooms with separate functions was connected to developments in technologies for heating, cooking, and lighting. As long as houses had only one fireplace, life necessarily revolved around that fireplace—at least during the colder part of the year. The open hearth contributed to heating the living space; enabled the preparation of meals; and provided light in an otherwise dimly lit environment. The fireplace was a truly multi-functional technology.

The masonry stove emerged as a serious alternative to the open hearth. Radiating heat at a lower temperature for a longer stretch of time, it fostered a more homogeneously heated domestic environment conducive to a larger variety of social activities. In the late Middle Ages, the first stoves had already appeared in the Alpine regions, initially in the houses of rich families who could afford the investment. Knowledge about the masonry stove's basic principles soon circulated to other parts of Europe, where different models developed. Originally, stoves were fueled by wood, but in the nineteenth century, coal became an option. The most obvious advantages of masonry stoves are their energy efficiency and heat-storage capacity: rather than escaping through the chimney, much of the heat is absorbed and retained by the masonry (usually clay, stone, or bricks and mortar). The heat then steadily diffuses into the surrounding room. In Continental Europe, such stoves were covered with colored or ornamental ceramic blocks and came to be known as tile stoves. An early-sixteenth-century copper engraving by the Bavarian artist Barthel Beham shows how a beautifully designed tile stove was used to heat a fairly large parlor, allowing people to undertake a number of activities in all corners of the room without freezing in the winter. Typical of the southern part of the German lands are benches fixed directly to the wall.[19]

Masonry stoves could be lit in one room and configured to emit most of their radiating heat in an adjacent room. German immigrants brought the concept with them to the United States, where they installed simple cast-iron stoves that were used for both cooking (in the kitchen) and for heating the so-called stove room (*Stube*, the German word for living room, has the same root as the English word "stove"). The outcome in New England and in large parts of Europe was the separation of the kitchen from the

Fig. 2.3 The Stove as Centerpiece: *It was the early 1800s when Europeans began to live in separate rooms, with the parlor at the heart of the home. Those separate rooms were warmed by individual stoves dedicated to heating rather than to cooking as well. Tile stoves—like the one in this circa 1900 photograph of a Budapest home—were known to heat more evenly than the open hearth. As a technology, the tile stove fostered the popularity of the parlor, an elaborately decorated room for reading, conversation, and music-making.*

parlor, a process with social and cultural implications. First, with the separate kitchen in place, wives or servants could prepare of meals without bothering other members of the household. Second, activities such as conversing, writing, playing, and mending— once carried out in the kitchen—moved to the parlor or the stove room. There is a direct link between developments in heating technology and the layout of houses and apartments. In Northern and Continental Europe especially, the masonry heater simplified the spatial separation of functions realized in the nineteenth century.[20]

Heating technologies played a third important role: creating homely comforts. In the seventeenth century, British visitors to the German lands had already noted the positive health effects of masonry stoves—and the enjoyable social life that unfolded in the *Stuben*. Nevertheless, the British in general favored the open hearth. In the Victorian period, members of both the aristocracy and the middle classes had fireplaces in the dining as well as in the drawing room. Why were closed stoves and central heating rejected? The reasons centered on the users' values: the open hearth embodied ideas of domestic convenience and coziness. Many British descendants in the United States perpetuated the open hearth. One nineteenth-century U.S. commentator wrote of this habit: "The old-fashioned fireplace will never cease to be loved for the beautiful atmosphere it imparts to a room, and the snug and cheerful effect of an open wood-fire. ... For the home, nothing except the fireplace would do."[21]

In England, as well as New England, then, the open fireplace was more than a heating device. It also served social purposes. During the colder part of the year, no reception or dinner was complete without the ritual of gathering around the fireplace. Robert Kerr, in his 1864 architectural manual directed at the British upper and middle classes, wrote: "...for a Sitting-room, keeping in view the English climate and habits, a fireside is of all considerations practically the most important. No such apartment can pass muster with domestic critics unless there be convenient space for a wide circle of persons round the fire."[22]

Coziness was an integral part of the British obsession with domestic comfort. Coziness was not only associated with snugness and warmth, but with decorative items and ornamentation, soft upholstery, and cushions. An illustrative example is an elaborately designed sofa that furniture manufacturers advertised under the name and headline "Cosy Corner, complete with drapery of cretonne." The furniture also came with buttoned upholstery and fine carvings. To the Victorian mind, such a corner provided physical comfort as well as protection from the outside world. It was a perfect place to immerse oneself in a good book and forget the struggles and tensions of daily life.[23]

In the Victorian setting, creating coziness and comfort was the responsibility of the woman of the house. It was her duty to turn the house or apartment into a home, a shelter from the harsh arena of business and industry. The middle-class ideal of the full-time housewife—a person tasked with working in the household and not with carrying out any form of wage labor outside the home—developed in tandem with an ideology and "cult of domesticity." This cult emerged in the English-speaking world of the Victorian era. Members of the Continent's (primarily urban) middle class immediately appropriated some elements belonging to the cult of domesticity. Breaking away from patriarchal authority, women reformers, in an alliance with the clergy, promoted the ideal of the breadwinner-homemaker household as a way of counterbalancing the ills of industrialization with a morally superior home life. The emergence of the cult sharpened the divide between private (female) spheres and public (male) spheres. As the seventeenth-century Dutch Republic and other consumer-driven societies exemplify, industrialization supported a process that one could call the feminization of the home. As bourgeois husbands in cities increasingly worked outside the home, they left—reluctantly, at times—the private realm to their wives, for whom the specter of working in a factory or an office was to be avoided at all costs.[24]

German Chairs & Persian Carpets

The middle-class passion for cushions and deep-pile carpets, for heavy draperies and knickknacks: this was an international phenomenon that came to characterize European domestic life. The middle class emulated what they regarded as aristocratic style elements, a process that the better-off members of the working class in turn repeated. The allegedly aristocratic "ruling taste" circulated across Europe and the Atlantic through handbooks,

Fig. 2.4 Home Sweet Victorian Home: *This painting showcases the Victorian obsession with heavy curtains, deep-pile carpets, oversized pillows, and frilly dresses. Toward the end of the nineteenth century, members of various reform movements attacked this predilection for what they called cluttered, stuffy, and dark homes. Appropriating insights from the new science of bacteriology, reformers advocated easy-to-clean furniture, walls, floors, and clothing. They demanded that apartments be designed with windows that were larger and more plentiful to let in light and air.*

magazines, and teaching materials; this sense of aristocratic style was enhanced by travel and international contacts. For example, most late-nineteenth-century guides on home decoration and "genteel behavior" published in Hungarian were translations from German. Even the Hungarian words for "living room" and "dining room" used at the time were modifications of German terms. Especially influential in the Austrian-Hungarian double monarchy was a book entitled (in translation), *The German Room of the Renaissance*. At the local level, upholsterers took on the role of mediators between the ruling taste and the interior decoration of individual apartments and houses.[25]

The "Oriental" influence on the aristocratic-turned-middle-class taste informs our understanding of how Europeans viewed their own culture. To this day, Persian carpets signal at least a certain degree of wealth, and ottomans can be found in every well-equipped furniture store in Europe. In the late 1820s, French artistic circles began focusing attention on the Arabic, Ottoman, and Persian worlds. Even before the heyday of imperialism, French novelists, painters, and decorators constructed a picture of the Orient that was more "legendary and dreamlike" than real. This romantic view of the Eastern World co-existed with a view of the West as modern and rational. In practice, the oriental ideals materialized in playful forms, elaborate ornamentation, and heavy textiles. Toward the end of the nineteenth century, Asian as well as African handicraft items came to signify originality and uniqueness in a period increasingly dominated by mass-produced goods. Both in European countries struggling for colonial supremacy and in the United States, consumers readily acquired baskets, embroideries, and paraphernalia from other continents. To surround yourself with oriental and exotic products became a political statement that signaled openness toward other cultures—without relinquishing the idea of Western supremacy.[26]

As furniture manufacturing mechanized and prices fell, the ruling taste of what turn-of-the-century critic Thorstein Veblen disdainfully called the "leisure class" found its way into broader segments of the middle class. Characterized by Veblen as the harbingers of *Conspicuous Consumption*, these classes crammed their homes with innumerable objects. On the basis of inventories from upper-middle-class apartments in Budapest, it is possible to reconstruct what awaited visitors in a typical drawing room: a couch or divan, several armchairs and a large carpet, one centrally placed table as well as "small, decorative occasional tables with chairs grouped around them," a cupboard with tableware and a mirror, oil paintings and vases, table lamps and a chandelier hung from the ceiling (fueled by paraffin or gas), clocks, and sometimes a piano. Windows and doors were framed with drapes. One inventory catalogued a "rocking chair with oriental fabric."[27]

The Budapest inventories do not differ substantially from descriptions of upper-middle-class homes in other European cities at the time. A lavishly furnished home became a symbol for a pan-European, middle-class domestic culture. Swedish sources from the second half of the nineteenth century provide a picture of home interiors that resembles those of Budapest. Still, the domestic culture was not uniform: a British parlor ordinarily contained a fireplace, while most Swedish and Hungarian drawing rooms included a tile stove. This stove was only heated when guest were expected. Perhaps surprisingly, this happened rather rarely—at least in Budapest. A contemporary Hungarian source suggests that "there were only a handful of families who could be said to enjoy on a regular basis the seasonal joys of social intercourse in their own homes."[28] Paradoxical as it might seem, the centerpiece of the apartment was only rarely used and, most of the time, was left unheated. The drawing room, or the parlor, obviously served non-functional purposes. It was a space where the middle-class family occasionally demonstrated its wealth and taste to others. This taste was clearly international in the sense that it embraced a wild mixture of French, German, Austrian, British, and oriental style elements, thus contributing to a sense of communality across national borders.[29]

The rooms of the typical middle-class home were filled with objects that had often been haphazardly lumped together. These artifacts were to be admired rather than used, and the feather-duster was always within close reach. Children were warned to keep their hands off tableware and figurines. It was a world that called for order and discipline. Although the cult of domesticity constructed a dichotomy between the warm, comfortable home and the cold, rational outside world, both realms were governed by the same values of discipline and self-control. Coziness and comfort could only be guaranteed if family members conducted their lives in an orderly manner. It is no coincidence that historical anthropologists who study the home frequently make reference to Norbert Elias's notion of *The Civilizing Process*.[30]

Considerable (mostly female) energy was mobilized to realize the ideals of domesticity. In line with similar publications in other languages, the magazine *Svenska husmodern* (The Swedish

Housewife) in 1877 explained to its readers that the "constant work to maintain [order]...is *her* task, which the husband *cannot* understand or perform at all."[31] Assisted by a maid, the professional housewife had to make sure that the many items filling the rooms were kept in their correct places. Chaos had to be avoided at all costs.

Toward the end of the nineteenth century, a new threat appeared on the home front: dust. The "war" against dust became an additional burden for women in both Europe and North America. As male doctors and bacteriologists popularized germ theory, sanitation and hygiene reformers acquired millions of new allies. Persuaded that it was no longer a mere aesthetic problem but a potential health hazard, citizens mobilized against dust. After all, the reasoning went, a dust grain could contain deadly bacteria. Only a few years after Robert Koch and Louis Pasteur formulated their ideas on the sources of contagious disease did the authors of household guides react by casting suspicion on heavy draperies and upholstery. Even long dresses came under attack: they were said to transport dirt from the streets into the home. The mop and the carpet beater became the main weapons against dust.[32]

Just as *style* ideals circulated throughout the Continent and across the Atlantic, so did the *sanitation* ideals. One outcome of this sanitation gospel was the introduction of straight lines and easy-to-clean materials into the middle-class home. Household handbooks published across Europe advised women to furnish the home with dust- and dirt-control firmly in mind. A Swedish household guide of 1891 made it clear to its female readers that heavy textiles were dangerous sources of "bacteria, bacillas [*sic*], microbes, and what else they are called, these invisible entities, that in recent times have begun to embitter our existence."[33] Those who sympathized with the broader turn-of-the-century reform movement were most likely to do away with the dust-attracting carpets and cushions in their own homes. These users were also more likely to dress in a lighter and more natural way. This *Lebensreformbewegung*—the reform movement as it was called in German—circulated more or less radical ideas in areas such as clothing, nutrition, housing, and leisure. To some extent the ideals of this early user movement squared with those of the aesthetic Arts and Crafts Movement. The furniture that Morris and Co. manufactured in its London factory was less ornamented than the previously popular classical style. Reform chairs and beds were designed to simplify cleaning.

Demonizing Dirt

Working-class households could afford neither oriental carpets nor reform furniture. To make ends meet, most working-class women either took in work or labored outside the home. Workers' apartments were smaller than those of the upper classes; housed more people per room; and were more sparsely furnished. The term "worker," however, could refer to any one of a number of categories, each with a different living standard. For example, some foremen and skilled craftsmen lived in reasonably comfortable conditions and developed a lifestyle that emulated that of the middle class to some extent. Day laborers and unskilled workers, by contrast, had to accept simpler circumstances.

It is difficult to differentiate between workers' living conditions in various parts of Europe; Reliable images that reveal working-class living standards are scarce. Most literary sources are based on observations made by benevolent representatives of the middle class, who for political reasons tended to dramatize the situation and apply foreign examples to create stark contrasts. The descriptions of a contemporary Hungarian social scientist are a case in point. Investigating the apartments of Budapest dockworkers in 1900, Manó Somogyi noted in a typical reformist style: "We arrived in smoky kitchens where the women were at the washtub, whose [sic] steam filled the entire flat. In the room, ropes were extended with wet underclothes hanging down."[34] At the same time, Somogyi was compelled to admit that the workers' conditions were not as bad as he had expected:

> In many places the interior and furniture of the flats surprised us because we were not prepared to find any traces of comfort and congeniality in the poor-looking flats. Most of the flats were maintained with exemplary cleanliness and tidiness…. The furniture was poor-looking, but decent. There were a few flowerpots in the windows. In other places, the blindingly white lace curtain [sic] made the workers' flats cozy.[35]

Such self-reflective comments are scarce in both historical literary texts and contemporary sources. Instead, the dominant discourse uses melodrama to describe the hopeless situation of destitute workers and peasants. The talk of dirt and filth found in Somogyi's report and Checkhov's *Peasants* started at least one hundred years

Fig. 2.5   The Working-Class Dwelling:
This photo's archival description—like many others of the time—belies middle- and upper-class attitudes toward the working-class life lifestyle: "Interior of a slum dwelling, c. 1930s." Starting in the mid-nineteenth century, benevolent commentators incessantly used negative words like slum to describe working-class living quarters. Judging from this boy's face, though, contentment could reign in apartments without electric stoves, washing machines, or hot-running water.

earlier and is by no means limited to Continental Europe or Russia. Ideas of cleanliness and hygiene emerged well before the sanitation reform movement mobilized in the second half of the nineteenth century. In Britain, Belgium, and other early-industrializing countries, benevolent commentators developed a patriarchic language of pity and assistance that would shape the political discourse well into the twentieth century. Vocal members of the aristocracy and the upper middle class invented the so-called social question, a topic that later led to the formation of public housing as a political field in its own right. Although the implications of such policies varied between countries, many of the proposed technical solutions were strikingly similar.[36]

This political discourse about poverty reflected larger concerns about the impact of industrialization. Environmental concerns—then referred to as sanitation problems—were also part of the discourse. By the late eighteenth century, Manchester in Northern England had already become a symbol for everything associated with the factory system: industry and ingenuity paired with social misery and slums. In the first decades of the nineteenth century, politicians and economists successfully obscured the obvious injustice of the system by referring to individual freedom and economic necessity; but in the 1840s, the critique took on new dimensions. Edwin Chadwick's *Report on the Sanitary Condition of the Labouring Population of Gt. Britain* ushered in an era of sanitation reform. The unvarnished description of *The Condition of the Working Class in England* by Friedrich Engels signaled radicalization of the political scene.[37]

Despite different outlooks and convictions, the reformer and the revolutionary agreed on at least one point: the situation of industrial workers had to improve. In Manchester, Engels visited Irish workers living in "ruinous cottages, behind broken windows, mended with oilskin, sprung doors, and rotten doorposts, or in dark, wet cellars, in measureless filth and stench."[38] And Chadwick, secretary to the Poor Law Commission, concluded that "the various forms of epidemic, endemic, and other disease caused, or aggravated, or propagated chiefly amongst the labouring classes by atmospheric impurities produced by decomposing animal and vegetable substances, by damp and filth, and close and over-crowded dwellings prevail amongst the population in every part of the kingdom."[39] Engels and Chadwick created a political discourse where filth, stench, overpopulation, and deterioration were central concepts. Its power was to be felt across Europe—all the way to Chekhov's Russia.

Reform-minded doctors and scientists put their faith in improved sanitation and hygiene. These buzz words became important rationales for the development and appropriation of new technical solutions. Chadwick argued strongly in favor of centralized water supply systems and sewerage networks, a proposition that garnered many followers. In Germany, the scientist Max von Pettenkofer became the most outspoken advocate of such sanitary solutions in the 1860s. William Lindley, a British civil engineer, brought Chadwick's ideas of the health benefits of water closets and kitchen sinks to the Continent. In the 1840s, Lindley started off in the fire-struck city of Hamburg and would then move on to construct water and sewage infrastructures in a number of European cities: Budapest, St. Petersburg, Frankfurt am Main, and Warsaw. In Berlin, a British-led company began installing a water supply system in 1853.[40]

The transnational circulation of engineering knowledge is impressive; it spanned the subjects of water supply and sewage treatment, gas and electricity, public transportation, street design, and traffic planning. In the mid-nineteenth century, engineers and companies were already transferring technical know-how among cities and among countries. City authorities set up expert commissions and sent them on study tours, while engineering journals reported regularly on the changing state of the art. The first international sanitation conference was held in Brussels in 1852, and the 1876 International Congress on Sanitation and Demography initiated a series of meetings. Whereas the sanitation conferences issued general recommendations to improve the living conditions of the urban working class—for example, by means of legal reform—the international congresses became a forum for scientists and engineers to exchange knowledge and experience.[41]

Although Chadwick had originally launched sanitation as a means to solve the problems of the poor, the solutions that scientists and engineers presented affected the penniless groups only marginally; it was the middle classes that profited. Workers could not afford water closets, and they benefited only indirectly—if at all—from various sanitation measures undertaken by municipal authorities. Instead of making the proletariat cleaner, running water increased the comfort of the better-off classes. Attempts to convince members of the working class to visit public baths met a similar fate. Admission was prohibitively expensive, and, apparently, many workers regarded the public health discourse as middle-class hype. Throughout most of the nineteenth century, sanitation and public-health technologies remained largely the privilege of an urban minority.[42]

City authorities' attempts to improve living conditions for the masses were more successful in another area. In the second half of the nineteenth century, a number of European cities and countries issued laws and statutes aimed at regulating building construction and land use in urban areas. In Copenhagen, an 1856 edict forbade the installation of basement cesspools and prescribed ceilings of at least 2.5 meters (ca. 8 feet) high. Three years earlier, the Berlin "building code" (*Bauordnung*) had specified a minimum size for backyards and the maximum heights for buildings. The city authorities' motto was "More Light! More Air!"[43] A closer look reveals, however, that many actions did not aim to improve public health but rather to protect the interests and investments of the middle class. For example, centralized water systems were installed not to provide the working classes with drinking water and indoor sanitation, but to protect private homes, public buildings, and factories from the risk of fire—a primary concern of the middle- and upper-class owners.[44]

Modernizing the street network was a further measure intended to improve the urban environment. This modernization began in the 1850s. Baron Haussmann, who worked under Napoleon III, was prefect of the *Seine département* that wiped out old houses and narrow lanes in Paris. With the stroke of a pen, Napoleon III replaced these structures with multi-story stone buildings and wide avenues. Throughout the 1850s and 1860s, this program continued in the name of the "Struggle against Filth and Germs." For Haussmann's superior, Napoleon III, this urban cleaning-up operation was meant to turn Paris into a well-ordered symbol of the Second Empire. Neither he nor Haussmann had discussed this with the residents, who would never be able to afford the rents in the new residences. As in the United States, European measures taken in the name of sanitation by reformers contributed to the displacement of the poor.[45]

Modernizing the street network, in the name of the struggle against filth and germs, also took place in the far southeast of Europe. In Constantinople (later Istanbul), creating new roads was a way of controlling fires as well as a means for the city authorities

Fig. 2.6   "Cleansing" the Cities: *In the name of modernization and sanitation, authorities radically changed the Parisian urban landscape during the 1850s and 1860s. With scant regard for residents of the old quarters, the city officials tore down thousands of old and allegedly unhealthy houses. In their place emerged broad thoroughfares and representative public buildings, ordered by Napoleon III. These structures connected to a new-fangled, centralized water-supply system. This 1856 engraving by cartoonist Emile Marcelin illustrates the dramatic transformation of the urban renewal process— from the perspective of those who demonized dirt.*

to manifest their power by seizing political and economic control of the so-called *mahalles*, the locally governed sections of the city. The political fight was especially pronounced during the *Tanzimat* period (1839–76), when new political elites seized power. Although Paris was an explicit role-model for the modernization of Istanbul, the outcome of these measures was not as drastic as in the French capital. Whereas most street projects in Paris forced poor inhabitants from their neighborhoods, most of the poor in Constantinople were allowed to remain in the same areas.[46]

When it came to modernizing Constantinople's water infrastructure, the local authorities turned to France once again. In the 1880s, when they decided to erect a modern water-supply system, they brought in a French consortium, *la Compagnie des eaux de Constantinople*. Not Turkish but West European engineers planned and managed this huge project that—not surprisingly—began in Pera (today Beyoğlu), the wealthiest part of town, with the highest concentration of foreigners. Pera had in fact been created as Constantinople's equivalent to the *6e arrondissement* in Paris. Those who could not afford running water in their own houses or flats had to continue carrying water from the nearest public fountain.[47]

Cooperatives & Home Inspectors

The decision to install a Western-inspired water network in Constantinople was far more than a measure to improve public health. It was part of a strategy to modernize the capital and solidify the power of the state. Also in Western Europe, modernization implied increased state control and surveillance, especially of the poor. In line with paternalistic notions, commentators assumed that the working class was unable to improve its lot on its own. To solve what came to be known as the housing question, parliaments throughout Europe took surprisingly similar actions, and a large number of well-networked organizations provided workers and the lower middle classes with reasonable and comfortable homes. In Northern and Southern Europe alike, cooperative societies were entrusted with the task of mobilizing and civilizing the working class.

The first laws concerned with dwelling standards and living conditions were accompanied by an array of inspection measures.

Fig. 2.7 West Meets East: *In the 1880s and 1890s, German and French companies were hired to create a centralized water-supply system in parts of Constantinople (Istanbul). Given that few could afford to connect their homes to this system, the authorities decided to retain public fountains, which gave the poor free water access. The elaborately decorated public fountain in this 1898 postcard looks like a traditional Ottoman artifact. Actually, though, the fountain's water system integrates Ottoman practices with West European engineering, which aimed to connect individual homes to a central water source. German Emperor Wilhelm II inaugurated the fountain in 1901.*

Enforcement relied on intense home surveillance, a task traditionally associated with the police. It is not surprising that the so-called "building police" (*Baupolizei*) was given the task to enforce the *Bauordnung* in Berlin. When the British Parliament in 1851 issued the Lodging Houses Act, enforcement fell to the local police. In Britain, a new role was invented to guarantee the implementation of sanitation acts: the medical officer of health—a professional who was assigned the power to inspect "not merely the exterior but penetrating into the interior, of the dwellings of the poorer classes." Housing and sanitation ordinances allowed inspectors to carry out surprisingly intrusive and demeaning investigations. Two police inspectors, who in the mid-1880s regularly visited lodging houses, explained their approach: "We compel [the residents] to scrub the bedroom floors once a week and we compel them to sweep them daily."[48]

Slowly but surely it became clear to concerned observers that the conditions of the working and the lower middle classes would never improve substantially as long as legislation focused on cleanliness and sunlight only. Considering the sheer number of impoverished families, it would have been an impossible task to inspect all of their homes on a regular basis. But what was the alternative? The "benevolent" representatives of the middle class who tried to improve the workers' lot by means of education and propaganda were also not particularly successful. Initiated to making cities more attractive to visitors and middle-class inhabitants, the sanitation measures often produced even more precarious living conditions for the proletariat: Haussmann's Paris was not the only city in which regulation forced poor inhabitants to leave their familiar neighborhoods.[49]

By the end of the nineteenth century, a new approach emerged. Governments now decided to support the construction of tenements for the working class, thus breaking the power of private landlords. Schemes ranged from the promotion of cooperative societies and public building companies to providing direct financial assistance—loans with low interest rates, for example. As in the area of urban engineering, solutions to the housing problem circulated among European countries as well as between Europe and North America.

Belgium's 1889 "Law on Workers' Tenements and the Establishment of Patronage Committees" was one of the first comprehensive European housing laws, and it came to serve as an exemplar for concerned groups in a number of other countries. The law offered workers and their organizations a range of supportive instruments. One of the first industrialized countries in Europe, Belgium had a tradition dating to the 1820s of subsidizing workers' apartments. Crucial to this initiative was the founding of the General Savings and Pension Bank (*Caisse générale d'epargne et de retraite*) in 1865. This national institution was mandated in 1889 to provide housing loans to people of modest means. To ensure that credit was given to the right projects, the government established local *Comités de patronage* in towns and boroughs.[50]

The new ideas for housing solutions were discussed at a number of international conferences. As part of the World's Fair in Paris, the first international congress of housing organizations for the working class convened in 1889. Whereas the Belgians in mid-century had often copied French legislation, their own act of the same year became a role model for others. Reformers in many countries took up the notion of Patronage Committees as mediators between the state and homeowners. Perhaps more important, housing acts throughout Europe now came to emphasize financial and organizational skills. Hoping that workers and members of the lower middle class would take matters in their own hands, proponents of the new schemes made sure that cooperatives were given due support. Residents were to be transformed from passive, power-less tenants to activists who designed and owned their houses collectively. In Germany, legislators gave cooperative societies a secure legal framework in 1889; many cooperatives were founded by upper-class philanthropists rather than by workers. In Britain (1890) and the Netherlands (1901), laws allowed towns to create communal building companies.[51]

Governments in Southern and Southeastern Europe also took legal initiatives. In 1903, the Italian Senate passed a law on Popular Housing (*case popolari*) that gave building cooperatives and welfare organizations the right to acquire favorable loans; eight years later, the Spanish Parliament regulated the construction of affordable apartments. In 1910, the Rumanian government paved the way for the creation of communal societies for cheap apartments. The fact that the Belgian law of 1889 had served as a direct role-model for the Italian initiative testifies to the strong transnational ties in those years. The Italians also actively participated in a number of international congresses on residential housing in the first decade of the twentieth century, appropriating technical and legal know-how from countries north of the Alps. Despite active housing policies in northern-Italian cities like Milan and Turin, the number of newly built apartments still remained low. Because of its strict rules requiring residents' financial security, the law hardly improved the lot of the industrial proletariat. Mussolini's seizure of power also did not alleviate the situation. Suburban areas built under Fascist rule were of mediocre quality. Some of these barrack settlements—pejoratively known as "Abyssinian villages"—were even designed and built according to plans that deprived inhabitants of running water and water closets.[52]

Social housing programs did not take off on a large scale until after the First World War. Perhaps most famous in these efforts were the progressive activities orchestrated by the Social-Democratic political majorities in interwar Vienna, Amsterdam, Frankfurt, and Stockholm. From this point on, housing became a central concern of the social-democratic movement, at the international, national, and local levels. Projects in cities like Amsterdam and Frankfurt were carried out by cooperatives or publicly owned companies, while the authorities of 'Red Vienna' undertook to build and finance apartments for the working class on their own. Compared to post-Second-World-War figures, the 59,000 apartments that the city erected between 1920 and 1933 may not sound impressive. But the Karl-Marx-Hof and other estates—complete with communal laundries, reading rooms, playgrounds, and nurseries—generated enormous interest throughout Europe. The inhabitants themselves also were apparently happy with the comparatively high technical standards. A "dream come true," one dweller recalled: "Directly lit rooms with a hallway, W.C. and running water and a small store-room in the flat, storage areas in the basement and in the attic, as well as a laundry with a drying area accessible to everyone."[53]

Dutch society chose a different path from Vienna. Supporting a tradition of local and collective initiatives rather than state sponsor-ship, the Dutch parliament created a legal framework that fostered cooperative housing. The 1901 housing law helped consumer groups, user-based communities, and other civic-society organizations to shape housing that was unique in Europe, if not in the world. With provisions to fund public–private partnerships, the law expanded civil society and, by 1940, helped to create about one million houses, most of them social housing. Catholic, Protestant, socialist, and liberal groups took a number of initiatives. In Amsterdam, where liberals and social democrats ruled, members of a food consumer cooperative named *De Dageraad* established a housing cooperative. As part of H. P. Berlage's city plan for Amsterdam, *De Dageraad*'s housing asso-ciation built the Pieter Lodewijk Tak Street between 1919 and 1923. Through a process of learning-by-doing, the cooperative's board acquired skills of good governance, while experimenting with forms of collectivism. *De Dageraad*, led by skilled workers and educated employees such as teachers and bookkeepers, also mediated between the city government, builders, and tenants. The houses were of high quality and equipped with hot water, central heating, and electric light—technologies that were seldom available even in middle-class houses. The board chose the residents, selecting those applicants who exhibited a "distinguished level of civilization" and, moreover, a strong commitment to "a cooperative way of home-making."[54]

The first tenants in the P. L. Tak Street were diamond cutters, salespersons, shopkeepers, public servants, traders, and skilled industrial workers. The cooperative board installed an attractive collective radio receiver system and a popular common library. To further reinforce the project's cooperative and social-democratic character, tenant committees encouraged the purchase of vacuum cleaners and boilers through installment plans; collectively bought

coal to save costs; and organized concerts, exhibitions, and festivi-
ties—perhaps most enthusiastically on the First of May. Although
residents appreciated the apartments' innovations, some of these
technologies were used in ways that were not intended by the association. For example, tenants tinkered with the heating system to the
extent that the cooperative's operating costs went through the roof.
To the dismay of the board, tenants used landings to park bicycles
or to let children play. And, given that they could not afford to buy
the requisite bathroom fixtures anyway, many residents stored their
tools and work materials in the space designed as a "bathroom."
When the Depression hit, the cooperative sought to help the
unemployed to settle unpaid bills. De Dageraad also lobbied the
municipal government to lower rents. Despite the initial building
setbacks and management woes, the apartments were a resounding
success. One tenant deemed the housing a "joy in stone," recalling
"I thought: tomorrow everyone will have an apartment like this.
Then we will all be living more or less in a utopia."[55]

Muckraking & Social Engineering

For many decades, Germany and Britain had been the inspiration
for progressive politics in the transatlantic world. And yet, the
British Labour government of 1929 to 1931 failed to find inspiration anywhere in Europe or beyond, proving helpless in facing
the Depression. Since the time of Bismarck, Germany had been
the role model for social policy; the Nazis crushed all progressive
policies in 1933. France went through more than twenty governments between 1929 and 1939. Only in the United States, where
the Depression hit hardest, and in Sweden, where a zealous
Social-Democratic government came to power, did an era of social
investment ensue. Like the communist Soviet Union, both countries appropriated earlier European experiments in social housing,
only to place them on a larger scale. The United States had been
lagging, but the scope and scale of the 1930s New Deal legislation
was peerless; the Swedish experiment became the harbinger of
the European post-Second-World-War welfare state. The Swedish
case illustrates how housing finally became a central issue on the
European political agenda—an issue that involved momentous
decisions of a technological nature.[56]

Directly after coming to power in 1932, the Swedish Social-Democratic government established a committee to investigate how to improve housing for the working class. Experience in Stockholm and elsewhere had shown that cooperative initiatives would not solve the problems facing many unskilled and poor workers. For example, the houses built by the cooperatively organized Tenants' Savings and Construction Society (*Hyresgästernas Sparkasse och Byggnadsförening*, HSB) came to be populated mainly by a working-class aristocracy of skilled workers, civil servants, and other niche members of the lower middle class. High technical standards and comparatively spacious floor plans made the HSB flats prohibitively expensive for the unskilled laborers the reformers had in mind.[57]

Muckraking journalism supported the social-democratic case. In 1938, author and journalist Ludvig Nordström toured Sweden, from the very south to the far north. The Swedish Radio Corporation (*AB Radiotjänst*) had commissioned him to investigate the living conditions of ordinary people in small towns and the countryside and to broadcast his experiences on a regular basis. To simplify access to interviewees, the Royal Board of Medicine (*Kungl. Medicinalstyrelsen*) and the Stockholm Deaconry Board (*Diakonistyrelsen i Stockholm*) asked local doctors and priests to accompany Nordström. "Lubbe" Nordström was already a well-known figure in Swedish public life. His radio documentaries attracted a substantial audience. When the Cooperative Society (*Kooperativa förbundet*) published his reports under the title "Filth-Sweden" (*Lort-Sverige*), the book became an immediate bestseller. In the well-established muckraking tradition on both sides of the Atlantic, Nordström narrated the grievances in a vocabulary reminiscent of Chekhov's: poverty and dirt abounded. In *Lort-Sverige*, we read about damp flats without bathrooms, insufficiently insulated cottages without central heating, and undernourished children deprived of appropriate clothing. To inspire the listening and reading public to take action, Nordström painted a picture of the 1930s Swedish countryside that invoked the Russia of the 1890s. For example, the house of the Finnish-speaking Alonen family, located in Pajala, north of the Arctic Circle, was portrayed in similar terms as the Chikildeyevs' hut. Two brothers and their families—a total of eleven people—lived in the small dwelling on the very edge of Europe. One kitchen had an iron stove built into the fireplace; the decaying floors showed holes; and the windows were insulated with rugs. One of the wives voiced her anxiety about the coming winter: "We do not understand how we are going to make it."[58]

Like reform-minded journalists in other countries, Nordström offered stories of Sweden's needy, comparing their plight to the ghastly conditions of "the Balkans or a Turkish village in Asia Minor." This was not only to shock, but to remind the public that there was a way out of the misery. If the government and society as a whole realized the gravity of the situation, then Filth-Sweden could give way to a modern country: "Bungalow-Sweden" (*Villa-Sverige*). Throughout his book, Nordström praised the diffusion of novelties such as "water, sewage, W.C., central heating," and "electric light." He shared the desire for technological modernization with his audience. One of his interviewees, a doctor from Eskilstuna, a small town not far from Stockholm, went so far as to claim that central heating would solve all sanitation, social, and economic problems and lighten housewives' daily burden. Indeed, such professional optimism was widespread among Swedish experts at the time, including engineers, teachers, and doctors.[59]

Modernization was an economic matter, and the Swedish state assisted those who wanted to repair or improve their houses. Subsidies did not, however, always reach the most underprivileged groups. Since the late nineteenth century, one widespread survival strategy for poor Swedish peasants and workers had been to move to the United States. Some returned home, in many cases bringing with them considerable savings and an outspoken willingness to invest in modern gadgets. A much publicized household item in those years was the AGA cooker, a large stove that Swedish Nobel Prize laureate Gustaf Dalén had invented in the 1920s. The intricate and expensive device had varying hot plates and different ovens for simmering, roasting, baking, and warming. Praised for its fuel efficiency, the stove had also found its way into Dutch and British country houses. According to a local politician whom Nordström had met in the Western part of Sweden, AGA cookers were originally bought by people of means. One returning emigrant bought an AGA, known as the Mercedes of stoves, because, "he thought it was so beautiful," not because he wanted the kitchen range for making meals. Rather than using it for cooking and baking, he had placed the stove—a symbol of his wealth—in the living room instead of the kitchen and only operated it to brew his "morning coffee."[60]

It was not a coincidence that the Swedish Radio Company, the Royal Board of Medicine, and the Stockholm Deaconry Board

Fig. 2.8 The Modernist, Feminist Home: *For decades, Alva Myrdal, pictured here, was a leading figure in what the Swedish Social Democratic Party referred to simply as "The Movement" (rörelsen). The Movement's vision of the future was defined by Myrdal, Sweden's Social Democratic Party, and an even wider transnational movement. Modern technology and modernist housing were central among their themes. In this utopian society, labor-saving household technology was meant to ease the burden of women's chores.*

exhibited such an interest in the population's poor living conditions. Demographic change—resulting from emigration and low birth rates—had been on the political agenda since the early twentieth century. It received new attention in 1934, when Alva and Gunnar Myrdal published the book, *Crisis in the Population Question*. Here, the young sociologist-economist couple detailed their plan for a modern welfare state in which women would be able to join the labor force without having to give up their desires to have children. Their plan mobilized a nineteenth-century international socialist-feminist tradition, dominated by French and German thinkers, who explored collective rather than individual solutions to help women negotiate home life and work, including laundry facilities, food-delivery services, and labor-saving household technologies. In the Myrdals' blueprint for Sweden, families had to be supported financially; nurseries and kindergartens had to be built and made economically accessible to working-class families as well; school children were to be given a free meal in the middle of the day; and housing standards had to improve to enable families to enjoy a reasonably comfortable life. To achieve all this, the state, local governments, and the cooperative movement had to act in concert.[61]

Housing took center stage in the couple's program. Four years before Nordström wrote "Filth-Sweden," the Myrdals had made it clear that the dire living conditions of the impoverished were detrimental to children. Their vision for the future did not focus on a country with scattered bungalows but newly constructed residential areas—a mixture of "cooperative services, apartments, and suburban developments," as Alva Myrdal later put it in *Nation and Family*. Their ideal was a modern home in a modernist apartment complex. Such multi-family houses, inspired by ideas of the American Christine Frederick, the German Erna Meyer, and the French Paulette Bernège, would include individual flats with small, rationally organized kitchens filled with "labor-saving gadgets," or designed in a way that certain tasks could be done collectively or by employed personnel. In fact, the Myrdal family lived for some time in such a collective house in Stockholm.[62]

The Myrdals envisioned a modern society based on a new gender contract. Their view of the future was also a modernist utopia. Their collective house in Stockholm had been designed by Sven Markelius, one of the leading Swedish modernist architects. Markelius was co-author of *acceptera*, a book that propagated international modernism in architecture, design, building construction, and urban planning. Published in 1931, *acceptera* was the outcome of a collective effort reflected in the international functionalist movement of the *Congrès International d'Architecture Moderne* (CIAM). Its authors included the most active champions behind the well-attended Stockholm Exhibition of 1930. Markelius helped found CIAM, a group of progressive architects and planners driven by a wish to solve social problems through urban planning and new technologies. Like the Myrdals, the modernists blended a utopian fervor with a dash of social engineering. Their goal was to create a new society by means of technology and social-reform measures.[63]

Among the most vocal planners connected to CIAM was Ernst May, city architect of Frankfurt am Main. Educated in London under the tutelage of Raymond Unwin, a pioneer of the garden-city movement, May belonged to the international avant-garde of modernist architects and urban planners. He actively contributed to circulating modernist concepts between Great Britain and the

Continent, as well as between the capitalist and communist parts of Europe. Two years before *acceptera* was published, May announced his ideas for "The New Frankfurt," a city that would care for all of its residents, including the poor. The document accompanied a number of social housing projects built under May in the second half of the 1920s, most famously the *Römerstadt*. In 1929, May proudly hosted CIAM's second meeting, an event that addressed "the minimum subsistence dwelling."[64] And one year later, he accepted an invitation from the Soviet Union to plan several new industrial complexes, the steel city of Magnitogorsk among them.

As in the case of *De Dageraad* and HSB projects, modern technological solutions were part and parcel of "The New Frankfurt." Central heating and running water were installed in every house. The city of Frankfurt proudly announced that the *Römerstadt* estate relied on electricity for cooking and water heating: "The main thing is electricity.... In the new houses it is the 'housemaid for everything': It makes the soup, fries the meat, bakes the cake, heats the bath and the dish-water—and, of course, it also brings light into the house.... We have America at our door-step."[65]

Despite its fame, though, *Römerstadt* was not a technological utopia. Nor was it a social utopia. Because of the housing's high utility costs, the vast majority of residents were not workers, but belonged to the lower middle class. Although they criticized the housing project in which they lived, the tenants did not question the ideology of modern technology and progress. May's political opponents phrased their concerns in this paradoxical manner: "Our present way of construction goes along with progress, but this kind of progress seems to us too expensive.... Perhaps it would be possible to save somewhat if we switched off modernity a little bit and made our buildings cheaper."[66] In the 1930s, May's opponents wanted progress, but they wanted it European-style: at the pace and for the price that reflected European values.

The situation would change again after the Second World War. The devastated European cities needed urgent rebuilding. The heirs of Haussmann would erase what was left of medieval street networks, creating U.S.-style "car-friendly cities." May and his colleagues across Europe seized the opportunity to erect modernist suburbs along the cities' outskirts, thus creating similar experiences for citizens on both sides of the Iron Curtain. And, as before, the lower classes would find themselves in housing structures—created by the state—whose design they could only selectively influence.

Cook's

CONTINENTAL TIME TABLES

TOURIST'S HAND BOOK

&

ECIAL · REFERENCE · TO · COOK'S · DIRECT · & · CIRCULAR · TICK

AND WITH

SECTIONAL MAPS

3
Crossing Borders—in Style?

Just as in fashion and housing, Europeans' experience of new transportation technology was governed by class during The Long Twentieth Century. Early in the 1800s passenger trains and steamships came on the scene as exciting innovations. Europeans across the Continent—and across class lines—made use of trains and ships for a range of reasons. The working class used the new transportation to migrate: more than 25 million people left Europe during the 1800s to start a new life overseas. Countless more migrated from the impoverished countryside to the industrializing cities. The middle classes boarded trains and ships for tourism. And the upper class deployed trains and ships to rule colonial lands, transplanting European imperial lifestyles. Indeed, trains and ships allowed passengers from practically all walks of life to enjoy the new travel freedoms. To the disdain of the middle and upper classes, though, the new technologies promised to blur class distinctions and co-mingle classes. But, in the end, one's travel experience of Europe and beyond depended almost entirely on the variables of class, gender, religion, and ethnicity. The very experience of crossing a border could be grueling or simple, depending on one's station in life. Train and steamship tickets came to be sold at various prices. The train and the ship thus reproduced class differences. And working class freedoms became tightly controlled. Each class encountered its own problems and anxieties. And for first-class travelers, the preoccupation became how to cope with the lower classes....

It was five o'clock in the morning on December 7, 1860; the train from Mulhouse had just pulled into the *Gare de l'Est*, in Paris. The body of the French Chief Justice Poinsot was found lying in a pool of blood, the victim's brains splattered across his first-class compartment. News of the murder spread almost instantaneously, setting off shock waves throughout Europe. All major European and American newspapers reported the brutal killing. Four years later, Thomas Briggs, a British chief banking clerk, met a similar fate as a first-class train traveler. Theft was the motive in both cases. In polite society's international press, the train entered the imagination as an ideal crime scene. For decades afterwards, commentators routinely identified European first-class train compartments as dangerous places. A passenger traveling alone, "[could] be exposed, by that solitary other fellow passenger, to all kinds of disagreeable things, possibly even robbery and murder, as has been shown, unfortunately by certain well-known events," a Swiss engineer wrote in 1866. A German colleague warned, in 1870, that a passenger could run into a shady figure, "who robs him in his sleep, or perhaps ever murders him, and then ejects his body from the compartment piecemeal, without attracting the train personnel's attention."[1] An American journalist pontificated in 1886: "Everyone knows beforehand that a sleeping man [in a European train] may be butchered in a railway carriage."[2] Inspired by these and other similar events, Émile Zola explored the human condition in *La bête humaine* (1890), typical of stories set in a train compartment. The wide circulation of such train-murder stories cultivated a deep sense of insecurity in Europe's wealthy passengers about the train as a mode of transportation. This unease extended to the new social mores that the mingling of classes, genders, and ethnicities in such public spaces required.[3]

The two iconic murders took place halfway between the 1840s and the 1890s, a period when railroad travel increased dramatically. These years marked a period of experimentation with standardizing an emerging public space to incorporate class, gender, and ethnic issues. Trains—rather than horseback, post, or private

Fig. 3.1 First-Class Anxiety: *Train passengers in first-class compartments faced many hazards, both real and imagined. Physically isolated in their private luxury cars, these travelers felt like "sitting ducks" for harassers, thieves, and even murderers. The European press readily publicized news of such crimes. Front-page news of a 1899 French paper, this report illustrates a passenger found dead in his first-class quarters. News stories like these fed upper-class fears about the consequences of "mingling" with the lower classes on trains. Debates raged over how to boost safety. To remedy travelers' isolation—while maintaining first-class standards—engineers offered plans to redesign train cars.*

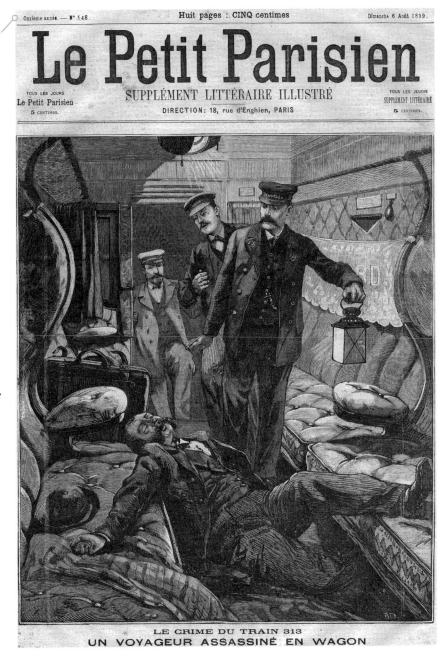

carriage—were designed for mass transit from the outset. This new public space prompted the establishment of classes in a practical and metaphorical sense. The class structures of society were reproduced in the four different "classes" of ticket and the amenities that tickets provided. On offer were various levels of comfort, service, and interiors in train cars, waiting rooms, public bathrooms, ticket offices, restaurants, and general train station facilities; each class was also connected with a different bureaucratic regime. In the pioneering years, railroad companies had served industry and business needs primarily, catering to the transport of goods rather than passengers. The situation changed as new user groups recognized the advantages of rail travel: speed, comfort, and a certain degree of anonymity. In particular, the explosion of organized tourism pushed railroads to consider travelers' needs. When acting as individuals, passengers could hardly shape the railroad system. Yet, customers managed to modify train designs to some extent; they tweaked travel regulations and improved passengers' rights through intermediaries. Travel agents, engineers, and legal scholars spoke up on behalf of passengers. So did emigrant self-help organizations and religious advocacy groups. Tourist clubs and commercial tour operators were most effective in mediating between the practices set by the railroad sector and the needs of passengers. These new professions and organizations had the collective power to negotiate with monopolistic railroad companies and state authorities.[4]

Here, we explore how trains were shaped to better fit users' needs and wants. The emerging tourist movement played a pioneering role in both reshaping Europe's geography and in ushering in innovations for railroads—as well as on ships. This is a paradoxical narrative: on the one hand, steam-driven vehicles and vessels contributed to the creation of transnational connections throughout Europe. On the other hand, these modes of transportation prevented passengers of different "classes" from having a common travel experience. The expansion of tourism within Europe and beyond depended on both trains and ships. These means of transportation were also essential in an area that may appear to be the opposite of tourism: migration. In this chapter, tourism and migration are juxtaposed to contrast the experiences of wealthy tourists and businessmen with the plight of poor farm daughters and Jewish emigrants looking for a better life in European cities or in the United States. Material gaps related to class, religion, gender, and ethnicity produced diverse technical infrastructures and social experiences. Transportation—like fashion and housing—was originally dominated by an aristocratic model that other classes tried to emulate. Accordingly, transnational railroads and steamships made promises when connecting distant places—promises of individual freedom and international fraternization that were at times fulfilled, and at other times denied.

Class & the Train Compartment

The train murders of the 1860s put the safety needs of (first-class) train passengers on the front pages of Europe's newspapers. Particularly frightening to the public was the isolation of the European first-class train compartments. The well-to-do judge Poinsot of the Imperial Court in Paris had been murdered without anyone hearing cries or gunshots. *The New York Times* quickly pointed to what observers saw as the problem with European trains: they were built without any "means of calling for aid."[5] The great public anxiety over the Poinsot and Briggs murders prompted a call for changes in train design. According to their critics, the design of European passenger trains (unlike its American counterparts) left first-class travelers insular and alone. Each train compartment had a separate exit door onto the platform without any connection between them. In an emergency, passengers could not communicate with other compartments and the railroad crew. To calm the public, the French and British governments mandated committees to consider numerous alarm systems that could solve the problem of first-class passengers' confinement. The French government's investigation of 1863 generated thirty-two schemes from engineers and inventors. One suggested running a voice tube throughout the train; another proposed a rope with a bell signal leading to the locomotive; and yet another inventor put forth a complicated system of mirrors to help guard all the compartments. The French government committee called for peepholes to facilitate visual communication between the compartments. Finally, one solution recommended mounting a footboard outside the carriage along its entire length so that crew members could move from one compartment to the next while the train was in motion. Eventually, the British government's inquiry into the Briggs murder led to an amendment to the Railways Act of 1868 that demanded proper devices for communication between passengers and train crew.[6]

In discussions surrounding the French and British murders, U.S. train interiors emerged as a new standard of comfort and safety. Six suggestions submitted in 1863 to the French railroad authorities after Poinsot's murder called for adopting the American example. Instead of secluded first-class compartments, U.S. trains sported seats on both sides of an aisle that ran the length of the entire wagon—allowing passengers and conductors to walk from one wagon to the next. Some carriages were even arranged as a parlor, where passengers could move around freely. The design of long, open, American-style coaches with their eight-wheel, double-truck car resulted in longer wagons, smoother motion, and fewer broken axles. These coaches were also cheaper to build. Promoters of the American exemplar believed that the design embodied democracy because travelers moved around freely and could approach each other. Despite notions of a class-bound Europe and a classless America, this idea was partly fictitious. As one U.S. commentator observed rightly in 1897, while in Europe there were first-, second-, third-, and even fourth-class carriages, the differences were mainly cosmetic: "Here [in the U.S.] we have the corresponding divisions without such harsh names: The Pullman is first, the day-coach second, the smoker third-class; and perhaps a seat in the caboose of a freight train may be called fourth-class."[7] American trains became increasingly stratified.

U.S. trains were also divided along lines of gender and race. Railroad companies installed special smoking cars for men, where joking, drinking, and spitting ruled. Catering to middle-class women and their male chaperones, designers created separate, private cars with refined accommodation: homes on wheels equipped with comfortable sofas, ice-water dispensers, and toilets. But the physical segregation into (working-class) smokers and (middle-class) ladies' cars proved problematic for at least two groups of travelers: genteel men traveling alone and middle-class African-American women. Both groups took to the courts to demand their right to travel in the more comfortable carriages instead of the filthy smoking cars. In practice, middle-class women, using Victorian codes of male gentility and female vulnerability to their advantage, increasingly displayed a confidence that dismayed social commentators. Most irritating to critics was women's expectation that men assist them in negotiating the discomforts of train travel. The idea that men should forfeit their seats to women when the train was full particularly peeved critics. In response, American railroad companies began to offer more secluded and comfortable places for men of polite society. These passengers could expect upgraded smoking cars with cigars and liquor, barber shops, divans, lounge chairs; library cars provided newspapers and writing desks.[8]

Fig. 3.2 Class Structure On & Off the Rails: *From the start, European trains reproduced society's class structures. First-class wagons [Top], in this British engraving from the mid-1840s, catered to the nobility and upper middle class. In contrast, the rest of the middle class [Center] settled for less costly, second-class compartments. The working class endured the discomfort of wooden seats of the third class. In the early days of train travel, some passengers even had to accept open wagons [Bottom]. Despite this, trains were vital to workers and farmers, day-laborers and immigrants: train travel afforded them new mobility and wider opportunity.*

European commentators came to appreciate the open spaces in U.S. trains as the best structural solution to the safety hazards of first-class passengers. Still, European railroad companies insisted that their first-class customers would never accept an American design. In the end, however, Europe's long-distance trains did adopt many U.S. design features. This freed European trains from their class-based roots of the coach era, which had avoided mingling with the masses. The American-style design allowed the conductor and train personnel to move more easily throughout the train in order to check tickets. The train murders had also highlighted the newspapers' power (if only fleeting) in publiciz-ing—and forcing authorities to respond to—travelers' concerns. The solution took the form of elite parlor, club, and smoking cars; private cars as well as exclusive sleeping and dining cars were less discrete and isolated, but still separated from other classes. This was the European variation on American safety and comfort that the company *Wagons-Lits* introduced in the 1880s. To the ordinary European transnational traveler, however, such luxury trains were far, far out of reach.[9]

Fighting for Passengers' Rights

Examining the rights of international train passengers, the young Swiss legal scholar Alfred Bonzon suggested in his 1896 disserta-tion that European travelers taking chartered, luxury, parlor, and sleeper trains were "hardly to be pitied." The first-class passengers in charter trains "[did] not notice changes of tracks," nor did they have to buy tickets along the way; wait for connecting trains; nor explain themselves in a foreign language. Bonzon observed: "Their only effort is getting onto the train and, once conveniently installed, killing time."[10] However, he noted, traveling in luxury chartered trains like the famous Orient Express *Wagon-Lits* was an exception. For most passengers, traveling in Europe was anything but smooth. During much of the nineteenth century, most ordinary interna-tional train users encountered multiple private railroad companies and an array of national rules and restrictions. This made booking international tickets complex if not confounding. More important, when travelers encountered problems, they had little legal ground to stand on. Bonzon was one of the few who tried to articulate and assert the rights of transnational train users.

Most of the time, of course, there were no murders to thrust the problems of train travel into the spotlight and thus attract support for train users. Left to their own devices, individual consumers faced railroads' spatial monopolies. And, notoriously, the railroads focused their legal efforts on cargo rather than on people. Under the lead of the Association of German Railroads (*Verein deutscher Eisenbahnverwaltungen*), rail companies throughout Europe in the 1870s began to harmonize rules for the transportation of cargo. Their goal was to ease cross-border traffic by coordinating intercompany agreements and by standardizing the seventy-five different administrative domains that covered eight countries. In the 1890s, the Swiss government took the initiative to unify legal arrangements across Europe. In any case, the agreements failed to protect international passengers when they transited countries and crossed borders.[11]

The typical international railroad traveler faced a mind-boggling variety of problems, all of which stemmed from the use of non-standardized systems. There were, for example, different track gauges and power grids; different regulations and tariffs—all impeding pan-European fraternization. Rail travel demanded a complex set of user skills, including knowledge about how to tweak rules and restrictions. For example, companies did not treat a cross-border trip from the Dutch city of Amsterdam to the Belgian town of Laeken, near Brussels, as an international journey. They lacked the administrative mechanism to issue direct international tickets from and to all existing stations (like Laeken). In the case of the Lausanne (Switzerland) – Paris (France) connection, only first- and second-class tickets were considered direct—and therefore international—journeys. This rule made it impossible for third-class passengers to book a direct trip. Passengers themselves had to figure out the local rules. In Switzerland, tourists could stop along the way for sightseeing, but in Germany, France, and Italy, the railroad companies maintained many restrictions on when and where customers were allowed to get off for a meal or some rest. Railways required passengers to register their baggage; fill out forms; obtain visas; and navigating national laws even when transiting a country without getting off the train. Planning train routes using notoriously complicated train schedules; calculating the many time zones along the journey—all of this required a new form of literacy. Reading foreign timetables and finding cross-border connections also demanded language skills.[12]

Luggage was another headache for transnational travelers. Most European railroad companies allowed customers to bring along small, "easy to handle" items that would "not annoy fellow passengers."[13] In principle, trains offered more room for luggage than before. During the era of the horse-drawn carriage, throughout the journey trunks were by necessity transported atop the carriage or at the back, outside the carriage; only smaller bags were carried along with passengers, inside the carriage. In contrast, railroad travel allowed passengers direct access to a larger number of personal belongings during the journey. In fact, the suitcase was an artifact of the railroad. Compared to carriages, passengers had more space in train compartments, but also had to change to make transfers more quickly; to negotiate these new opportunities and demands, travelers found carry-on suitcases advantageous. This meant immediate access to books and writing materials, extra clothing as well as food and drink. One traveler, August Hare, counted twenty-eight items of hand luggage among his seven fellow passengers in their train compartment; this excluded items in the baggage wagon. Manufacturers offered suitcases in various sizes, along with specialized products such as "patented ladies' hat-cases."[14]

The tightest regulations limited passengers to carrying one "small" suitcase aboard. For passengers changing trains and crossing borders, it was unclear how each railroad company defined "small," "easy to handle," or "inoffensive." No standard maximum weight existed for hand luggage; in Switzerland the limit was 10 kilos, while in France and Belgium it was 25 kilos. Sometimes, one had the right to bring along dogs and small birds in the compartments; at other times, no pets were allowed. The rules for registered luggage (checked baggage) transported in separate wagons also differed from one country to the next and from one company to the other. The Prussian-Hessian Railroads allowed tourists to register wheel chairs, baby buggies, musical instruments, working tools, and bicycles, as long as they were for personal use only. British companies used stricter criteria for carry-on luggage; the Swiss were obstructive when it came to bicycles: they did not accept bicycles as personal items and refused to register them as luggage.[15]

Most persistent was the utter lack of international consumer protection, as legal scholars discovered. In his dissertation, Bonzon, who would go on to work for a French railroad company and the League of Nations before becoming a notable Swiss politician, found that while an integrated transnational agreement governed the transportation of goods, passengers enjoyed no such international protection. Goods were apparently more important than people. The average international-train user was in dire straits. As Denis de Leeuw, Bonzon's Dutch colleague argued, it was quite paradoxical that goods being transported from Paris to St. Petersburg fell under the 1890 Berne Convention and could cross the Continent without disruption, whereas passengers traveling the same route had to submit to demeaning—and often contradictory—national and corporate regulations. Legal loopholes could result in remarkable differences in how people and things were treated. Commercial goods and mail packages shipped over the same railroad line often physically shared the same railcar. While mail and goods traveling from Geneva to Amsterdam were thus exempt from customs inspections and border controls, passengers' registered luggage failed to enjoy the same freedoms. Indeed, De Leeuw implied that, legally speaking, train companies treated a bag of potatoes better than a human being. He cataloged the many difficulties stemming from this regrettable legal void. Throughout the decades, Swiss, Belgian, and Russian authorities tried to improve the situation, eventually putting passengers' rights on the international agenda. They maintained that passengers' rights rules should be modeled after the international cargo and mail agreements. France and Germany routinely sabotaged their efforts.[16]

Many obstacles for passengers persisted well into the interwar period; even then, protection was substandard. Train users remained at the mercy of both national and corporate rules and customs and border-crossing regulations. In some places, passengers were required to be present personally for baggage inspection on the platform; in other places they were not. For the lucky traveler, customs control and the inspection of luggage took place en route in train compartments. Less fortunate passengers—even those with direct tickets in hand—had to get off and have their luggage inspected inside the station out of sight of the train, risking missing their connections. Authorities took no responsibility for delays and lost connections that arose from border inspections. In case of loss of life, accidents, or delays, the burden of proof lay

Fig. 3.3　Crossing Europe's Borders: *For luxury-train passengers, border crossings were indeed routinely easy; inspectors boarded trains to streamline the process. For less fortunate passengers, border crossings were disruptive—demeaning, even. At inspection centers located inside train stations, customs officials rifled through personal belongings, rooting out prohibited and taxable items. Ever-present was the threat of missing connecting trains and ships—and being left behind in a traveler's no-man's-land.*

with passengers, who were left on their own to fight in local and national courts. Given that there were no strong user movements or consumer organizations in place at the time, plaintiffs were compelled to employ their own lawyers. Even those travelers who had the time, energy, financial resources, and language skills to navigate local jurisdictions often lost their battles in court.[17]

As late as 1936, international travelers were advised to take precautions to avoid legal battles that they were bound to lose. The warning came from Charles Ackermann, a transport consultant and diplomat in Geneva, the world capital of transnational agreements. At a minimum, Ackermann recommended that passengers check to see whether their trunks and packages were still in the right wagon after customs inspections and when train companies swapped railroad cars. The best protection tourists could buy, he insisted, was the help of a travel agency, which provided accurate information, negotiated the best deals with transport companies and hotels; and offered travel and baggage insurance. He insisted passengers had to protect themselves as much as possible, because rail companies refused to take responsibility. Luggage insurance offered another vital form of protection for the international traveler, until proper

Fig. 3.4   The Liability of Train Travel: *Journeying through Europe, train passengers encountered a host of problems, from missed connections to accidents and lost luggage. Train companies tried to maintain the legal status quo—and limit their liability for international passengers. International laws of the 1890s protected freight first— not people. For protection, passengers had to rely on third parties such as insurance companies. The Budapest-based European Goods and Travel Luggage Insurance Company (referred to in this ad as "l'Européenne") offered its services in almost all countries across the Continent.*

Avant de partir en voyage
n'oubliez pas de conclure une
Assurance des bagages
et contre les **accidents de voyage**
de „ **l'Européenne** "

En cas de dommages, vous pourrez adresser votre réclamation
à l'une quelconque des

„EUROPÉENNES"

à Amsterdam, Bâle, Belgrade, Berlin, Bruxelles, Bucarest, Budapest, Copenhague, Helsingfors, Lisbonne, Londres, Madrid, Milan, Munich, Oslo, Paris, Le Pirée, Prague, Presbourg, Stockholm, Varsovie, Vienne ou Zagreb.

user protection was in place. From 1907 on, it was possible to buy an insurance policy for the international transportation of personal luggage. For those who could afford it, the "European Goods and Travel Luggage Insurance Company" offered its services in several countries. This Budapest-based company opened its first offices in Berlin, Milan, Sarajevo, and Monaco. It soon acquired a monopoly position in many countries. The cost of the insurance depended on the distance traveled and, of course, the luggage's estimated value; typically, it amounted to less than one per cent. Ackermann warned transnational travelers to be well prepared for rail companies' shamelessness: they showed no good faith whatsoever in settling claims, ruthlessly engaging "in genuine attempts at blackmail" if they believed that the traveler might be at fault. Ackermann hoped that the courts would clamp down on any railways failing "in their duty and compromising the life or interests of their clients."[18] Indeed, throughout the nineteenth and the early twentieth centuries, rail companies succeeded in limiting their liability longer than other transport industries when it came to international passengers and their luggage. While the shipping and the airline industries recognized their collective responsibility respectively in 1910 and in 1924, not until the 1980 Berne Convention did railroads guarantee passengers full compensation for loss of life and luggage when travelling internationally.[19]

Middle-Class Infrastructures

Members of the aristocracy and the middle classes grew more and more frustrated with railroad companies' inability to operate and cooperate across borders. To help solve the problems, private tour operators and travel agencies, as well as insurance companies, positioned themselves as mediators between individual travelers and train and shipping companies. In fact, it was shipping companies and entrepreneurs connected with shipping that pioneered organized tourism. Their activities co-evolved with the steamship. Most likely the first person to offer tours with a paddle steamer was a certain Henry Bell. Intrigued by the possibilities of steam traction, Bell had *The Comet* designed in 1812 to transport customers from Glasgow down the river Clyde to his hotel and spa in Helensburgh. In the following decade, Londoners learned to appreciate regular and fairly predictable trips along the river Thames and even beyond. By 1840, more than one million tourists per year arrived by boat in Margate on the North Sea, about 70 miles east of London. Increased speed brought travel time down from 12 hours in 1815 to an hour and 45 minutes, which allowed a round-trip to Margate in one day. The option was especially important for workers, who usually only had Sundays for leisure.[20]

The business of organized tourism contributed to the formation of transnational bonds and, in part, a sense of shared experiences among Europeans. The British were especially active at the outset: in the 1860s an estimated 300,000 travelers crossed the Channel annually; by 1900 the figure had risen to almost one million. Paris and parts of Switzerland were their most popular destinations. For one thing, vacationing on the Continent was less expensive for the British than taking a holiday at home. Moreover, compared to Britain, which had industrialized at a frantic pace, daily life on the European Continent was slower and less "modern": to vacation there was to experience a deeper authenticity. In the eighteenth century, sons of the British aristocracy and American *nouveaux riches* had already begun to tour Continental Europe—a more or less mandatory journey of discovery and learning. Their grand tour began with the White Cliffs of Dover, where Channel boats transported passengers to the Flemish port of Oostende. Then it was on to the culture capital of Paris, followed by visits to the sublime nature of Switzerland, the Renaissance art of Florence and the antiquities of Rome. Along the way, the scions of the British and the American well-to-do purchased souvenirs; collected art; picked up a language (usually French); and socialized with local peers. Such proto-tourism provided instant gratification and made the most of their leisure time. Even before the advent of trains, British middle-class men and women passed through Paris to reach Switzerland on shorter trips. By the mid-1800s, their tours took two months rather than two years, and new experiences were included in the usual itinerary. Following in the footsteps of artists like Turner, authors like John Ruskin, and various mountaineers, the well-heeled enjoyed the alpine landscape at Chamonix and Zermatt or stopped at the Dolomites. Still, such leisure travel still had to be validated socially by a more convincing pretext: a search for spirituality, culture, or health justified visits to churches, ruins, and spas.[21]

The tourist exploitation of the Rhine Valley and Switzerland had begun after Waterloo, practically as soon as the guns fell silent; this was before the era of the railroad. In its own way, tourism contributed to healing the wounds of the war-torn Continent. In 1816, citizens on the river bank in Cologne were dumbfounded by their first sighting of a British steamship. Inspired by the "twenty-four highly finished and colored engravings" published by Baron J.J. von Gerning, and influenced by the writings of Lord Byron and Mary and Percy Shelley, British tourists cultivated a romantic view of the Rhine, its surrounding castles and ruins, and the famous Lorelei rock. By 1832, passengers could travel all the way to Basel by steamer. In contrast to the difficulties that would later plague trans-border railroad traffic, shipping along the Rhine went smoothly. Immediately after the Napoleonic Wars, an international commission had been set up to regulate Rhine navigation, and in 1831, seven abutting states signed The Act of Mainz. The document removed local privileges, reduced duties, and simplified passage for freight and passenger boats. British tourists in particular profited from these arrangements.[22]

In the Austrian-Hungarian Empire, the combination of steam and rail enabled upper-class tourism. The Austrian railroad company, Lloyd, had opened the first line along the Dalmatian coast in the late 1830s, taking passengers from Trieste to Split and further south to Montenegro. Still, tourism in the area did not take off until almost half a century later. Contributing to its breakthrough was the inauguration of the Vienna–Trieste rail line in 1857, which made it possible for the capital's upper classes to reach the coast within a reasonable amount of time. But train tracks alone did not allow tourism to flourish. Only after *Lloyd Austriaco* and the so-called Southern Rail Company invested in luxury hotels, fancy restaurants, sanitation facilities, and parks, did a rich clientele arrive in greater numbers. The town known as Opatija in Croatian, Abbazia in Italian, and Sankt Jakobi in German emerged as an attractive destination for well-off international visitors. In 1867, then, Lloyd followed suit and erected a luxury hotel in Dubrovnik, fittingly named "The Imperial."[23]

Back in Britain, Thomas Cook & Son had, in the meantime, pioneered domestic rail tourism for the working classes. Thomas Cook was a cabinet maker who was anti-alcohol. He devised organized tourism for workers and their families as a form of social uplift. In 1841 he negotiated with the Midland Counties Railway to secure affordable tickets for five hundred Leicester temperance activists. In the process of arranging for the activists to attend meetings in Loughborough, Cook discovered that railroad companies would allow groups to travel more economically. Building on this initial success, Cook started to organize train excursions to English

**Fig. 3.5 Early User
Manual:** *Issued under
various titles,* Cook's
Continental Time Table
*began to simplify life for
middle-class European
travelers and tourists in
1873. Early editions of
the handbook contained
timetables plus maps
and other valuable
information. Planning
itineraries, plotting
train connections, and
negotiating international
logistics required
considerable skill on the
part of consumers. Cook's
"user manual" helped
people to navigate the
complex travel corridors.*

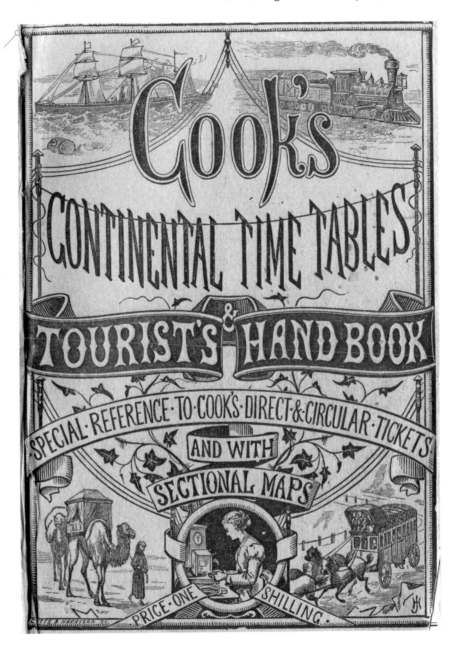

coastal resorts and to Scotland. To make sure that his working-class
clientele noticed the sites along the way, Cook provided guide
booklets written in easy-to-understand prose. His great commercial
breakthrough came with the tours he organized to the 1851 Crystal
Palace Exhibition, an event attracting six million visitors from all
over the world.[24]

In contrast to middle-class travelers, who preferred an individual, private culture that shunned group and communal activities, workers appreciated Cook's collective form of traveling. Cook
assisted working-class clubs in getting affordable deals through
advance bookings and bulk purchase of transport and accommodation. He also helped clubs to organize services, guides, and
excursions in the hope of enhancing the sense of familiarity and
community. There were other workers in search of like-minded
travelers. Intrigued by the prospect of meeting fellow socialists
from other countries, working-class clubs themselves organized
trips to the Great Exhibition. These pioneers of the tourist movement negotiated time off with employers and lower charter prices
with the railroad companies. Once in London, the parties needed
accommodation at a reasonable price as well as inexpensive food
that was familiar. The success of these working-class clubs and
enterprising operators like Cook (and his closest competitor, Henry
Gaze) showed that collective actors could wield considerable
purchasing power when buying in bulk. In fact, such passenger
groups functioned like latter-day consumer organizations.[25]

As Cook himself explained, expansion throughout Europe
was seldom easy and required a certain degree of flexibility. In
arranging deals, Cook's firm, "had a hard fight with continental
companies; and it required unceasing vigilance to keep on the good
side of hotel keepers, money changers, booking clerks, and others
with whom we had pecuniary transactions." Most mindboggling
were the "monetary perplexities" of Europe's fluctuating currency
rates; "the wretched and uneven appearance of coins and notes;
the conglomeration of francs, centimes, thalers, gold and silver
groschen, pfennigs, florin and kreutzers."[26] Similarly confusing
was Europe's linguistic diversity. Cook's employees negotiated
with partners in different countries and languages on behalf of
their British clients.

Although Cook was not a socialist, he echoed the labor movement's value of mutual understanding across national borders.
Nevertheless, his clients—deprecatingly called "Cookies"—
sometimes met with incomprehension, even disgust. In the 1860s
Charles Lever, a highbrow Irish novelist and vice-consul for the
United Kingdom to the Ligurian city La Spezia, attacked modern
mass tourism: "Tribes of unlettered British…come over, not in
twos or threes, but in scores and hundreds." Assuming the Italian

point-of-view, Lever wrote of tourists: "They deride our church ceremonies, they ridicule our cookery, they criticize our dress, they barbarize our language." The obvious butt of his outrage was mainly working-class tourists, whose "gross ignorance is the very smallest of their sins."[27] The liberal *Illustrated London News* took issue with such class-based sneering, claiming that Cook travelers did not differ from others, "save they were better behaved and more anxious to acquire information" than the "Traveling Gentlemen" from the "stuck up" classes.[28]

In the 1860s, Cook moved beyond working-class trips. He began to organize tours to Continental Europe for a middle-class clientele: doctors, lawyers, accountants, clergymen, shopkeepers, and professors. His company took visitors to the established travel sites of the British privileged classes: Paris, Cologne, Luzern, and various Italian cities. The tours were especially popular with middle-class women. His success was based on the principle of turning over the product as quickly and cheaply as possible, while incurring minimal transaction costs based on economies of scale. The Grand Tour had been a rite of passage for young aristocratic men; it was a ritual designed to cultivate individual maturity and foster political networking.[29]

Cook upended the aristocratic notion of the Grand Tour: he integrated accommodation, meals, and itineraries through unknown cities into a pre-arranged package. Under his guiding hand, travel became a closed system of standardized experience. To smooth the bumps in the road, his company mobilized all kinds of transportation and communication technologies and invented such novelties as traveler's checks and hotel vouchers; and it pioneered the publication of international timetables. The company reached more people from different class backgrounds, and went further geographically than other tour operators. As *the* package-tour pioneer, Cook effectively began to modify the topology of Europe by means of an array of social and technological innovations.[30]

Imperial Tracks

Compared to Thomas Cook & Son, Georges Nagelmackers' *Compagnie Internationale des Wagons-Lits* served a much wealthier clientele. The luxury train-travel company, founded in the 1870s, introduced what would come to be known as the Orient Express. At the time, most rail companies still operated cars without lavatories, restaurants, bars, vestibules, heating, or artificial light, and sported only hard seats. In contrast to this bare-bones approach, Nagelmackers offered aristocratic and wealthy customers a package deal of dining, lodging, and entertainment. This was one seamless experience—without the hassles of changing trains or waiting. *Wagons-Lits'* famous kitchen and dining cars came to symbolize the efficiency and luxury of frictionless travel. Like Cook's travel products, the *Wagons-Lits* package was standardized to give clients a predictable experience; unlike Cook's deals, the upper- and middle-class package was based on the individual journey rather than a communal travel model. Both companies thus helped to internationalize train travel and promote the feeling of a shared experience (along class lines) among their clients. It is questionable whether these efforts helped to dismantle prejudices and foster international understanding, as tourist operators claimed and advocates of Europe's infrastructures hoped.[31]

Nagelmackers was a mining engineer and the son of a wealthy Belgian banking family that was connected to the royal family. In 1870, he traveled extensively in the United States. In fact, the example for the Orient Express—the long-distance parlor trains that incorporated sleeping cars and restaurant cars—came from the other side of the Atlantic. There, railroad services linked the formerly disconnected U.S. territories from coast to coast, forming a new, continental nation state. Impressed by the luxury Pullman carriages, Nagelmackers decided to transfer the model to Europe. Having tried and failed to mobilize American capital and know-how, the entrepreneur experienced a breakthrough only after he exploited his family's business connections with King Leopold. On behalf of *Wagons-Lits*, the Belgian Head of State opened the doors to Europe's royal houses and politicians to secure concessions in the numerous railroad jurisdictions. This international network of royals, aristocrats, and bankers was the key to Nagelmackers' success in acquiring rights of way and providing the access to various destinations, as well as in attracting a wealthy clientele. In the 1870s, the company headed for the Ottoman Empire and in the 1890s expanded beyond Europe's borders, before consolidating its operation in the period after 1905 until the First World War.[32]

Fig. 3.6 Tools of Empire: *This 1890 poster glamorizes Europe's famed "Orient Express" experience, the travel extravaganza from London to Constantinople in true imperial style. But the train, as a technology, did much more than sate the appetite for luxury travel. Along with telegraph lines, railways enabled faster communications, plus the transport of personnel and goods. This meant simpler control—and exploitation of—colonial lands. Railroads also brought businessmen, soldiers, and tourists to terminal harbors; here they embarked on ships bound for Africa, Asia, and South America. Acknowledging the technologies' power, historians have named telegraph lines and railroad tracks "tools of empire."*

Wagons-Lits was a charter company; the firm did not own any tracks. When setting up international routes, Nagelmackers had to negotiate with up to twenty different individual states and railroad companies. For one single line, it could take up to ten years to arrange for rights of way, transits, and border crossings. Success hinged not only on Nagelmackers' pan-European aristocratic network, but also on Europe's relative political stability between 1871 and 1914. In the absence of an international regulatory framework for international passengers, *Wagons-Lits* representatives participated in international railroad conferences and diplomatic negotiations. After the First World War, the company even took part in the Versailles negotiations.[33]

Wagons-Lits' international negotiating power was legendary. For example, until the early 1880s, because Pullman held exclusive rights to Italy, British passengers traveling through that country to the port of Brindisi had had to transfer in Bologna from *Wagons-Lits* to Pullman carriages. By deploying British diplomatic pressure on the Italian rail company, *Wagons-Lits* in 1883 managed to push out its competitor. From then on, its customers could travel without interruption to Brindisi, where ships waited to welcome passengers to India. Through the mediation of *Wagons-Lits* and Cook, luxury train passengers gained a range of traveling privileges that ordinary international passengers were denied. For example, customs officials came on board luxury trains instead of tourists disembarking at borders. *Wagons-Lits'* first Orient Express poster in 1888 boasted that travelers could go all the way to Constantinople without changing trains—and without a passport between Paris and Bucharest. When headed for Egypt, tourists could buy tickets for the boat service to Alexandria either in advance from a travel agency or on board from the train steward. Escaping cold and foggy winters at home, British tourists traveling to the French Riviera or further south, to Rome, could avoid the interruptions of customs. "Border patrols enter the wagons for inspection," another poster advertised in 1884; and when traveling from London to Lisbon to connect with the steamship service to South America, passengers' luggage could be registered through to the final destination.[34]

Wagons-Lits' long-distance express services as well as Cook's transcontinental tours were fully embedded in colonial ventures radiating outwards from Europe. Cook helped transform European railroad networks and shipping lines into corridors of empire, all the way to Egypt and the Middle East. When faced with competition in Europe, the founder's son, John Mason Cook, led the company's expansion in Egypt. He transformed the family business with its mission of social uplift and small profits into a modern global corporation intertwined with the rise and fall of British imperial expansion. In the Egypt of the 1880s, Cook faced competition from French and Swiss travel agents who had already set up a resort infrastructure, including lodging facilities, promenades, and an army of waiters, chefs, nurses, doctors, and tour guides. To provide his clients with more comfort, Cook went further. He integrated continental train and Mediterranean steamship services with Nile transportation, accommodation, guides (by the hundreds), porters, and servants all over the country. By 1890, one third of Cook's employees were working in Egypt and Palestine. Approximately half of the profits came from the Nile fleet, which included six passenger steamers, four mail steamers, and fifteen *dahabeahs* (sailing houseboats). Cook's enterprise and British colonialism became intertwined in other ways. For example, the company made great profits from shuttling the British army to the Egyptian-Sudanese border in 1884; and the tourism business there expanded once the British consolidated their control over Egypt and Sudan. Cook built its own fleet of steamers and a chain of hotels, including the Cataract in Aswan, south of Luxor. In turn, the many British tourists who had been to Egypt on Cook tours helped to sway skeptical public opinion in favor of British rule on the Nile. European tourists' interest in seeing Egypt had been piqued by the popularity of Egyptology as a scientific field; by travelogue publications; and by the wide circulation of images (photographic and stereoscopic) of ancient Egyptian treasures. Cook's customers went to Egypt to experience a place they had already glimpsed in their home parlors; they sought an ancient world, which screened out modern Egypt and the lives of local people.[35]

In addition to Thomas Cook & Son, *Wagons-Lits'* long-distance express services also linked Europe to faraway territories. The trains shaped the colonial experience and framed passengers' ways of seeing. From 1884 onwards, the luxurious Orient Express departed daily from the Paris *Gare de l'Est* and sped through Strasbourg, Vienna, Budapest, and Bucharest to the Bulgarian port of Varna—the only interruption being a ferry trip across the Danube. In Varna, passengers would transfer to the ferry to Haydarpassa over the Black Sea on their way to Constantinople. The Orient Express served as Europe's gateway to the Middle East. Similarly, the *Nord* Express connected Paris with St. Petersburg, the capital of the Russian Empire, and the *Sud* Express linked Paris with Lisbon, the Portuguese Empire's Atlantic capital. Lisbon provided European travelers with a gateway to ocean liners serving

Madeira, the Antilles, Cape Town, and South America. *Wagons-Lits* operated as a *de facto* travel agency for most European royal houses in the decades before the First World War (which devastated the aristocratic way of life). The world that Nagelmackers created for his royal and aristocratic patrons was a Europe without borders. Before the Treaty of Versailles created new nation states and generated firm borders with more rigid passport controls, this world served the needs and interests of transnational aristocratic elites.[36]

Strategically placed hotels at the edges of Europe's continent served as gateways to the colonies. Between 1891 and 1914, *Wagons-Lits* built luxury Terminus Hotels at key locations. The company's *Hôtel de la Plage* in Oostende, Belgium offered a stop on the way to boats across the Channel and linked British travelers to train connections deeper within Europe. The Avenida Palace in Lisbon connected the railroad with Atlantic shipping lines to South America. The Grand Hotel International in Brindisi was the stopover for British travelers waiting for ships to Bombay. *Wagons-Lits* also built hotels in the Atlantic port of Bordeaux and the Mediterranean port of Marseilles, connecting the French motherland with its colonies. The company helped shape extra-territorial places outside Europe's geographical boundaries, linking many "Little Europes" around the world. In the European Quarter of the Ottoman capital Constantinople, for example, the stately Pera Palace Hotel overlooked the Golden Horn, the border between Europe and Asia. In Cairo, the Ghezireh Palace served as a pre-boarding meeting place for European elites who were to take trans-African trains to the Cape in Southern Africa. And in Beijing, the train company's *Grand Hôtel* offered the only Western lodgings in the Foreign Legations quarter —the extra-territorial entity which Western powers created on the edge of the crumbling Middle Kingdom, beyond the reach of Chinese law. At this hotel, against the backdrop of the weakening Chinese Empire, German, French, Belgian, and Dutch residents built a European identity, at times forgetting national differences. The railroad hotel offered a European ambiance in the Forbidden City, a few minutes away from the train station. Another outpost, the *Grand Hôtel de Pékin*, was known for its fine French cuisine and wine. The orchestra played classical music, and the Thomas Cook travel office was situated in the lobby to help tourists negotiate train schedules, day trips, and accommodations. The station and the *Grand Hôtel de Pékin* best symbolized Europe's claim on Western access and investment in China. The Terminus Hotels likewise represented a connection between international train and shipping services.[37]

The *Wagons-Lits* created a uniform social space and experience where the *beau monde* could glide from one European capital to the next without interruption from—or confrontation with—other classes. No smells from local farms, no sweaty workers, no restive Balkan revolutionaries. Undesirable sights could be screened out literally and figuratively: on the train from Paris to Constantinople (Istanbul), *Wagons-Lits* instructed the conductor to draw the blinds at one point in Serbia because country women were bathing in the nude.[38]

People from different classes traveling along the same train tracks nevertheless experienced different versions of Europe. The same lines the charter luxury trains followed also carried the second-class and interregional trains that allowed East European intellectuals to travel in the opposite direction. Those voyagers sought to sample West European culture or to discover a multi-ethnic Europe. Most could not afford *Wagons-Lits'* comforts and had to endure a slower, more fragmented rail system. In their social space along the journey, they experienced another Europe: one of diversity. Mingling with other groups, these intellectuals may have undercut the intentions of the Austro-Hungarian authorities, who slowly but surely began to consolidate the different multi-ethnic parts of the monarchy. The Czech intellectual Václav Staněk, for one, enjoyed railroad travel as a way of seeing the newly-connected Slavic world. Specifically, Staněk appreciated meeting uneducated German-speaking Czechs and Magyarized Slovaks who, while they spoke Slavic languages with great difficulty, were becoming a part of this new Slavic world. The world of *Wagons-Lits*, by contrast, was a Europe of monumental capitals built in the image of Paris and its urban planner, Baron Haussmann. This was a Europe in which the countryside could be consumed at a safe distance, through the filter of a window pane. The *grandes lignes* of the francophone *Wagons-Lits* and the package tours of the British Thomas Cook Company—together with their chains of international and terminus hotels—created a seamless system of comfortable corridors. These corridors ushered selected groups of tourists to—and through—faraway colonial lands, offering a transnational experience of Europe. Even more fundamental, when arriving from areas plagued by cholera and other epidemics, first-

class (and second-class) travelers learned how to avoid the humili-
ation of medical detentions. Well-heeled international passengers
could avoid the intrusive methods that officials used to disinfect
lower-class passengers and their luggage. Those crossing Europe's
borders in third or fourth class could not.[39]

Emigrant Corridors

The conflict-free world of the aristocracy and the *haute bourgoisie*
had their opposites: the stream of young men and women from
rural Europe lured to the cities in search of better lives. Europe's
great internal migration generated its own infrastructures. In
1877, religious groups founded the International Union of Friends
of Young Girls, a branch of the International Federation for the
Abolition of Prostitution. In countries throughout Europe, their
"Station Assistance" staff (*Bahnhofshilfe*, *Oeuvre des Arrivantes*,
Stationswerk) offered special support to young women traveling
alone. To earn a living, these young women came to the cities from
farms, villages, and rural towns. Arriving at train stations and
ports, they were easy prey for pimps looking for new recruits. The
religious organizations built missions at stations and ports; these
sites were the gateways for those in transition. Equipped with
booklets, badges, and armbands, the women of the Protestant,
Catholic, and Jewish associations provided a wide-ranging support
network for young travelers. On offer were practical advice, tips,
and free lodging to prevent them from "falling off the right track"
and drifting into prostitution. Through their posts at train stations
and ports, missionaries tried to forge what they believed to be a
safe future for travelers migrating from Europe's impoverished
regions to the city.[40]

The stream of transatlantic emigrants—and their experiences—
represented yet another version of Europe. One Russian-Jewish
teen-aged girl, Mashke Antin, described how, in 1894, she, her
mother, and her sister managed to flee from their hometown of
Polotzk to the western border of Poland's Russian section. With
valid passports and prepaid third-class train tickets, they set off
for the port of Hamburg, en route to America to join Mashke's
father. The first hurdle was the transit through the German Empire.

At the German-Russian border, in the town of Eydtkuhnen, they were stopped by German gendarmes and forced off the train. Their passports were confiscated. Only after the intervention of a Jewish relief organization and an act of bribery was the Antin family able to board a cramped fourth-class immigrant train. But first their luggage had to be "steamed and smoked" at the railroad depot at Keebart—at their own cost. Then they were rerouted to Ruhleben in a deserted area just outside Berlin. The station had been built in 1891 as Germany's central hub to keep transmigrants out of sight of Berlin's busy train stations.[41]

Ruhleben train station was part of an intricate distribution system that ensured immigrants were channeled through the

Fig. 3.7 Enforcing U.S. Immigration Law—in Europe: *En route to Ellis Island, European emigrants were subject to U.S. law— while still on European soil. This German engraving from 1895 depicts one of Europe's networks of stations built to disinfect emigrants and their belongings. The station interior shown here was in Ruhleben, outside Berlin, hidden from the city's commuters. The pipes pictured in the background emitted the noxious disinfecting vapors. The German text reads: Disinfection Location for German Lloyd [Steamship Line]. This image represents the standard operating procedure for all U.S.-bound third- and fourth-class emigrants.*

Vor der Desinfektionsanstalt des
Norddeutschen Lloyd.

country without defecting or coming into contact with "regular" train passengers. Here, German officials in white coats rushed the migrants off the trains and into a small building. They separated men, women, and children, and threw their luggage on a massive pile. The terrified migrants were forced to undress and undergo disinfection in a shower installation before they were allowed to board another (even more crowded) fourth-class train to Hamburg. Like other Russian families, the Antin family was involuntarily quarantined for two weeks in prison-like conditions. Held in an area with high walls and barbed wire, the detained travelers were forced to take baths and to be vaccinated—again at their own cost. Upon arrival at New York's Ellis Island, the immigrants had to endure yet another medical, financial, and social screening. The German disinfection stations at Ruhleben and Hamburg were part of an international system of cleansing procedures and checkpoints. At any one of these hubs, families could be broken up; as family members could be deemed unfit and turned back.[42]

In the course of the nineteenth century, the movement of people and goods meant freer international travel and, for some, even mobility without passports. Train and steamship companies offered new opportunities for mobility, but administrative and physical walls were erected to control migrants, with delousing stations to disinfect them. Emigrants faced many obstacles. A head of household from feudal lands who wanted to emigrate had to get an official letter to free him from his obligations, usually paying a substantial fee. He also needed a recommendation letter from the local church stating that he was in good standing. A permit certified that bills were paid; affairs in the community were settled; and he and his family were free to leave. Passport in hand, the family was allowed to cross district, provincial, and country boundaries. Sometimes, the permit and the passport were combined into a single exit visa issued by the authorities. Yet these administrative hurdles were only the beginning. As many emigrants found out along the way, the fourth-class and third-class train and shipping tickets they had bought in their hometowns failed to guarantee a trouble-free ride. At every step of the way, emigrants had suffered demeaning cleansing procedures. Countries constructed medical inspection and disinfection stations to cope with mounting numbers of germ scares, and private companies did the same to conform to American immigration restrictions.[43]

The German *Kaiserreich* succeeded in preventing immigration from Eastern Europe before the First World War. Instead, the empire became the primary transit country for emigrants on their way westward to the Americas. From the 1890s until 1914, migrants from Eastern Europe traveled by special, sealed trains through Germany to harbors in that country (Bremen, Hamburg), the Netherlands (Amsterdam, Rotterdam), France (Le Havre, Cherbourg), Belgium (Antwerp), and Britain. Migrants from Czarist Russia could avoid Germany by departing from the Baltic port of Libau for the ports of Britain (Hull) and Scandinavia (Copenhagen). The Austro-Hungarian authorities tried, without much success, to promote the ports of Trieste (Slovenia) and Rijeka (Croatia) for Habsburg subjects going overseas. Trains first brought emigrants to Atlantic port cities for ocean-liner connections to receiving countries like the United States, Canada, and Argentina. Between 1880 and 1924, more than 25 million Europeans crossed the Atlantic in order to reshape their lives.[44]

In collaboration with state authorities, shipping and rail companies established a new infrastructure of containment and medical routines further to control migration. The regime confined emigrants to sealed trains, quarantine quarters, disinfected ships, transit camps, railroad junctions, and ports. These transit points were staffed with policemen, clerks, inspectors, medical workers, and interpreters. The Europe that the emigrants experienced was one of isolation; its topology was one made up of straight lines and transitory junctions. To overcome the difficulties, emigrant groups established self-help organizations to come to the travelers' rescue along the way, at posts near all transit points. The relief organizations operated transnationally. As the modes of transportation expanded internationally, so did the ruling classes' fear of diseases and the numbers of "undesirables" crossing borders. An exodus of East Europeans in response to harsh politics and pogroms and the epidemics (typhus and cholera) of 1892 (in Hamburg and New York) resulted in stricter immigration laws, especially in Germany and the United States. Tougher medical controls at Europe's transit and exit points—as well as the U.S. entry points—were also put into effect.[45]

Faced with the anti-immigrant policy in Prussia and tightened U.S. immigration laws, the German shipping company, *Hamburg-Amerikanische Packetfahrt-Actien-Gesellschaft* (HAPAG) established a public–private partnership with the Prussian state to protect its dominance of the East European passenger market. To meet the strict demands of the U.S. immigration authorities, the company built a transnational corridor for migrants on their way to the other side of the Atlantic. After the first wave of five million, when German emigration began to ebb in the 1880s, HAPAG and its competitor, *Norddeutsche Lloyd*, found profitable new markets in Eastern Europe and Russia. The second wave of emigration mobilized approximately 2.3 million Russians and Poles; three million citizens from the Austrian-Hungarian Empire; and roughly half a million Rumanians. Most were guided through the German ports of Hamburg, Bremen, and Bremerhaven. In the 1890s, the biggest rush came from Jewish emigrants escaping pogroms in Eastern Europe and Russia. A coalition representing German shipping lines, the state-owned railroad companies, and relief organizations, first lobbied the Prussian authorities to reverse their policy of sealing off eastern borders. The coalition also offered a system of control stations for the transiting migrants. Simultaneously, U.S. immigration officials watched closely and inspected the European network. U.S. port health officials and the American public demanded rigorous medical inspections and quarantine stations to screen out "the diseased, defective, delinquent and dependent."[46] Rules became ever more stringent. In 1905, immigrants were checked (and rejected) for infections of trachoma and favus. In 1906, U.S. federal law demanded six days of quarantine before embarkation for all immigrants. From 1907, immigrants with disabilities were banned in order to weed out "idiots," paupers, polygamists, criminals, anarchists, prostitutes, tubercular patients and epileptics. Later, illiterates were added to the list.[47]

The severe 1890 U.S. sanitation laws for immigrants had a ripple effect across the entire European system. These laws created a separate regime for the emigrants *en route* from border crossing of Myslowitz-*Oświęcim* (Auschwitz) via the port city of Hamburg to Ellis Island. Specifically, U.S. and European governments held the steamship lines responsible for enforcing U.S. immigration policies and regulations. For example, U.S. authorities demanded from the steamship lines a fine of one hundred dollars for every immigrant who did not pass inspection on arrival. U.S. laws forced the companies to create an advanced medical screening and disinfection system on Europe's shores. Thus shipping lines and railways opened special border stations and "Immigration Cities" to channel and contain the migrant flow and to meet the stringent regulations. The transmigration system had two crucial hubs at borders—at Myslowitz-*Oświęcim* and Hamburg—and a hub in the Berlin suburb of Ruhleben. The system included delousing stations at the Russian-German border (Batajohren, Tilsit, Insterburg, Eydtkuhnen, Prostken, Illoo, Ottloschin, Posen, Ostrovo), the Austrian-Hungarian border (Myslowitz, Rutibor, Leipzig); the German-Dutch border (Bentheim-Oldenzaal); and the central hub of Ruhleben. It was a concerted effort to weed out the infectious diseases feared by the U.S.[48]

For disinfecting the emigrants, their clothing, and their luggage, a great variety of substances were used, including steam, hot air, carbolic acid, Lysol, paraffin, and creosote. Ships were disinfected before leaving for the United States. Such cleansing procedures were the responsibility of the train and shipping companies at borders, railroad junctions, and ports, rather than state agencies. German authorities, in fact, were happy enough to leave the control and organization of the transiting migrants to private companies—as long as the poor travelers did not escape, cause trouble, or become a financial burden for the state.[49]

The steamship companies were willing if not eager to comply with U.S. regulations. Collaborating meant maintaining—and growing—the lucrative emigrant travel market. The steamship companies participated in helping the U.S. government create a medical system of inspection to screen emigrants to ensure their fitness as workers. The intrusive medical inspection came to function as a way of qualifying emigrants for industrial work: the companies, governmental organizations, and aid organizations devoted substantial resources to screening those diseases and conditions that could compromise an immigrant's capacity for work—and, by extension, citizenship in the industrializing world. More broadly, the medical inspection was a rite of passage in which peasants became industrial workers.[50]

The German transit corridor was as efficient as it was cruel. Other companies that covered Fiume (Rijeka), Trieste, and Odessa tried to get into the profitable emigration trade as well. Their prices were higher, their facilities more primitive, and their services limited. For emigrants passing through, Fiume's laxer rules meant that the chance of being stopped on Ellis Island and sent back to Eastern Europe was much greater than for those who endured the ruthless rigors of the German centers. Moreover, in collaboration with the German police, migrants who tried to circumvent Germany were harassed and stopped at every turn. Under pressure from the powerful German shipping lines, Belgian, Dutch, and British competitors signed a cartel agreement that guaranteed each a fixed share of migration market from Europe.[51]

Parallel with the first- and second-class world of Cook and *Wagons-Lits*, transnationally operating Jewish, Roman Catholic, and Italian relief organizations assisted poor emigrants navigating baffling sets of rules and bureaucratic systems. Since 1865, Jewish communities, in particular, had established an informal transnational support network to help poor Jewish emigrants when states failed to protect them. Jewish community leaders acted—and were accepted—as the spokespeople for those fleeing Russian pogroms; they entered into negotiations with local officials and commercial interests to lobby on behalf of arrested or deported migrants. Together with the entrepreneurs, these non-governmental actors shaped an emigration infrastructure separate from the corridors that *Wagons-Lits*, for example, had created.[52]

The delousing stations and lodgings at the end of the line were the 'photo-negatives' of the *Wagons-Lits* terminus Grand Hotels. Many of these commercial buildings for processing emigrants were carefully designed and had attractive facades, but the interiors of the emigrant hotels were only minimally equipped. In 1901, Hamburg harbor opened a new complex (*Auswandererhallen*) that included disinfection stations and quarantine quarters, sleeping and dining halls, churches and synagogues for emigrants awaiting their ships to America. Elsewhere in Europe, similar emigration infrastructures developed in ports like Rotterdam (HAL, 1901–8) and Amsterdam (Lloyd complex, 1921–38). Often, the delousing stations and lodgings were connected to the main railroad lines by a separate set of tracks. All such emigrant stations led to Ellis Island in the harbor of New York City—and to other entry points overseas in North and South America.[53]

At sea, class segregation continued to create sharply different travel experiences for those on the upper decks and those in steerage. The second-class, transatlantic traveler Alfred Steiner offered a glimpse of this on the elegant ocean liner *Kaiser Wilhelm*

Fig. 3.8 Pre-Selecting Immigrants: *After 1890, U.S. immigration laws and practices became increasingly harsh. Cholera and other epidemics were feared. And only those considered healthy and fit were granted entrance to the U.S. In addition to disinfection and delousing, intrusive physical examination was part of the procedure. In this photo, taken at Ellis Island in 1907, immigrants end their journey with the final medical check. Note the eye chart—in Hebrew— on the wall.*

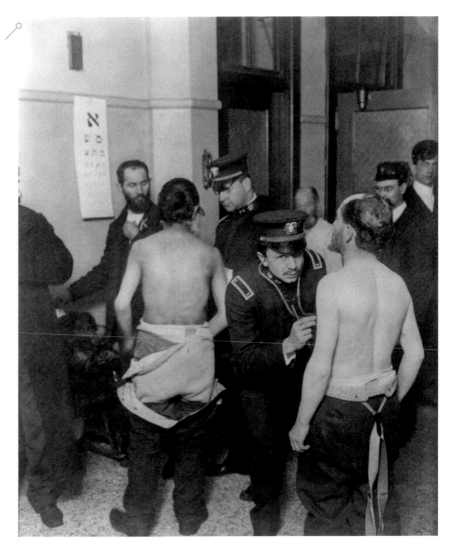

II of the North-German Lloyd line. In 1906, Steiner, a successful German emigrant who had come to the United States in steerage years earlier, looked from his second-class deck with empathy at the emigrant world below—a world that he had left behind. Steiner described how, in third-class, people were "packed like cattle," "the stenches [became] unbearable," and "the food … [was] miserable." He concluded, "On the whole, the steerage of the modern ship ought to be condemned as unfit for the transportation of human beings." Steiner also accused the shipping companies of dishonest practices. Although second-class passengers paid only twice as much as third-class passengers in steerage, the second-class traveler enjoyed disproportionally more space and better services, including a well-furnished cabin and access to a dining room. Emigrants slept 200 to 400 per section, on bunk beds stacked one above the other. "The barriers which keep the classes apart on a modern ocean liner are as rigid as in the most stratified society," Steiner maintained. "A matter of twenty dollars lifts a man into a cabin passenger or condemns him to the steerage; gives him the chance to be clean, to breathe pure air, to sleep on spotless linen and to be served courteously; or to be pushed into a dark hold where soap and water are luxuries, where bread is heavy and soggy, meat without savor and service without courtesy."[54]

The divergent experiences between upper and lower decks extended to arrival procedures. First- and second-class passengers—whether immigrants or not—were the subjects of only perfunctory customs checks conducted aboard the ship; there were no medical procedures, intrusive or otherwise, for these passengers. In contrast, once they reached their destination, immigrants below second-class were physically separated and had to go through entire cleansing procedure all over again. Steiner remarked: "Let no one believe that landing on the shores of 'The land of the free, and the home of the brave' is a pleasant experience; it is a hard, harsh fact, surrounded by the grinding machinery of the law, which sifts, picks, and chooses; admitting the fit and excluding the weak and helpless."[55]

The European states were overwhelmed by the number of people who, unprotected by their governments, chose to emigrate. The emigration infrastructure that evolved was a response to European states' failure to, for example, patrol their borders and protect migrants from commercial exploitation and political violence. Ironically, the German-sponsored forwarding stations—put in place by powerful commercial interests in service of U.S. immigration policy—offered the emigrants some protection against swindlers and lax officials. To meet the strict U.S. immigration requirements, emigrants passing through the German transit system benefited from the efficient cleansing sites at train stations, ports, and borders: the system prepared them to be accepted as immigrants to America. But the process was indeed cruel: once East European migrants left their homes, they "stepped into an extra-legal space where they were deprived of agency" until after they were admitted to the

U.S.[56] For emigrants and their advocacy organizations, the relative ease that the steamship and train companies offered produced a bittersweet experience.

Stuck at the Border

If the francophone *Wagons-Lits* presented a transnational Europe in which all rail lines radiated from Paris, the German *Kaiser* wanted his own railroad empire with Berlin as the central node in the system. The First World War offered him an opportunity. In 1915, while the soldiers were still stuck in the mud and dying in the trenches, the Prussian Minister of Public Works, Paul von Breitenbach, worked out plans to replace Paris with Berlin as Europe's center of infrastructure. The Kaiser wanted to reach the Far East through Constantinople on the edge of Europe. Here, his railway would hook up with the so-called Baghdad line that the Germans were about to build through the Ottoman Empire. With financial and logistical help from the Ottomans, the Austro-Hungarian Empire, and the Principality of Bulgaria, Germany founded the company called Mitropa (*Mitteleuropäische Schlafwagen- und Speisenwagen-Aktiengesellschaft*) in the midst of war. The name Mitropa, which stood for Middle-European Sleeping and Dining Car Company, was chosen deliberately to denote that part of Europe slated to be under German control. The company began to operate with railroad cars and railroad lines confiscated from its competitor, *Wagons-Lits*. In a grand symbolic gesture of the war, the Mitropa first ran a luxury train, the *Balkanzug*, from Berlin via Vienna and Budapest to Constantinople on New Year's Day, 1917. From a German perspective, the Balkans was the political bridge to the Middle East, all the way to the Euphrates and Tigris rivers. The Balkan train, symbol of the alliance, did not last long, however. Weekly service commenced only in summer 1918, but ended in October, a month before the defeat of Germany. The war had shown that trains were not simply a means of transportation: they were indeed tools for building and maintaining empires.[57]

The First World War and the ensuing collapse of the Russian, Habsburg, and Ottoman empires put an abrupt end to the aristocracy's transnational world of conflict-free and comfortable travel.

The humiliating signing of the armistice took place in *Wagons-Lits* car no. 2419, stationed just outside Paris. Allied victors huddled in the smoking rooms and hallways of Versailles to create nation states out of the ruins of multi-ethnic empires: the Baltic States; Finland; Poland; Ukraine; Czechoslovakia; the State of the Slovenes, Croats, and Serbs; and a number of separate Balkan states. The new nationalism that the Versailles treaty dictated meant that even the transnational *Wagons-Lits* had to give up its monopoly and accept the German competitor, Mitropa. At Versailles, *Wagons-Lits* demanded that Mitropa be dissolved, but the Allies left the German company intact. As a compromise, the Versailles Treaty's Article 367 ruled that Mitropa operations would be restricted to Germany and its foreign lines returned to *Wagons-Lits*. The victors reconfigured international rail transport between the Western powers and the new Balkan countries by creating a route that avoided Germany and Austria: the Simplon-Orient Express. Not surprisingly, *Wagons-Lits* was to organize and operate it. After the First World War, international travel became a smoother experience but with more formalized procedures. New state authorities—forged from the collapsing empires—issued passports, carried out border controls, and instituted hygiene regimes. The end of the First World War also meant an end to the hermetic transit corridor through Germany. New borders and anti-immigration laws stranded millions of migrants who had nowhere to go: many ended up stateless and unprotected.[58]

The collapse of empires; the fear of an expansive Germany; the belief in a system of nation states along ethnic lines—all these factors contributed to the decision to put a hold on plans to build new transnational railroads. They also halted the international standardization of railroad hardware and again made crossing borders more complicated. The new circumstances prevented the international labor movement from establishing automatic couplers to protect railroad workers from fatal accidents. The blocked transnational railroad development did not prevent various players from making proposals for a new system. Some plans to build railroad lines stemmed from the geopolitical power moves of old imperial powers. The French sought to curb German expansion by connecting France to the newly established Balkan states or to its colonial possessions in Africa. This vision had all international railroads radiating from Paris—and circumventing Germany.

Other plans, which circulated in the halls of the League of Nations, came from pan-European visionaries with loftier ambitions. They believed that the new nation states could be drawn together peacefully by building railroad interconnections to stabilize the new political order. Nevertheless, the dominant trend between 1919 and the 1950s was for governments to nationalize railroad lines.[59]

Despite the problems arising from the freshly drafted map of Europe, international tourism again flourished from the mid-1920s onward. In fact, tourism now began to attract an ever-increasing number of European citizens. Until the First World War, *Wagons-Lits* services, with the exception of dining, catered solely to first-class travelers. With most of the aristocratic amenities now obsolete, middle-class travelers forced train companies to offer comfortable surroundings in second-class compartments. Unions negotiated paid holidays for their rank and file. Battlefield tourism blossomed, creating yet another experience of what Europe meant: the same tracks that had delivered troops to battles now ushered in the former soldiers as tourists. While the rich Russian nobility was gone, demand for long-distance express trains was nonetheless

Fig. 3.9 Tourism at Home: *The First World War ended with the Treaty of Versailles, which created new nation states but also imposed restrictive borders. Now, even first-class travelers faced more problems, including struggles with new borders and changing travel infrastructures. A wave of nationalism ensued, reflected in this advertisement from a private British railroad company. Travelers were urged to explore their own countries instead of going abroad. This 1925 ad claims that a visit to Cornwall could be just as exciting as a trip to Italy—traditionally one of the most popular tourist destinations.*

high, driven by the Europe of new nation states, unprecedented diplomatic activity, and large wartime profits. Traditional travelers, accustomed to the all-teak interiors of the sleeper cars, now found all-metal designs. In 1922, *Wagons-Lits* refashioned its wagons in the modern Art Deco style and changed its corporate color from brown to royal blue. The company also invested in *Étoile du Nord*, the trains connecting Paris with Brussels and Amsterdam. During the Weimar Republic, Mitropa renegotiated the Versailles terms, winning back its foreign lines in 1925. One of its most famous routes, winning would be the *Rheingold* along the majestic river Rhine, connecting the Dutch well-to-do with Swiss mountain resorts. Services were revamped on the traditional Cannes-to-Rapallo route that serviced, for example, members of Prussian high society on their way to the French and Italian Riviera.[60]

After Versailles, migrants without the right papers, however, were rendered stateless or stranded at borders. For *Wagons-Lits*, the collapse of Europe's aristocracy and royal houses meant that the former first class became "single" berth accommodation, second class became "double," and third-class with three berths became "tourist." From 1925, triple-berth sleepers connected the newly established states of Poland, the Baltic region, and Czechoslovakia with Western Europe. Moreover, the former rule prohibiting third-class passengers from entering the dining car was abandoned. From then on, for an extra fee, third-class passengers could get a taste of upper-class luxury, if only for the duration of a meal.[61]

Revolutions had toppled regimes. The labor movement had challenged class boundaries. Democracies had emerged. Nation states had been reassembled out of old empires. Yet, class distinctions had not disappeared. In the worlds of travel, home comforts, and fashion alike, class distinctions had only been reconstituted through encounters with the new technologies. The emergence of new nation states created further streams of immigrants who now had to grapple with the new political realities of the Continent. Countries tried to contain and ethnically cleanse populations. Closely policed borders in East Central Europe were one problem; the end of transatlantic migration as a result of U.S. restrictions was another. The 1921 and 1924 U.S. immigration laws put a nearly-complete stop to the influx of emigrants from Southern, Central, and Eastern Europe, as well as those migrating from Asia. When the door slammed shut for European immigrants, rural American migrants took their place: from the American South, millions of African Americans began to migrate to the industrializing North to escape racism. They made this escape using three major railroad lines. These migrants found factory jobs, which the new laws now

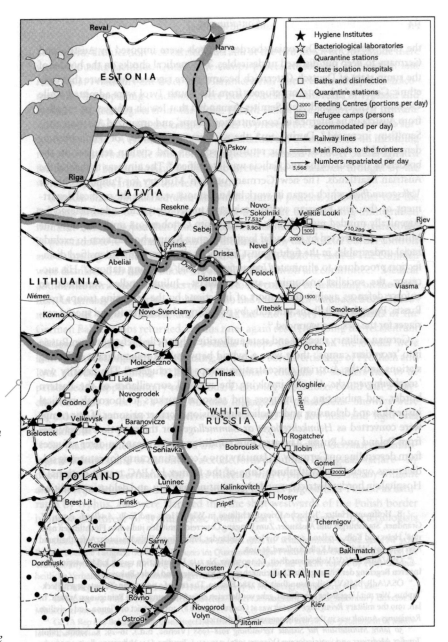

Fig. 3.10   A Europe of Boundaries: *This 1922 map documents the network of authorities and private companies tasked with controlling the movements of Russian emigrants and refugees. Note [Legend, Upper Right] the quarantine and disinfection stations, the feeding centers and refugee camps. All were connected by railroad lines and roads; they were also supported by sanitation institutes and bacteriological laboratories. Since the 1890s, this system had evolved into a region-wide network that restricted mobility.*

prevented European immigrants from reaching. Everywhere border control systems and fortifications were put in place. Countries instituted severely restrictive immigration policies, making emigration and transmigration increasingly difficult, if not impossible. During the interwar period, steamship lines had to seek other lucrative enterprises. They focused on the cruise ship business for first- and second-class passengers, to make up for lost revenues. From a commercial point of view, tourism now outstripped migration, and individual means of transportation began to challenge the prevalence of collective travel. But the new and active role that governments had assumed would have a profound impact on the ability of user movements to shape the new technologies in the decades to come.[62]

II

AFTER THE GREAT WAR: WHO DIRECTS TECHNOLOGY?

DE LOS
NTORNOS DE BARCELONA

por
J. Bordons.

Vendrell 65 K.
Tarragona 97 K.
Villanueva 66 K.
Villafranca 66 K.
Igualada 66 K.
Montesa 116 K.
Tarrasa 116 K.
Berga 118 K.
Puigcerdá 21 K.
Vich 68. K.
Granollers. 29 K.
Mataró 28 K.
Arenys 38 K.
MAR

Provincia

Region Catalar

c. á s. Culsat

Tibidabo

Valldaura

Montañas de
S. Pere Martir

SARRIA

Pedralbes

Bonanova

Las Corts.

S. GERVASI

Horta

Velódromo

SANS

Poble
Sech

FC. de Granollers

S. AN
de Pal

Castillo
Monjuich

SAN MARTI   de Provensals

c. á la Frontera

Bes

Rio

Poble Nou

FC. de Francia

Signos convencion

1 Sociedad de Velocipedista   Ferro-Carril
2 Club Velocipedico   Travesias generalmente en mal e
3 Peña Ciclista   Via Transitable
4 Cyclist's Club   Via no siempre transitable

1 K   2 K   3 K   4 K

4

Bicycling & Driving Europe

As transportation technologies, the train and the steamship had united Europe in some ways and divided it in other ways. Class, gender, and ethnic divisions had indeed been challenged. But overall, the train and steamship had simply recreated the social order in the form of separate corridors. And user groups in each of these separate corridors—from steerage to first-class—experienced a vastly different version of Europe. As for the technologies of *individual* transportation, it is the automobile that has traditionally been the focus of attention. Likewise, the motorist has been celebrated as enjoying the ultimate expression of middle-class individual freedom. But a closer look at who shaped Europe reveals that it was bicyclists who actually paved the way for the automobile's success. Cyclists and cycling organizations turned the bicycle into the bourgeoisie's main mode of individual transportation. Likewise, bicyclists contributed to yet another mapping of rustic Europe. Liberal Britain served as the model for the middle-class organizations that championed cycling. Symbolically, the bicycle itself was squeezed between an aristocratic world of privileges and the growing demands of working classes everywhere. After the First World War, the bourgeoisie discarded their bikes and turned to cars, while the working classes began to cycle in overwhelming numbers. Clashes in the streets resulted in

a full-fledged conflict over how European cities should be reconfigured in the wake of war. Despite these contests, a transnational
group of traffic engineers and urban planners—working alongside
local policy makers—projected a seamless, car-governed future, in
which cyclists would not necessarily have a place.

In the late 1950s, a generation of Swedes developed a vision of
Europe's future—a vision they found quite convincing. The Swedes
made their predictions based on what they had observed during
numerous study tours to the United States. Their delegations included
traffic politicians, engineers, and urban planners. One of the delegates,
Sweden's social-democratic commissioner for municipal building,
Helge Berglund, reported that on the other side of the Atlantic, automobility represented "an irresistible development, to which urban
planning has to adjust." In the same vein, construction engineer
Sven Lundberg published an article with the telling title, "What to
Learn from the U.S.A. in the Area of Traffic." And, despite a certain
degree of ambivalence toward "the Americanization of our way of
life," architect Sven Tynelius regarded the increasing importance of
the automobile as *"inevitable."* These politicians and self-proclaimed
experts in the postwar Europe believed that the future belonged to
the United States and to the car.[1]

Half a century earlier, the situation had been completely different:
the bicycle was then seen as the vehicle of the future. In 1899, an
American guide gushed that "Europe offers the best field for ...
the bicycle." Nowhere but in Europe was there "such a variety of
nationalities, manners, customs, languages and natural scenery so
compactly joined." For an American, a European cycling tour represented "the crowning pleasure."[2] Two years earlier, A.M. Thompson
of the British (and socialist) Clarion Press captured the key role that
cycling had assumed in modern life: "The man of the day is the
Cyclist," who was hotly debated everywhere, in discussions about
"his health, his feet, his shoes, his speed, his cap, his knickers, his
handle-bars, his axle, his ball-bearings, his tyres [sic], his reins,
and everything that is his, down unto his shirt." Thompson wrote,
"the cyclist is the man of the Fin de Cycle—I mean Siècle. He is
the King of the Road."[3] A woman columnist for an Irish women's
society journal concluded the same year: "We are all cyclists
now."[4] In the 1880s and 1890s, thousands of enthusiasts—men and

Fig. 4.1   The Bicycle & the War: *Cycling surged during the Second World War—in all European countries and the U.S. Primarily to save resources, members of both the middle and working classes eagerly appropriated the bicycle. The demand for repair services boomed along with the bicycle. In this image from occupied Paris, a repairman hawks his services. Apparently, he is operating from a makeshift, street-side shop.*

women of all ages—celebrated the sense of freedom that cycling gave them. They contrasted that sense of freedom with the oppressive grind of train schedules. Devotees from commercial classes, liberal professions, and clerical occupations praised the bicycle as a cheap and reliable alternative to riding and maintaining expensive horses. Pedals were stirrups, handlebars were reins, and seats were saddles. Like aristocratic equestrians, bourgeois cyclists rode with their backs properly rigid and straight, not bent over like reckless working-class teenagers. Despite such references to aristocratic norms, cyclists of the 1890s cultivated an unabashedly middle-class culture around their innovative bicycles.[5]

Hundreds of thousands of urbanites shared the thrill of riding the novel "machine," as contemporaries called the bicycle. Their role model came from Britain. As a Portuguese enthusiast observed: "The Saxon race, which gives us the example of its physical finesse, by cultivating corporal exercise, of its healthy morality through a culture of hobbies that cleanse the spirit, like excursionism, whether rich or poor, in yacht or automobile, bicycle or foot, have always traveled, experiencing at a close distance the natural beauties of its country."[6] Forming a pioneering user movement, middle-class cycling enthusiasts all over Europe and the United States established clubs and lobby groups; they wrote bicycle guides, articles for magazines, even poetry and novels. Bicycles offered an exhilarating feeling of self-directed, rapid mechanical movement, which the aficionados compared favorably with collective forms of transport such as diligences and trains. One British aficionado summed up the "charms of a well-conducted bicycle tour." Cycling freed the traveler from "consulting and conforming to time-tables." It offered "the delicious sensation of constantly having a free ride"; generated a freedom of movement "at once swift, easy and independent"; ideally combined "exercise and rest"; opened up an "unceasing series of exhilarating…mild adventures"; and "provided an unrivaled opportunity for seeing the picturesque in nature and observing the life of the people out of the beaten path of travel." Touring the Continent on a bicycle—individually or in a group—was the ultimate antidote to the "well-worn circle of European travel," of package tours, which the commentator likened disdainfully to "ready-made clothes which may or may not fit you."[7]

Middle-class cycling had not always been associated with touring and travel. Professional racing—sponsored by manufacturers to boost sales and by newspapers to generate sensational copy—offered lower-class young men the chance to win prize money, sudden fame, and stardom. At the outset, bicycles represented danger and pleasure—a plaything rather than a means of transportation. The high-wheeled bicycles of the 1860s were macho machines ridden by young men who cherished the sense of danger and of freedom that they evoked. Sudden braking regularly caused cyclists to flip over the handle bar and crash—"headers" as these accidents were called at the time. With no standard brakes, coasting downhill with feet on the handle bars was common practice. For the manly thrill-seekers, such hazards were part of the appeal; they enjoyed testing their user skills in the balancing act of keeping control over the machine. As late as 1898, one bicyclist observed how young men coming down the Alps without brakes backpedaled with the right foot, while pressing the lifted left foot against the front tire to control the dangerous downhill acceleration. Tweaking the machine and flirting with death went hand in hand. Amateur racing and competitions in balancing skills in cycling *sur place* attracted young middle-class men. The early cycling machines became symbols of irresponsible male modernity.[8]

Between 1880 and 1900, the meaning and design of bicycles went through a process of differentiation. Demanding more safety, many rejected the working-class penchant for racing. Among these dissenters were older men, urban couples, single middle-class women, and the upper classes. They preferred the pace of bicycle touring, first on the easier-to-mount tricycles and later on redesigned bicycles. Even members of European royalty adopted the vehicle of the bourgeoisie for their outings. The new safety models incorporated two sets of user needs into the design: sportsmen's relentless search for increasing speed and lighter materials, and tourists' interest in safety and user-friendliness. The demands of the new user groups, fascinated by touring rather than racing, drove safety-bicycles' sales.[9]

The low-wheeled vehicle with the diamond-shaped frame, pneumatic tires, and a chain drive on the rear wheel is the bicycle as we know it today. This design delivered greater speed and its identically-sized wheels provided better balance. The bicycle's dropped frame allowed for easier access, and air-filled tires conferred greater comfort. Compared to earlier models, "the safety" was more user-friendly and required less technical skill. But before this modern machine for mobility could be exploited to the fullest, it had to evolve further to suit female riders. Clothing innovation was another area in which the mutual shaping between the bike and its riders took place.[10]

Like so many consumer technologies of The Long Twentieth Century, the bicycle originated in Europe but only became an industrially mass-produced innovation in the United States. As in the case of both the sewing machine and the automobile, American entrepreneurs transformed European inventions into products for a mass market, making use of aggressive advertising campaigns in which images of young men and modern women dominated. In fact, many sewing machine manufacturers turned to the mass production of bicycles. And when the market in the United States collapsed because of overproduction, companies entered the automobile business. European tinkerers created and produced innovations successfully, but U.S. entrepreneurs were particularly strong in mass producing and mass marketing them.[11]

Bicycles became vehicles of exploration, both at home and abroad. They helped constitute a middle-class, individualistic tourist "gaze"—the word used to describe the tourist's way of framing the travel experience as authentic. British and American tourists crossed the Atlantic and the Channel to cycle around the European Continent. The Germans, Scandinavians, and Dutch followed, while French, Italians, and Swiss pioneers concentrated on exploring picturesque places in their own countries, especially in regions that trains could not reach. France, the most beloved destination of transnational bicycle tourists, offered superbly maintained roads and a spectacular variety of landmarks and views. Historians typically associate freedom and personal mobility with the arrival of the automobile, but these luxuries began with the bicycle. Historically tucked in between railroad travel and touring by car—but mostly ignored by scholars—bicycle riding paved the way for individual, middle-class mobility in the twentieth century.[12]

To promote their interests, cyclists established powerful clubs and associations, both nationally and internationally. Such organizations became successful spearheads of a middle-class user movement centering on one specific technology. Between 1880 and 1900, urban and liberal middle-class cycling clubs were responsible for creating a touring apparatus and rural service corridors. This was comprised of a system of bicycle support services, guidebooks, maps, hotels, railroads, and signs, in addition to clean beds and good food. Internationally oriented from the start, the clubs contributed to creating a genuine transnational and pan-European feeling of collaboration, generating a tourist infrastructure and a touring experience that automobile lovers and their organizations would expand and perfect. Connections also extended to Europe's colonies. Working independently of governmental channels and across borders, cyclists lobbied forcefully for better roads and established networks based on the idea of mutual assistance. In many instances automobile users later absorbed cyclists' organizations and emulated their institutional structure.[13]

Fig. 4.2 The Bicycle as Main Transportation: *From the First World War until the early1960s, the bicycle was Europeans' dominant form of urban transportation. This chart compares bicycle use to cars, buses, streetcars, and mopeds. The numbers at left indicate bicycles' percentage share of the total "rides." Blank spaces represent the Second World War, a time when bicycle use was not studied—but actually soared. Only after traffic planners engineered the way for motor vehicles did the number of bicyclists in Europe's cities decline.*

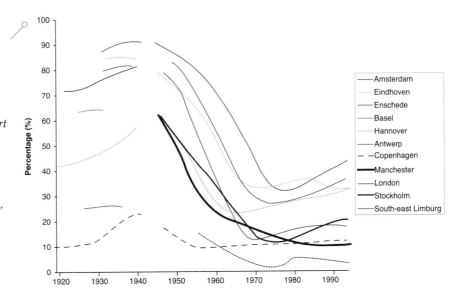

Bicycles—their numbers soaring in the interwar period—and automobiles ultimately became competitors on the roads. Policymakers and experts came to cast cyclists as unruly and uncivilized, translating such representations into far-reaching policy. For policymakers and experts like Berglund, Lundberg, and Tynelius, bicycles came to symbolize Europe and the past; automobiles came to be associated both with America and the future. Given that the historical literature has, until recently, overemphasized the relative importance of the car in the first decades of the twentieth century, this chapter underscores the contemporary impact of the bicycle. It also focuses on how the state and experts like Berglund began to act in the place of users when shaping new innovations after the First World War.[14]

Liberal Britain & Iron Horses

Many marveled at the great strides cycling had made and the speed with which the machine had become widely adopted. By 1900, European and American pioneers had cycled to virtually every country on the European Continent. They had mastered Europe's mountains and explored its edges, pushed into the colonies, and even pedaled around the globe. In 1886, Thomas Stevens, the British-American son of a working-class man, completed a two-year-long endurance trip around the world on a high-wheel ordinary bicycle. He was sponsored by a newspaper and cheered on by local cycle clubs along the way. A decade later, pedaling across Europe and around the world, British travel author John Foster Fraser and two friends crossed the Caucasus over the Kasbec Pass from Vladikavkas to Tiflis on their user-friendly safety bikes. The British friends ventured onto a newly constructed Russian military road, which they described as "the most wonderful road in the world, 130 miles long, a marvel of engineering." Smaller than the majestic Alps, the Caucasus still seemed awfully forbidding at first: "They frighten you," the men admitted. But the spectacular feeling of coasting down from the highest point at the junction of Europe and Asia was the greatest reward: "From that altitude, on the roof of the world as it were, began the descent," they exclaimed. "Instead of twirling round sharp corners we might have flown off into space... Twisting like a corkscrew, the road dropped and we flew like the wind."[15] They rode their cycles through colonial lands and cities like Calcutta and Shanghai, where local expat cycle clubs welcomed the three British explorers with parades and celebrations. In their way, cyclists contributed to the stretching of Europe's borders and to the stabilization of colonial relations.[16]

For others, the thrill of cycling was not about mastering mountains, conquering lands, and setting records. At a more leisurely pace, middle-class cyclists toured parts of France, traveled along the Rhine, and climbed the Alps in order to coast down into Northern Italy. They enacted the tradition of the bourgeois *Bildungsreise* as the crowning achievement of becoming a cultured European. On their trips, cycling tourists sought to learn about people and customs. They also yearned to experience the landscape of still lesser-known and unspoiled corners of Europe, even as trains sped by. The cyclists also evaluated Europe's colonial lands for their tourist potential. From Cairo an enthusiastic report celebrated "a delightful spin to the Pyramids" on a "good macadam" road.[17] Cyclists hailed the bicycle as the ultimate vehicle for off-the-beaten-path individuality and juxtaposed their experience to those of the many Cook tourists traveling in groups by carriage, train, and boat. The American artist-writer couple, Joseph and Elizabeth Pennell, shared their transnational cycling pilgrimages with millions of readers through evocative prose and brilliant illustrations in dozens of magazine articles and books. First on tandem tricycles and later on their low-wheel safety bicycles, they toured England, Italy, the Alps, Transylvania, and Andalusia in the years between

Fig. 4.3 The Swiss Alps, American-Style: *In the 1880s and 1890s, the American couple Joseph and Elizabeth Pennell, artist and writer, toured Britain and the European Continent—by bicycle. Their "tourist gaze"—expectations of what they would find—informed their well-illustrated books with romantic depictions of landscapes and people. In this illustration, taken from a typical Pennell book, the artist portrays himself and his wife on an Alpine tour. Note that even the steam locomotive—a beacon of industrialization—is peacefully integrated into the landscape. The only thing missing here are the throngs of other travelers crowding tourist routes, as humorously described by Elizabeth in her travelogue.*

THE RAIL AND THE ROAD OVER THE ST. GOTTHARD.

1884 and 1900. At one point, they teased their readers, claiming to have found the most enchanted and picturesque Italian town, but refusing to divulge its name or location for fear of having to share their authentic experience with masses of followers. Many other progressive and liberal couples cycled around Europe: the Fabians Beatrice and Sidney Webb; the physicists Maria Skłodowska and Pierre Curie; and the feminist doctor Aletta Jacobs and her husband, liberal politician Carel Gerritsen, to name but a few.[18]

Most cyclists lived in cities. Members of the urban middle class initially used the bicycle to explore the countryside in their spare time. From 1913 on, Spain's Pedal Festival (*Festa del Pedal*) united urban cyclists from sporting and touring associations; cyclists ventured out on collective trips to towns on the outskirts of Barcelona. The bicycle festival, which grew into a decade-long tradition, included a banquet, competitions, and a raffle with cycling-related prizes. The profits went to charity. Other towns that were further from Barcelona organized their own festivals. Like clubs elsewhere, those in Barcelona also designed special bicycle maps charting tourist trails and indicating terrain gradients, road conditions, and places to rest.[19] The Portuguese Velocipedic Union, when organizing a trip to Alenquer in 1903, followed the French touring club format in choosing the itinerary and offering cost analyses, places to eat, railroad transport expenses, and booklets with technical advice and tourist sites to visit. As Duarte Rodrigues, the founder of Portuguese cycling tourism explained, the tours were scheduled for Sundays, when "those losing vigor in a store, at the counter, or at a desk, in an office" had the opportunity to get out into the countryside and "say hello to the shepherd we run into." In planning the itinerary, the club created an experience of the land-scape set in a rustic rhythm in contrast with the pace of the train and automobile. From the British, Rodrigues suggested, the Portuguese ought to adapt the practice of "touring...touring...to ride a bicycle in a fresh summer morning..."[20]

In a deeply Romantic way, bourgeois urbanites of both sexes viewed the picturesque aspects of rural life at a close—yet safe—distance. Away from the noisy city and the excesses of industrial life, they marveled at the countryside's blissful silence and tranquility. Bicycles—combined with secondary railroad networks—facilitated easy rural travel. Bicycles also reconnected smaller villages bypassed by the train lines, causing old inns to close, post offices to move, and diligence connections and coach lines to fall out of fashion. Before the arrival of the bicycle, "did one ever see these caravans covering the countryside of our beautiful France?" a trade journalist asked rhetorically in 1895. "For every tourist you used to see on foot, you see a hundred cyclists these days roaming the road." These cyclists rode "at good speed, slow enough to admire the countryside but fast enough to escape the boredom of the same scenery."[21] Railroads had transformed the countryside into a quieter and more isolated place, but trains also brought scores of city folks physically closer to nature. In fact, trains and bicycles became mutually supporting technologies. In 1894, American railroads carried more than 430,000 bicycles to places where riders could start tours. The Portuguese bicycle club negotiated train rates for its members to tour the country. Both the French petit bourgeois and skilled workers took railroad daytrips to the beach, bringing along their bicycles to tour the coastal area. Affluent cyclists with time to spare enjoyed longer trips in the countryside, idealizing peasant life, yet also passing through the landscape fairly rapidly. Social critic of industrial life John Ruskin naturally disapproved of such a pastime, advocating contemplative walking, instead.[22]

Bicycles and cameras maintained a symbiotic relationship. For a successful country trip, cyclists considered a camera (at the 1900 Paris Exhibit, one in six visitors carried one) essential gear. Thus outfitted, urban cyclists created authentic experiences by framing rural vistas, scenic ruins, and picturesque peasants. Responding to the trend, Kodak manufactured purpose-built cameras that could be carried on the bicycle in a special pouch: "Nothing so fits the pleasures of cycling as photography," Kodak advertised in 1897, claiming that their light, bicycle-ready Kodaks were the "most compact cameras made," and, "with our perfected bicycle carrying cases, are entirely out of the way, yet instantly available for use."[23] As they sought to escape the chaotic city and the effects of industrialization, the bicycle, the camera, and the notebook helped capture a vanishing world, one which—ironically—tourism had a hand in destroying.[24]

The Bourgeoisie & their Bicycles

Cycling generated a form of middle-class consumer activism organized around one specific technology that was unparalleled at

Fig. 4.4a, b   Twin Technologies: the Bicycle & the Camera: *It was the U.S. Eastman Kodak Company (founded 1889) that popularized photography. Europe's wealthy middle-class cyclists appropriated that technology for their own use: mobile cameras became all the rage. On cycling tours, here illustrated in the magazine of Touring-Club de France in 1897, male and female bicyclists snapped shots and drew images of rural vistas for friends and family at home. Kodak provided consumers not only with a special camera [Below], but with special carrying straps and other paraphernalia for bringing cameras with them for the ride.*

the time. By associating in local, national, and international clubs, bicycle users in all industrializing countries participated in democratic governing practices. As the previous chapter showed, individual train passengers often found themselves on the losing end in conflicts with railroad monopolies. Forming a pioneering user movement, cyclists were more successful in lobbying municipal, local, national, and foreign governments for better roads and services and pressuring manufacturers for better designs. The *Touring Club de France*, for example, played an important role in promoting research for better touring bicycles. The club organized chain and brake tests on mountain roads in the Alps and the Pyrenees, persuading companies to design lighter, stronger, adjustable, and easier-to-repair touring bicycles for users without special training. In their search for user-friendly solutions, the clubs shaped road infrastructures, pushed for higher hotel standards, and built up experience-based knowledge and competence around tourist infrastructures. Their user advocacy had a deep impact, establishing the example for automobile tourism, which followed.[25]

National clubs lobbied for better roads and road-building legislation; published the latest road-building research; and even co-financed new roads through specialized organizations. British cyclists pioneered in 1878, establishing the British Cyclists' Touring Club (CTC). Soon after, a host of other organizations all started: The League of American Wheelman (1880), *Union Vélocipédique de France* (1881), the *Dansk Cyklisk Forbund* (1881), *Royale Ligue Vélocipédique de Belge* (1882), the Dutch *Algemene Nederlandse Wielrijders Bond ANWB* (1883), *Deutscher Radfahrer Bund* (1884), the *Union Vélocipédique Algérienne* (1884), and the Swedish *Hjulförbundet* (1888). National organizations in Italy, Switzerland, Austria, Russia, Luxembourg, and the Czech lands followed. Cyclists also established numerous clubs in Europe's colonies, including British India, French Algeria, and the Dutch Indies. Federating hundreds of local organizations, bicycle associations appropriated existing foot- and towpaths for bicycles and fought for legal access to main arteries. The CTC convinced Parliament to allow cyclists access to all roads through an 1899 amendment of the Local Government Act. The British Road Improvement Association campaigned to improve the existing, often poorly surfaced roads and tracks. In the United States, where new laws denied cyclists access to main roads, the League turned to the courts to win access; brought suits combating harassment and aggression against cyclists; and fought restrictive state legislation. The American Good Roads Organization, established in 1880, pressured local governments. The Dutch ANWB in 1889 created its *Wegencommissie* to advise local governments on road building. In Germany, local bicycle organizations like the Hannover *Radfahr-Renn-Verein* and the Magdeburg *Verein für Radfahrwege* did the same. In Italy, the *Touring Club Italiano*, the country's strongest middle-class organization, sponsored the first cycle paths.[26]

Still, there were notable national differences. The French, for example, had a less urgent need for road improvement because the nation's road network was probably the best in the world. The *Corps de Ponts et Chaussées* managed a superior maintenance system based on a long tradition. While well-built, the British and German roads were increasingly neglected because authorities invested in rails rather than in roads. In the United States, road building and reform were locally organized and rather dismal. Cyclists loved the Dutch brick roads (*klinkers*), but despised the Belgian use of cobble stones (*kinderhoofdjes*). Local circumstances dictated different lobbying efforts.[27]

Clubs in country after country created a tightly woven support network of *hulpdozen*, *boîtes de secours*, or first-aid boxes stocked with

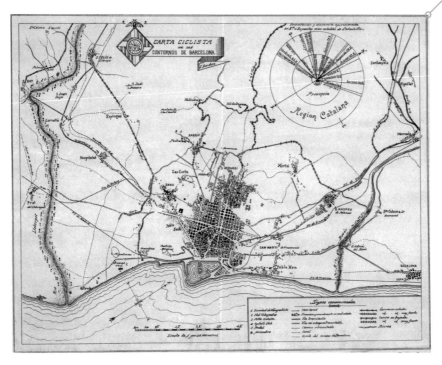

Fig. 4.5 Mapping Europe's Tourist Landscape: *All across Europe, middle-class bicyclists formed clubs that helped shape Europe in important ways. For example, this 1898 map, created by a coalition of Catalonian bicycle clubs, made it easy for any cyclist to explore the countryside of Barcelona. The map's legend details the gradients of the roads and calls out points of interest. In addition to socializing with one another and making maps, bicycle club members advocated for better roads and bicycling facilities—locally, regionally, nationally, and transnationally.*

emergency items for both the rider and the cycle in dire need. The boxes were placed strategically at cycling distances along the road and at busy spots in hotels, cafes, and garages. Invented in 1894 by a cycling entrepreneur from the Belgian city of Leuven, Joseph Delin, the public first-aid box combined an apothecary, bicycle repair set, and spares. Exported to the Netherlands, the upper half of the box contained special "wadding and other cotton, some sublimate pastilles to making bottled water, English and salicylic patch, a linen bandage, some safety pins, a bottle of peppermint oil." These items were "individually packed in parchment paper or in glass tubes sealed with a cork, in small quantities, so that no damp, dust or dirt can enter." The lower part included "several tools to repair bicycles and tires, and some bicycle parts, which are easy to replace like a key, pump, oil can, wooden hammer, screwdriver, awl, scissors, gas rod, burning oil wicks, copper wire, single chain links for different sizes, rings of different sizes, a locking nut, beads, and everything needed to repair tires."[28] Reflecting a larger social commitment, the boxes in Leuven also included an alms box for charity.

By establishing a system of representatives, "consuls," as they were called, local members were given the task of guiding stranded fellow cyclists by offering advice, roadmaps, and technical support. Finally, the associations negotiated discounts for railroad fares and accommodation. In 1898, the British CTC claimed 55,000 members, who had privileges with 4,500 contracted hotel keepers. With 15,000 members, the Dutch union boasted a network of 250 hotels and cafes and 180 repair stations. The American club contracted with over 7,000 League hotels to guarantee good food and meet middle-class expectations of quality and cleanliness. Similarly, the French and Italian clubs negotiated with hotels to raise standards of hygiene, amenities, and food. As user organizations, the clubs built competencies and experience-based knowledge about bicycle design, innovations, road conditions, and tourist infrastructures. The associations broadened their self-help systems and urged politicians to institute safety measures. The associations went on to pioneer the traffic-sign conventions that governments would later adopt.[29]

Given tourists' penchant for exploring ever-more destinations, it is little wonder that the national clubs began to cooperate across borders. Under Dutch and Belgian leadership, the bicycle clubs of

Fig. 4.6 Europe's 1890s "Help Desk" for Bicyclists: *A "modern machine," the bicycle was a breakthrough, but it needed care and repair, as did the cyclist. To support their fellow cyclists, bicycle clubs designed* boites de secours *and* hulpkisten (help boxes) *such as this one [Right] invented by Joseph Delin of Leuven, Belgium. Kits typically contained bicycle-repair tools and first-aid supplies. Bicycle clubs such as the Dutch ANWB installed the help boxes at club-certified hotels such as this Dutch one [Left].*

the two nations went international in 1897. The editor of the Dutch cyclists' magazine explained why cycling needed a world union. International cycling had boomed across Europe. Within fifteen years, cycling had become amazingly common, the editor wrote, transformed from a sport for young men of moderate means in search of thrills, speed, and adventure into a "mode of transport for the family." Even more remarkably, cross-border touring had become routine: many "who went on their summer trips along the Rhine or to Switzerland, used to taking the boat or the train, are now faithful adepts of the bicycle."[30] In 1899, the journal generated statistics to support its claim. With a membership of 200,000, the Dutch organization responded to more than 2,500 requests about international travel and issued border permits and *permits de circulation* for bicycle-touring to Belgium, France, Switzerland, Italy, and Norway. Cycling had become a means of creating common cross-European experiences.[31]

To deal with the bureaucratic hurdles of transnational bicycle touring and standardized service, the Dutch and Belgian national clubs sent out a call to their sister organizations. In 1897, associations from Great Britain, Belgium, France, Luxemburg, Germany,

Austria, Switzerland, and Russia answered the call to meet in Amsterdam and Brussels. To address the explosive growth of cross-border bicycle traffic, they launched the *Ligue Internationale des Associations Touristes* (LIAT). Later, Scandinavian, Italian, and U.S. societies joined the League. Although in principle Europeans did not need travel documents, cyclists found that in practice, a passport prevented a lot of trouble from suspicious or uninformed local authorities and customs officials. Moreover, each country had its own rules and tariffs for border crossings. To protect their national industries, several governments treated bicycles as export goods instead of vehicles for personal use. When John Foster Fraser and his companion entered Austria, they had to pay customs duty equaling £2.10 for each bicycle.[32]

At the first international conference, the Dutch initiators were keen to achieve the "principle of reciprocity" between states—the idea that favors, benefits, as well as penalties granted by one state for its citizens should be returned in kind. The Dutch also mobilized the international trade convention of "most-favored nation," according to which countries agreed to give all signatories beneficial trade advantages. In support of an international bicycle alliance, they evoked a language of international relations and peace through cooperation. The organizers articulated a vision of transnational cycling as a vehicle of mutual respect and understanding among nations and advocated the establishment of a World Cyclists Union. Their proposal was a bridge too far for many, however. The British, American, French, and Austrian clubs expressed concern over possible breaches of national autonomy that an international union might entail. The Austrian club, sensitive to the domestic politics of the Habsburg Empire, prevented the Czechs from joining the international organization. Moreover, France was the Continent's most popular holiday destination for cyclists; French revenues increased in part because of foreign tourists who joined the French cycling club to profit from the many perks that membership offered. The *Touring Club de France* had little incentive to share its highly valued privileges with sister organizations. Nevertheless, the national organizations had to admit that there was something to gain from international cooperation. In particular, as the champions of cycling tourism in Europe, the British and Americans stood to benefit from services of the sister organizations in their quest for authentic European experiences. Members could get bicycle maps with profiles, discounts at hotels, and the most coveted perk of all: the privilege of crossing international borders without bureaucratic obstacles.[33]

Despite crushed dreams of securing a World Bicycling Union through a single, multilateral agreement, the outcome was still impressive. When meeting next in Amsterdam in 1897, Luxembourg in 1898, and London in 1899, twelve national clubs—with the exception of the French—signed bilateral agreements, allowing their members full access to the services of collegial organizations. Some national clubs even negotiated successfully with foreign governments. As a non-governmental organization, the British cycling union concluded pioneering agreements with national governments, allowing duty-free entry for members' cycles. Other clubs pursued similar agreements, with varying degrees of success. Two decades later, automobile advocacy groups followed the precedents set by the cycling organizations.[34]

The national cycling organizations played a key role in forging a middle-class culture of governance and leisure. Proudly liberal, anti-clerical, and committed to the free transnational exchange of goods, people, and ideas, club members practiced a political culture constituted through grassroots democracy. In France, the *Touring Club* allied with the anti-clerical civic culture as part of a Third Republic social contract that sought to appease the tensions between the right-wing conservative power of the *haute bourgeoisie* and the leftist heirs of the French Revolution. Dutch cyclists were similarly progressive: they were anti-confessional (against organized religion in politics). These cyclists were liberal elites who advocated for a secondary-school system that encouraged modern languages rather than a classical education. In Italy, touring was closely allied to the young republic's emerging bourgeois culture in northern industrial centers like Milan and Turin. In Germany and the United States, organized cyclists also came from liberal and internationally oriented circles. In his *Philosophie des Fahrrads* (*Philosophy of Bicycles*) published in 1900, the German writer Eduard Bertz identified cosmopolitanism as the vision that made every cyclist "a member of a great, world-encompassing party of reform."[35] Through personal, family, and economic ties, many cyclists, moreover, felt closer to Britain's entrepreneurial middle-class culture than to the aristocratic culture of Paris. This preference was best expressed in their attraction to the British-inspired sport fashion. Resistance to cycling came from anti-modernists, like the Church, farmers, and cultural conservatives, as well as some aristocrats. To the aristocracy, the editor of an Irish cycling magazine wrote, "cycling is vulgar under all circumstances, and the rational dress has not received the hall-mark of fashion, and is therefore, immodest and unseemly."[36] Others observed that tourist cycling emerged from the middle classes rather than from the old elites. It is hardly surprising, then, that the clubs formed national federations. As active grass-roots organizations, they cultivated a thoroughly civic and democratic political culture.[37]

Yet, riding the pathways of liberalism was always a balancing act. Despite the clubs' highly developed democratic culture of inclusion and commitment to a liberal worldview, they routinely excluded women as full members. The only option for women was to establish separate clubs of their own. Jewish cyclists were not welcome in some German clubs, while Émile Zola's progressive stance in the Dreyfus affair led the French club to demote him from full membership. Committed to free trade and to fostering international ties, touring clubs at the same time articulated national identities. In Germany, two organizations represented opposing ends of the political spectrum: citizens at one end celebrated the nation; citizens at the other end supported internationalism and peace.[38]

In France, the touring clubs helped reinforce the secular religion of the Third Republic, which was to know and love *la patrie*. The Dutch club's version of this was a nationalist ideology of "character building" and learning about one's own country in all its diversity and beauty. To simplify the access to landscapes deemed typically Dutch, the club designed an intricate national system of cycle paths, maps, books, gear, (youth) hostels, and repair services. The Italian Touring Club summarized the same mission with the motto *far conoscere l'Italia agli Italiani* ("making Italy known to Italians").[39] In 1915, the president of the Italian club sought to strike a subtle political balance between liberal internationalism and emerging nationalism. He argued that the club's activities were essentially pacific: "The most important tourist developments have found their best and most salient expression in fruitful international collaboration." In the middle of the First World War, with hopes of international peace shattered, he claimed that "touring is a powerful expression of national unity and of solid social concord."[40] The bourgeois

Fig. 4.7 Bicycle-Club Founders & Friends: *This photo, taken in Luxembourg in 1898, commemorates the meeting of the* Ligue Internationale des Associations Touristes *(LIAT), founded one year earlier. Transnational in scope and led by bicyclist groups, touring clubs simplified cross-border travel, primarily in Europe. Until the time of the Second World War, the bicycle tourism movement was hugely popular, mostly among the middle class.*

cycling organizations thus helped build both European national identities and transnational collaboration across borders.[41]

To summarize: cycling organizations built tourist infrastructures and practiced international governance—decades before the-much celebrated automobile clubs did the same. With corporate sponsorship from manufacturers and newspapers, the well-organized cyclist associations and their commercial allies were key players in forcing access to—and improving—existing roads. In a very literal sense, the cyclists paved the way for the automobile pioneers. Cyclists established a tourism apparatus and arranged international agreements and standards with thousands of federated local clubs throughout Europe and the United States. Britain boasted more than 2,000 clubs in 1898; France had fostered 800 different local clubs by 1910. And those many associations represented only their country's most politically well-organized segment of the bicycle-user movement. There were many more cyclists on the road than the already impressive national clubs' membership figures suggest. For example, while the British national touring organization claimed 44,000 members, the country already had 1.5 million cyclists by 1897. In the same year, the French national club reported 67,000 members, but tax records suggest that almost 400,000 French citizens owned bicycles; that number jumped to as many as one million within a scant three years. Italy, Belgium, and the Netherlands also became cycling nations, passing the one million mark around the First World War. And the numbers continued to rise.[42]

As a result of this continuous growth, the bicycle lost its avant-garde character and ceased to be a novelty. Expecting large profits, American industrialists jumped on the newly standardized bicycle with their mass-production methods. The "bicycle craze" or a "velomania," as it was called in the summers of 1896 and 1897, vanished from the front pages as suddenly as it had appeared.[43] Bicycles no longer generated sensational stories for a press hungry for new copy; cars had taken over as the Next Big Thing. But the end of media coverage did not mean the end of users' interest in the bicycle. During the interwar period, bicycle sales soared again. The new user group came from a lower social class. The shift also meant a change in user activism.

Bicycles as Proletarian Shoes

The 1900 Paris exhibition is often taken to symbolize the moment automobiles entered the twentieth century as the trailblazers of individual mobility and technological modernity. Yet it was cycling rather than motoring that came to dominate the streets of European cities like Amsterdam, Aveiro, Barcelona, Basel, Bremen, Rotterdam, Milan, Leuven, and York. The touring bicycle became a work tool, "a horse of the poor." Professionals (ministers, doctors, and midwives), shopkeepers (bakers and butchers), and civil servants (postmen and soldiers) began to mount bicycles for daily use: to conduct business, to commute, and to transport goods. Workers cycled to their jobs in the morning and back home in the evening, and on weekends they gathered to explore the surrounding countryside. French hairdressers organized cycle excursions on their day of rest with their *Club Sportif de la Coiffure*; so did French butchers, policemen, railroad men, workers in Parisian department stores like *Bon Marché*, and clerks at major banks like the *Société Générale*. Young rural folk, followed by farmers and their wives, began to cycle to reach the field, the next town, the church, and the cinema. Adapting to these new user contexts, manufacturers designed bikes for intensive daily use. They improved brakes and racks (front and back), and added sturdy stands for making short stops and quick deliveries. This new form of bicycle was a workhorse rather than a fancy touring machine. Next, the bicycle industry brought down its prices still further in the 1920s by tapping into the market of unskilled workers. To save costs, manufacturers got rid of extra accessories like chain guards, coaster brakes, and back racks, all of which had made cycling more user-friendly.[44]

Compared to bicycle sales, car sales increased dramatically percentage-wise but remained exceedingly small in absolute terms. In contrast, bicycles came within the financial reach of more and more groups. Bike sales continued to soar. By the mid-1930s, there were roughly 15 million bikes in Germany, 9 million in Britain, 7 million in France, 4 million in Italy, and 2 million in Belgium. The fact that every second Dutch citizen owned a bicycle prompted an American journalist to assert in 1934 that in the Netherlands, the bicycle had become "almost a part of the body," and to joke that, if evolution theory had any say in it, within a century one would see Dutch babies, "coming into this world on tiny bicycles."[45] Seven years earlier, a comparative study had shown how widespread bicycles had become. That year, every third resident in the Netherlands had a bicycle, closely followed by Sweden (every fourth), Denmark, Switzerland, and Belgium (every fifth), Germany and France (every sixth), Britain (every seventh), and Italy (every thirteenth). With one bicycle for every seventy citizens, the United States represented the exception in the industrial world. Whereas in that country there were seventeen cars to every one bicycle in the 1930s, in urban Europe there were seven bicycles to one car. As Figure 4.2 shows, bicycles remained the most popular means of transport well into the early 1960s compared to cars and public transportation.[46]

Despite these figures, the popular myth is that the car rather than the bicycle was the most important innovation on the road during the twentieth century. This distorted view of history comes from a misreading of data from the United States, where bicycle-production figures plummeted from sky-high levels and car sales increased dramatically. In the public imagination, in policy projections about the future, and in later studies on automobiles, America's present seemed to signal Europe's inevitable future. Even in bicycle-friendly Amsterdam, the City Department of Public

Works in the 1930s compared the number of cars in the Netherlands and the U.S. in terms of lagging behind.[47]

The shifting meaning of the bicycle—from a middle-class to a working-class means of transportation—proved politically precarious. After 1900, many national touring organizations, dominated by bourgeois elites, either renamed themselves automobile clubs or simply began to redirect their focus away from cycling. Other national cycling clubs collapsed. The bicycle-based *Ligue Internationale des Associations Touristes* (LIAT, 1898) and the car-based *Association Internationale des Automobile-Club Reconnus* (AIACR, 1904) at first collaborated in road improvements and at easing cross-border traffic flows pioneered by LIAT for bicycles. The newly established Permanent International Association of Road Congresses (PIARC, 1909), however, completely excluded the interests of horse-and-buggy drivers and cyclists. That exclusion was a sign of things to come.[48]

In the interwar period, bicycles boomed as every-day devices in urban settings, and bikes contributed to conferring urban Europe with a different character from the United States. The upper middle class "shifted gears" culturally from promoting cycling to promoting automobile touring. Many European city governments began to treat bicycle traffic as a problem to be solved rather than a solution to be embraced. Policy discussions centered on whether roads should remain accessible to mixed traffic or become a mono-functional space privileging motorized transit. Outside of cities, Italy and Germany carried out the first experiments with *autostradas* and *Autobahnen*. They excluded cyclists as a matter of design principle. Experts and policymakers in many cities linked the explosive growth of urban traffic and the increasing numbers of accidents to the cycling boom rather than to automobiles. Given that bicycles were now associated with the working class, governments and urban planners focused on regulating them.[49]

Fascist Italy exemplified this new regulation. In the industrializing Northern provinces in the large and smaller cities of the Po plain—Milan, Turin, Padua, Parma, Brescia, and Bologna—cyclists dominated the roads. When the Fascists created *L'Azienda Autonoma delle Strade Statali* (AASS) (The National Road Agency) in 1928 to modernize the national road network, the presidents of the *Touring Club Italiano* (TCI) and *Unione Italiana Velocipedistica* (UVI) sent a petition to the Minister of Public Works asking that cycling interests be included in the planning. The advocates argued that bicycles were immensely useful for scores of workers who pedaled on Milan's outlying streets and the plains of Emilia Romagna in order to find a day job. Moreover, bicycles were responsible for carrying much greater numbers of people than all other vehicles combined. Indeed, even in car-dominated Milan, city statistics of the 1920s showed that cycling was more prevalent than motoring. The Fascist Black Shirts established the Military Road (*Strada della Milizia*), which was followed by the new Highway Code (*Codice della Strada*) in February of 1929. Despite the petition, the Fascist road builders sought to discipline rather than to facilitate cyclists' interests. In the 1929 Italian Highway Code, there were three rules limiting cyclists' mobility. First, Article 53 prohibited cyclists from forming "groups that could impede the movement [of cars]." Often this rule became locally interpreted as requiring cyclists to pedal "single file" on the right side of the street or highway. Second, cyclists could not ride on sidewalks. Third, cyclists were obliged to use (often non-existent and poor) bicycle lanes. While motorists were given ample street and road space, cyclists were to occupy as little space as possible so as not to hinder vehicles. Facilitating the motorists' speed was most important. The cyclists' safety was secondary. Disciplining cyclists' unruly behavior was the third concern.[50]

Today, we associate bicycle lanes with safety and sustainability; but in the interwar period, many European cycling organizations either opposed or grudgingly went along with the construction of separate paths for cyclists within cities. The clubs seldom promoted a policy of separating the traffic streams of horses, carts, pedestrians, and cars. Instead advocacy groups focused on improving roads—especially surfaces—for everyone. At first, few if any of the European local and national bicycle societies took an explicit anti-automobile position. Some clubs, like the Dutch national tourist organization, did advocate—and even finance—separate cycle paths along rural roads for touring. Similarly, powerful local bicycle organizations in German cities like Bremen, Hamburg, Hannover, Lübeck, and Magdeburg promoted special road sections to foster bicycle tourism in the countryside, but such projects were not top priority. The retired municipal engineer and urban planner in Magdeburg, Carl Henneking, in the mid-1920s wrote that "as long as the road is built with noiseless pavement, asphalt, tar, concrete, or wood, the installment of special cycle lanes on the roadside is not essential."[51]

By this time, Henneking's thinking was somewhat old school. He had been a local personality who managed to push through the construction of 400 kilometers of cycle paths throughout the forest and farmland around the city. Henneking was a member of the Magdeburg Association for Bicycle Lanes, established in 1898. By the time he retired in 1927, the club had grown to 49,000 members, most of them local businessmen and civil servants, including city officials with influence in key agencies. This urban planner and bicycle advocate recalled how, in the pioneering days, club members cycled together "through the wide and pretty park of Magdeburg to places outside the city, drank their morning coffee, and went if possible swimming in the Elbe river." After their morning break they "returned after 1 or 1.5 hours back to their daily work, refreshed from their time in the beautiful countryside, cycling and swimming in the river." Such men "understood that if cycling was to be a complete pleasure and relaxing activity, suitable paths had to be created."[52] Yet the goal of "suitable," separate, or sectioned-off cycle lanes always centered on cyclists' comfort and not on their safety nor on sorting out traffic streams. In the decades between 1880 and 1920, the middle-class clubs had succeeded exceedingly well in shaping existing and emerging infrastructures for all. As civil-society organizations, they had developed user-based political expertise in municipal and state governance. Neither the state nor experts intervened systematically on behalf of cyclists. But the situation changed in the late 1920s.

After the First World War, states and local authorities took a more active part in transportation, urban development, and social planning. Particularly in the (war) economies of Communist Russia, Fascist Italy, and Nazi Germany, the state dominated urban planning and traffic policies, strongly favoring motorized mobility. For Benito Mussolini, cars were not only the engine of the future and of progress, but also a way to neutralize the proletariat's revolutionary tendencies: the fact of not owning a car could turn workers into revolutionaries. Although many governments propagated cycling as an emergency measure to reduce oil consumption in war-time, the First World War proved to be a defining moment for projections about the future. Motorized taxis in Paris; the omnibus at Verdun; trucks for the American soldiers—all lent an epic quality to motorized mobility, its decisive role in winning wars, and inevitable future. Mass-scale automobility was still a dream. In the hands of policy makers, urban planners, and traffic engineers, these projections turned out to be far-reaching.[53]

When motorized traffic began to challenge the bicycle in cities, local policymakers launched plans to reorganize streets; manage the exploding numbers of vehicles and people; and improve urban efficiency. City councils and construction departments began widening streets, filling canals, demolishing houses, building sub-surface roads, and regulating traffic speed to control what they considered to be the chaotic development of the city. Giving priority to speed, an Italian expert argued that the sole design purpose of the profession was to "allow the car to run, because....that's why it came to the world." He further argued that the motorist had to be the "king of the street, respected and despised, and also something feared (that does not hurt)."[54] Even though cyclists dominated the streets, the *Royal Automobile Club Italiano* (RACI) felt confident enough to announce in 1928 the "twilight of the bicycle," whose rider was just a "poor pariah of the street." The Italian cycling union (UVI) responded that cyclists had only become pariahs in the streets because motorists were not "admitting that the street also could serve others."[55] All over urban Europe, echoing automobile lobbyists, traffic experts and policymakers expected that, in the long run, motorized traffic would inevitably and rightfully substitute what they, by extension, cast as old-fashioned bicycle mobility. Car proponents began dismissing bicycles as hopelessly out of date, as being the "enemies of fashion" and as "irreducibly conservative."[56]

Indeed, bicycles and cars were on a political collision course in the interwar period. In the 1920s, Dutch socialist MP Florentinus Wibaut questioned the bicycle taxes that were levied disproportionally against workers and that failed to acknowledge that bicycles had become their basic need. The tax revenues, "mostly generated by proletarians," were an uneven burden "for many workers for whom the bicycle is the same as for others the soles of their shoes."[57] For the government, the bicycle tax turned out to be a cash cow when collectors discovered a surprising 1.7 million cyclists instead of the projected one million. In post-*Anschluss* Vienna, one businessman who lacked sympathy for socialists severely criticized the Nazi regime for unfairly taxing workers for whom the bicycle was a cheap means of commuting: "For...thousands of unemployed it was the only practical method of getting around to look for jobs...[and] for apprentices on very low wages it was almost indispensable."[58] In a speech to the Italian Parliament in 1938, Italo Bonardi, an active member of the Italian Touring Club, conceded that cycling lanes should be built if only to "liberate motorists of cyclists and cyclists from motorists," but he reminded his fellow politicians that instead of complaining about them, they should recognize that the millions of cyclists filled the government's coffers.[59] His was a lone voice in Fascist Italy.

Throughout European cities, bicycles, no longer bourgeois pleasure vehicles, became workers' indispensable "shoes." Working-class organizations like the German *Solidarität*, the Italian *Ciclisti Rossi*, and the British Clarion Cycling Club demanded more room for cycling infrastructure. In 1935, British Labour MP John Banfield put it sharply, asking "the Minister of Transport whether he is aware of the resentment felt by the large body of pedal cyclists, numbering some 10 million, against the restrictions imposed upon their use of the roads by recent regulations?"[60] Under threat of losing his seat to the Labour party, no doubt, even the Conservative British member of the House of Commons, Sir Wilfred Sugden, pleaded for cyclists' interests, saying, "the 9,500,000 cyclists ought to receive as much consideration in these democratic days as the 2,250,000 motorists."[61] The experts, however, took a different path entirely.[62]

Urban cycle-lane construction and road funding became a focus of class politics that now attracted the interest of European political chambers. Along with other momentous changes in the First World War era, workers had gained access to parliaments. But parliament was not the only stage on which the class conflict was played out: more often than not, the clash between bicyclists and motorists was negotiated by engineers and planners in drafting departments. As engineer Einar Nordendahl bluntly remarked in the Regional Plan for Greater Stockholm in 1936, constructing special lanes for "bicyclists and pedestrians [was] to a large degree motivated by the urge to free the main lane from such traffic elements."[63] In this political context, rules, regulations, and separate lanes took shape as measures to, quite literally, push bicycles aside. One Italian newspaper explained in 1934 exactly whose road rights were to be privileged: "Given that you cannot ban [motorists]," it was best to "manage their fury."[64]

Contesting the Urban Machine

Experts throughout Europe and the colonies were quite united in seeking to tame cyclists—regardless of whether or not car ownership in their countries was growing rapidly. Indeed, there was a weak causal relationship between the (small) number of automobiles on the streets and the kind of problems experts sought to solve in their urban visions. Their visions were based on their expectations of how traffic would develop rather than what they actually observed in the street. In the Netherlands, where the automobile arrived late and tourist organizations supported a classless image of cycling, policymakers tended to regard bicycles as a fact of life, and legislated accordingly. Yet in Denmark, where automobility arrived early and expanded quickly, policymakers also assigned equal traffic rights to cyclists and motorists. By contrast, Germany and Italy motorized comparatively late, but their policies sought to control cyclists early on. Thus the level of automobility does not account for the differences in traffic policy affecting bikes. What (nearly) all of the European nations did share was a vision of the future of cities. This vision was voiced by a strong coalition of motorist lobby groups, new engineering professionals, and policy makers.[65]

A new generation of experts—urban planners and traffic engineers—shaped the debate about the design of the street. Enabled by government agencies and influenced by U.S. developments, planners and engineers began to challenge the user expertise of civil-society organizations. As new professionals began to speak on behalf of cyclists—and successfully claim new knowledge in the arena of urban traffic—cycling clubs largely lost their influence on local decision-making. Increasingly, the new professionals framed cyclists' dominance of the road as a problem. Municipal traffic engineers portrayed bicycles as dangerous—and as obstacles to more modern modes of mobility. The engineers accused cyclists of causing traffic congestion and engaging in dangerous behavior. The bicyclist, asserted Stockholm's traffic engineer Nils Lidvall in the late 1930s, was "the road user who has the worst manners,"

Fig. 4.8 The Bicyclist—Sidelined: *In the 1930s, many European countries built dedicated highways. These roads for motorists physically separated cars from cyclists, horseback riders, and pedestrians. Cars were designated as "fast traffic"; bicyclists and everyone else as "slow." In this photo, taken outside the city of The Hague in the Netherlands, cyclists are marginalized, forced to ride on bicycle paths to make way for future motorized traffic. Cyclists, heavily taxed, paid disproportionally for this new plan that separated traffic—and banished the bicycle from main roads.*

who "could not be tolerated in civilized society."[66] Administrators and experts also blamed cyclists for the alarming growth of lethal accidents, even though statistics did not bear this out. Voicing a commonly held belief, a prestigious Italian engineering journal argued that the bicycle "is almost always the main cause of serious and fatal accidents" because cyclists "do not respect any of their duties."[67] Similarly, middle-class newspapers also fostered the idea that motorists and cyclists were incompatible by reporting on the alarming— and still growing—number of accidents. Road safety became the leading argument on policymakers' agenda. And in making the argument, policymakers blamed accidents on cyclists. Middle-class newspapers campaigned against unruly working-class cyclists. Statistics absolved cyclists by showing that, in both absolute and relative terms, cars were far more lethal than bicycles. Cyclists were nevertheless cast as erratic and undisciplined; they acquired the moniker "mosquitos of the road."[68]

To deal with increasing traffic flows, governments took many measures that singled out cyclists. Looking for ways to reduce accidents, officials sought to re-educate and discipline cyclists rather than motorists. The state even discouraged cycling altogether. German and Belgian cities subordinated bicycles to public transport and cars. In 1937, the mayor of the Belgian working-class city

of Antwerp proposed prohibiting cyclists from riding side by side;
he ordered his administrators to close off many roads to cyclists,
forcing them to use cycle lanes in other parts of the city. In Italy,
banning cyclists for several hours daily first happened in 1928 in
Florence, Rome, Palermo, Turin, and Parma; next, cyclists were
expelled altogether from the university city of Bologna, in 1932.
The same happened in Hannover, Germany.[69]

In the hands of engineers, bicycle lanes were another tool for
managing cyclists and allowing motorists the right of way. Cycle
lanes were reframed in terms of separating cyclists from motorists
rather than serving cyclists. Sponsored by the Italian automobile
club in 1933, a conference of traffic control experts strongly recommended constructing cycle lanes outside cities "as a privilege
and not an imposition" for cyclists.[70] In the manufacturing city of
Manchester, the Labour-dominated city council gave priority to
building an extensive public transport system over investing in
cycle lanes for its working-class constituency. And while the political elite of the Labour party often advocated public transportation
for its constituents, the rank-and-file kept on cycling because it was
much cheaper.[71]

For their part, cycle clubs continued to lobby for general access
to roads and higher-quality road surfaces. They did not advocate
separate bicycle lanes, nor did they take an active anti-car position.
Cyclists, they believed, were entitled users of the roads that they
had pioneered in improving. By the 1930s, the British National
Committee on Cycling (NCC)—an alliance of the Cyclists' Touring
Club (CTC), the National Cycling Union (NTC), and bicycle manufacturers—took a stand against bicycle lanes. They resented the
new, car-governed traffic policies that restricted cyclists' freedom
of movement. Since the 1890s, British cyclists had retained the
right to use the "King's Highways." The clubs were not about to
give that up easily. In 1938, the CTC opposed separate cycle lanes
because they would do away with cyclists' fundamental rights as
traffic participants: "It is wrong in principle that a cyclist should be
required to put himself to considerable trouble and expense to keep
from being run into by a driver who is not obeying the common
law of England."[72] For the same reason, the NCC opposed other
road-safety proposals like mandatory rear lights. In the ensuing
political battle, working-class cyclists and bicycles became closely
linked with stubbornness and disorderly conduct. In turn, this

led to new anti-cyclist policies. In Fascist Italy, cyclist associations responded to state rules in a somewhat more subdued way. Rather than holding motorists responsible for safety, the rules demanded that cyclists protect themselves. When rules required cyclists to use headlights and red retro-reflectors on the rear fender—accessories that most lower-class riders found prohibitively expensive—the Italian Touring Club sponsored a campaign in 1930 to provide them free. The club also offered to paint cyclists' fenders for free and lobbied large employers to do the same for their workers.[73]

In the Netherlands, the tourist organization did encourage separate and wider recreational bicycle paths along rural roads, which were increasingly reserved for cars. Meanwhile, the fight that continued concerning creating urban lanes was as political as it was class biased. The touring club, professional organizations, and the Dutch labor party vehemently and bitterly fought over the despised bicycle tax once reintroduced as a temporary measure to balance the nation's budget. When parliament debated a fund to finance and maintain the road system in 1926, the government proposed to use bicycle taxes to subsidize both motorways and separate (and smaller) cycling lanes to give motorists more room. The policies led to a situation where cyclists not only paid for roads benefiting motorists, but also financed their own marginalization. This was exactly what the British CTC had so feared—only it was happening to the Dutch, and not the Brits. When the bicycle tax had been levied the first time around, the Dutch touring club—which represented both motorists and cyclists—had protested on principle. But when the bicycle tax was proposed the second time, the Dutch touring club remained silent, even though the government's bill proposed to use bicycle taxes to build roads serving the interests of wealthier motorists.[74]

Nazi Germany and Fascist Italy witnessed a similar but more strategic and ideological shift. The Nazi regime viewed roads in terms of future military planning, favoring motorized traffic and separated traffic flows. German traffic engineers and city planners picked up the idea of cycle lanes but modified its meaning. Their designs no longer focused on creating good road surfaces for cyclists' pleasure; they now focused on segregating bicycles from cars altogether. In 1926, the government agency *Zentralstelle für Radwege* had become responsible for the design and construction of cycling lanes in cities. Under the Nazi regime, the 1934 *Reichs-Strassen-Verkehrs-Ordnung* regulated the separation of traffic and the *Radwegebenutzungspflicht* coerced cyclists to use bicycle lanes. Bicycle lanes were defined as the safe solution for preventing traffic chaos and congestion. Traffic safety, traffic separation, and bicycle lanes became locked in an iron triangle. However, reports surfaced about subtle forms of resistance to separating traffic flows into slow and fast lanes, as envisioned by the Fascist *autostradas* and the Nazi *Autobahnen*. In the summer of 1937, there were repeated complaints about groups of cyclists on their way to work: bicyclists were riding illegally on the highway between Milan and Turin and on the Milan–Brescia–Bergamo car route. The focus on separate traffic flows as a means of control was not an exclusive obsession of the Nazis and the Fascists. Traffic engineers and urban planners shared the modernist notion of uninterrupted traffic flows—a view that would guide urban traffic planning even more forcefully in the postwar era.[75]

There were local exceptions to these modernist engineering principles. In Amsterdam, where a coalition of social liberals and socialists governed, urban planners included working-class mobility practices in their designs. Embarking on ambitious projects of social housing and urban planning, the Dutch capital gained a reputation as the mecca of urban planning. The municipal engineer and modernist Cornelis van Eesteren, who led the *Congrès Internationaux d'Architecture Moderne* (CIAM) with Walter Gropius and Le Corbusier, played a key role as chair in the CIAM discussions on the "Functional City." First in Berlin (1931) and then aboard the cruise ship the SS *Patris II* sailing from Marseille to Athens in 1933 (Stalinist Moscow and Nazi Berlin were no longer hospitable sites to the internationalist visionaries), the Amsterdam plans were presented as a prototype for international developments. After investigations of street use and traffic streams concluded that as many as 400,000 cyclists commuted to and from Amsterdam's city center on a daily basis, Van Eesteren decided to incorporate the bicycle in his plan for the future. His team measured the distances from downtown to the newly designed working-class neighborhoods in terms of how long it would take to bike from home to work; the verdict was that urban expansion should not go beyond 30 minutes of bike commuting. Importantly, Van Eesteren received support from the socialist alderman F.M. Wibaut, who, in the 1926 Parliament had fought the bicycle tax that unfairly penalized working-class cyclists. Wibaut would now play a powerful role in

Fig. 4.9 The Bicycle Rules: *As this 1930s photo of Copenhagen shows, bicycles far outnumbered cars on Europe's urban streets. Workers, civil servants, and small entrepreneurs appropriated the bicycle as their means of transportation. Until well into the 1950s—if not the 1960s—cyclists dominated European streets. This photo documents workers and civil servants commuting to Copenhagen's central Nørrebro district via the Queen Louise Bridge.*

implementing the ambitious concepts that Van Eesteren and his group formulated. Within CIAM, the Amsterdam blueprint served as inspiration for a number of cities in Europe, the United States, and the colonies. The bicycle-related component of this urban vision was nonetheless abandoned in subsequent decades. When dealing with cycling, planners in other cities did not follow the Amsterdam example.[76]

War reversed priorities, if only briefly. From the occupied Netherlands and Nazi Austria to Fascist Italy, regimes abolished the much-despised bicycle taxes to appease the working class. Bicycling boomed in all European cities and the United States. While the Fascist regime revoked taxes on cars to stimulate consumption as part of their vision of modern Italy, ending taxes on bicycles was merely a political move to curry favor with the working class. Unsurprisingly, newspapers duly reported the "wave of satisfaction" among workers in the Ferrara provinces and in Turin, where there was one bicycle for every 2.6 residents.[77] Another journal announced triumphantly that "the bicycle is in vogue, is patron of the cities, reigns supreme on the beaches."[78] The Fascist leisure society for workers (The *Opera Nazionale Dopolavoro*) organized "Bicycle Day" to celebrate cycling's "utility and sporting excellence," or as someone said, "I like cycling because it is a sport of poets."[79] As soon as the oil shortages and the war were over, however, the temporary pro-cycling war measures were abandoned.[80]

"Death of a Cyclist"

What is most remarkable about the interwar period is that middle-class policymakers and traffic engineers were convinced that the car would inevitably become the dominant mode of transport. Policymakers and engineers seemed oblivious to the fact that few people could afford cars. They also disregarded the streets around them that were bustling with a growing number of cyclists. European-trained traffic engineers and colonial admin- istrators projected similar class-based views onto their colonial subjects in British India, French Indochina, and the Dutch Indies. While colonial officials in India recognized wealthy motorists, "Malabar" chauffeurs, and the rural poor as users with valid needs and rights, cyclists were no such thing; they were put in the same category of "ambling ox carts" and "wandering animals," that the police saw as "a source of annoyance and danger to all other classes of road users."[81] Like their counterparts in European cities, pedestrians and cyclists were the most likely to die from road accidents. Nevertheless, the blame (and the solution) centered on the victims rather than the perpetrators. "No single factor" would guarantee their safety better than instilling in them the proper "road sense," the Madras police wrote in 1937.[82] French colonial administrators in Hanoi, Saigon, and Tonkin published rules and regulations on how pedestrians and cyclists should behave in traffic. The cyclists' undisciplined behavior continued to exasperate officials and motorists. The middle-class and colonial projections of the future represented a bias on the part of public-opinion leaders, policy makers, and urban planners. During the interwar period, the experts' plans were still merely visions—visions that were highly contested by bicycle organiza- tions, at that. In the postwar period, these visions would become realities through the reconstruction programs of many European countries.[83]

Under the influence of Le Corbusier's hyper-modernist plans, interwar discussions were recalibrated to fit with a car-dominated vision of modern cities. The Swiss architect's idea for an urban machine—skyscrapers and steel-framed office buildings set within large, rectangular green spaces—featured at its center a

Fig. 4.10   In with the Car, Out with the Bicycle: *After the Second World War, motorists prevailed. Bicyclists became increasingly imperiled; road deaths spiked; and bicyclists were blamed. Juan Antonio Bardem's 1955 film,* Death of a Cyclist, *memorialized the bicyclist. The Spanish movie depicts an accident in which the protagonists hit a bicyclist with their car—and leave him to die. Explored in the film is the moral dilemma of an illicit affair between an upper-class woman and her lover of lower social status. The plot symbolizes class conflict, and with it, the clash of the upper-class car and the working-class bicycle in Franco's Spain.*

huge transport hub. The blueprint specified depots for buses and trains; highway intersections on different levels; and on the top, an airport. Le Corbusier segregated pedestrian circulation paths from roadways. He erased bicycle paths and glorified the automobile as the primary means of transport. Le Corbusier envisioned adopting Taylorist and Fordist strategies from the United States to reorganize society. After the Second World War, European modernist planning—following Le Corbusier rather than Van Eesteren and using U.S. rather than classical European cities as their models— inspired an entire generation and would dominate the reconstruction of devastated cities.

When it came to implementing plans for the car-governed city, Sweden became the lead European country. The new generation of urban planners included Sven Markelius, Carl-Fredrik Ahlberg, and Göran Sidenbladh. Backed by politicians like Helge Berglund, they proactively planned new cities by making predictions about cars and public transportation. One professor of urban planning was rather surprised in 1960 to discover that old practices did not die easily: "Pedal cyclists and mopeds [are] supposed to have disappeared from the streets and traffic routes."[84] In the context of the Cold War, visions of classlessness became further re-inscribed in American rather than Nazi or Fascist versions of car-governed cities. In these blueprints, the American automobile came to represent the nation's newfound global economic and technological power. Now, the car symbolized modernity and the future. Cultural critics believed it was only a matter of time before whatever happened in the U.S. would reach Europe's shores.

The power struggle that played itself out around the bicycle on the street inspired cinema enthusiasts. In Italy, Vittorio De Sica's wrenching film *The Bicycle Thief* (1948) showed a poor worker losing his bicycle—and thus his ability to land a job. The Spanish director Juan Antonio Bardem's realist film, *Death of a Cyclist* (*Muerte de un ciclista* 1955), explored through metaphor the class and political dimensions of Spain's painful civil-war past under General Franco. In *Death of a Cyclist*, two illicit lovers, a wealthy socialite housewife and a university professor, accidentally hit a cyclist with their car. Although they see that he is still alive after the accident, they decide not to call for help, out of fear that their affair will be revealed. The couple drives away, leaving the bicyclist to die. The moral dilemma and tensions resolve in the man's moral rebirth and the woman's demise. These moviemakers understood brilliantly the struggle that was soon to be played out. Despite numerous collective actions and interventions by cycling advocates—in tandem with the resistance shown by countless cyclists on the streets—bicyclists ultimately lost their case. Only in the 1970s, as the so-called oil crisis motivated governments in some European countries to arrange car-free Sundays, were cyclist organizers able to begin taking back the streets. It would take another forty years before European cycling would be placed at the heart of urban policy—and set an example for the world once more.[85]

5
Eating around the Continent

During The Long Twentieth Century, access to food was at times a struggle for Europeans. In the first half of this period, consumers often openly criticized the food they were provided—by the state, by scientists, and by companies. In particular, they rejected the emerging system of industrially processed foods. Vegetarians made up a strong section of the "life-reform movement" (*Lebensreformbewegung*) in many European countries. Consumer groups on both sides of the Atlantic rallied against high prices, food adulteration (poisonous food additives), and inhuman working conditions in the food industry. Trying to regain control by manufacturing their own foods, consumers developed various new food-related skills, from canning to yoghurt-making. Manufacturers reacted to consumers' wish to retain control of what they were eating by providing those who could afford it with household machines and do-it-yourself kits. Later, in the second half of The Long Twentieth Century, the industrialization of food production was accompanied by the centralization of food distribution. Heavily influenced by U.S. business models, European retailers after the Second World War seized on the idea of the supermarket—or, as it was often called in Europe at the time, the "self-service store."

In 1948, the Migros Cooperative Alliance (*Migros-Genossenschafts-Bund*) opened a self-service store in Zurich, Switzerland. The first of its kind, the shop allowed customers to take items directly off the shelves without having to consult a salesperson. Customers flocked to the supermarket, but the Swiss press was not immediately convinced that this novel form of retailing would catch on. Although the shop on the Seidengasse witnessed increased turnover, journalists doubted that the store's sales growth could be sustained once the charm and novelty of self-service had worn off. The *Neue Zürcher Nachrichten* reported on customers' initial reticence: "As they first entered the store—with its strange traffic rules, members of the public exposed a certain degree of timidity."[1] The founder of Migros, Gottlieb Duttweiler, admitted that U.S. supermarkets had been his principal role models. Anticipating European sensitivities to the threat of Americanization, Duttweiler claimed that in Switzerland, there would still be a place in the market for traditional, individually oriented retailing.[2]

Six years later, Duttweiler's somewhat apologetic tone had been replaced by a more definitive, expansive strategy. In 1954, Elsa Gasser, a prominent Migros manager, summarized her company's business philosophy for a U.S. audience. An economist with a PhD in hand, Gasser gave a talk to the Twenty-sixth Annual Boston

Fig. 5.1 The American Supermarket—in Europe: *European consumers and commentators were critical of the supermarket as an icon of American consumerism. By the 1950s and 1960s, though, the supermarket had pushed its way into most European countries. The Swiss Migros Company was one pioneer. This 1950s photo, of a Berne Migros store, shows housewives and others engaging in self-service shopping—a key promise of the supermarket concept. Self-service supposedly freed consumers from the control of mom-and-pop shopkeepers. But the supermarket also required consumers to know everything from how to read a label to how to compare prices. Note the woman on the right, studying the available selection.*

National Conference on Distribution. It was entitled "Europe's Mrs. Consumer takes to Self-Service." Gasser conveyed the message that, contrary to popular belief, women consumers in Europe and North America did not fundamentally differ: "Take Mrs. Warren in a New York skyscraper, Frau Trudi Schmidt in a small Swiss town and Signora Rapelli in a South Italian hamlet," she said. Their lives may seem totally different, Gasser acknowledged. But, "ask about their husbands and children, the small cherished circle that makes up their everyday life, their cares and joys, their wishes and disappointments," and one would discover that "the similarity of their impulses and reactions is unmistakable." Gasser expected that, in the end, the self-service system would create worldly "paradises" all over the globe. After all, the supermarket was a place where "the daughters of Eve will stroll about and pick up all those wonderful things of the earth—at not too high a price."[3] Alluding to Genesis, the Migros manager implied that the modern self-service store enabled women to combine economic rationality with consumer pleasures.

During the 1950s, in search of new consumers overseas, American policymakers and businesspeople sought to transform European retailing. Starting in 1953, a traveling exhibit funded by the Marshall Plan, "Modern Food Commerce," introduced Europeans in France, Belgium, the Netherlands, Denmark, and West Germany to the basic ideas of the supermarket. Fascinated by the prospects of higher turnover, European delegations subsequently toured the United States to study recent trends in food distribution and retailing. Such trips were part of the Marshall Plan. The West European and Scandinavian visitors became convinced that a trip across the Atlantic Ocean was a journey into what would become Europe's inevitable future. The members of a Migros delegation in 1961 wrote: "In the next era the conditions in the USA will to a larger degree establish themselves also in our country."[4] Although instrumental in importing the mass-scale distribution and mass-scale consumption system into Switzerland, the Migros managers, like their colleagues in other European countries, viewed modern America with mixed emotions. The Swiss visitors to America were taken aback by the sheer size of supermarkets, which they disapprovingly described as "retail factories" (*Verkaufsfabriken*); they wondered if the lavish character of the whole system would not lead to wastefulness. Also, they feared that the high degree of standardization would produce a kind of "punch-card mentality" that threatened the close relationship between customers and retailers. Although they were convinced that the supermarket system would inevitably conquer Europe, the Migros managers were not ready to accept all aspects of Americanization. They claimed that the can opener was the U.S. housewife's most-used cooking utensil, and that the husband's midday meal had been reduced to a hectic lunch in an anonymous, mechanized, fast-food establishment.[5]

Ambivalence about American consumerism is a common thread that runs through many Europeans' postwar accounts of the United States. Although many visiting experts were convinced that they had experienced the future when touring the U.S., most were reluctant to acknowledge American innovations across-the-board. Another 1956 Swiss delegation, which focused on the hotel and restaurant business, concluded: "It would be wrong to copy slavishly the developments and phenomena we observed in America." Instead, they looked for ways "to adapt the many valuable pieces of inspiration to Swiss circumstances in order that they do not fail."[6] Employees of the Basel Cooperative Society nicely summarized the approach that most Europeans in the hotel and restaurant industry shared. Managers of self-service stores and restaurants regarded the United States as a source for new ideas and principles but they were unwilling to accept the American approach lock, stock, and barrel. For the Europeans, the extraordinary size of the supermarkets and the speed with which food was dispatched to diners was a threat rather than a boon.[7]

The supermarket and the fast-food restaurant symbolized the Fordist consumption regime. The guiding principles of this regime combined high turnover and speed with standardization and economic concentration. If Ford Motor Corporation's assembly line has traditionally represented manufacturing, the shopping cart and the cash register were seen as its retailing parallels. A blueprint of the first Migros store showed flowcharts on how customers were guided to move through the aisles. On the one hand, it is as if Migros imagined the customer as a robot—similar to Charlie Chaplin's mechanized assembly-line worker in his 1936 film, *Modern Times*. On the other hand, Elsa Gasser offered a view of the self-service-store customer that represented a real human being. To her, "Mrs. Customer" was an active, well-informed shopper who made her own decisions about what and how much to buy. Gasser understood "self-service" in a very literal sense. The French equivalent, *libre-service*, strongly conveyed the customer's freedom. Boosters of the supermarket argued that, in traditional shops, the housewife was at the mercy of the merchant to, for example, choose which of the apples she was to buy and which of the chickens she would bring home. The modern female shopper was, in Gasser's contrasting view, the merchant's "collaborator and partner."[8]

The same year the Swiss store opened, the Dutch brothers Van Woerkom advertised their country's first self-service store in the city of Nijmegen. At *Gebr. Van Woerkom*, which bore the unapologetic tagline "little America in Nijmegen," customers could "service themselves" and did not have to submit to "endless waiting." Customers also had "enormous choice of all current brand names." They enjoyed "cheap prices," and had a "comfortable overview of all kinds of foodstuff." The brothers advertised, "self-service is earning money." British surveys of the time confirmed that many shoppers, particularly the younger generation and working-class customers, welcomed the freedom of self-service. Men reportedly were particularly open to the new system.[9]

In many countries, the food-cooperative movement—and not private dealers—was among the most active promoters of self-service retailing. In Stockholm, the Swedish equivalent to the Co-op chain (*Konsum*) opened several so-called "quick buys" (*snabbköp*) in the early 1940s. The first of them was, perhaps predictably, given the name *Yankee*.[10]

Always progressive, the cooperative movement has roots in the late nineteenth century. Scientists, educators, and other self-proclaimed experts tried to teach workers how to improve their diets and avoid unhealthy and adulterated foods. Their activities—exemplified by the formation of home economics as a field of research, propaganda, and teaching—were responses to the rapid changes that accompanied urbanization. Middle-class reformers often sought to alter the eating habits of urban workers and landless peasants, believing they did not know how to feed their families properly and had limited opportunities for growing their own vegetables and storing home-grown produce. In another response to industrially processed food, urban middle-class housewives developed strategies to provide their families with homemade products—either individually or as a part of smaller networks and associations. New technologies and user skills also played a role in this process. Home canning was one of the outcomes; buying in bulk and establishing member-based stores to reduce the costs of produce was another result.[11]

The history of food has a strong transnational character. For centuries, Europeans appropriated and modified products that originated in other parts of the world—not only in the United States. The food system was intimately linked with colonial regimes. Accordingly, the history of so-called colonial products—including chocolate, coffee, and sugar—cannot be told without citing the cruelties that accompanied their cultivation. For decades, these foods remained luxury items that only the wealthy could enjoy. The history of these colonial crops is one of economic centralization and exploitation. But, as far as consumers were aware, the

Fig. 5.2   **Rewards of Home Canning:** *The technology of canning was never the preserve of large-scale companies alone—despite the existence of global food manufacturers like the U.S. Heinz company. As a pastime—and for reasons of self-sufficiency, middle-class families often made their own marmalade and chutney, for example. This photo of 1949 Britain portrays canning as a small-scale, community business in Folkestone, Kent, where women produced up to 250 cans per day.*

colonial crops were a world away from the European food chain. In terms of the food produced in Europe, housewives in rural and urban areas reclaimed traditional technologies like canning in an effort to maintain control of how their families' food was produced and consumed. Reformers and the state, for their part, tried to feed the many by providing a system of healthy, efficiently produced food.

"Homemade" & Self-Esteem

Nineteenth-century Europe saw a great migration from the countryside to the cities, where laborers hoped to find work. To feed the urban dwellers, larger quantities of agricultural products had to be transported over longer distances. The railroad, as well as new food-conservation techniques, enabled farmers to supply far-away customers with products of reasonable quality. Entrepreneurs built factories to produce margarine, milk powder, meat extract, and sausage. Emblematic of these trends—more pronounced in the United States than in Europe—was the establishment of huge slaughterhouses near train hubs; the transportation of chilled meat over large distances; and the manufacturing and distribution of canned food. At the dawn of the twentieth century, as many as 25,000 railroad wagons equipped with cooling facilities rolled across the North American Continent. Britain imported substantial amounts of refrigerated meat from both North and South America. In Uruguay, investors established large-scale factories to produce meat extract based on a method developed by German chemist Justus von Liebig. Canning began to play an important role in conservation, packaging, and long-distance transportation of food—particularly in the United Kingdom. The First World War proved to be a watershed moment in the history of European food consumption: governments started to provide the troops with canned meals.[12]

Compared to the more industrialized sector of the food industry, individual farmers could not afford to undertake such capital-intensive operations; installing refrigeration equipment and setting-up disinfecting and canning machinery required substantial investments. Instead of selling their produce directly in the nearest town, many farmers and their wives supplied local canning companies with vegetables and meat. Rural women found seasonal work in the canning industry; this became an important supplementary source of income for their families. The outcome was a degree of industrialization of production and the geographical expansion of distribution networks—processes that also took place in the brewing industry. The improved reliability of mechanical refrigeration triggered a growing popularity of bottom-fermented *lager* beer on the European Continent. Even at the end of the nineteenth century, only the very largest breweries were able to serve regional or national markets because of the prohibitive costs of ice-making machinery.[13]

The corporate world's industrialization of food production was less than well-received, however. Consumers were uneasy, the public protested, and the state intervened legally. For example, under public pressure, Prussia issued a comprehensive law in 1868 that allowed detailed inspection of food producers and distributors. Although demand increased as the price of canned meat and vegetables declined, many customers were not convinced that these products were safe and healthy. Scandals arose over old and rusty cans. The press eagerly reported fatalities resulting from poorly sealed cans and poisonous additives like copper and arsenic. Manufacturers failed to calm the public's fears. Even after autoclaves were introduced, and manufacturers' products were reliably sterilized, the public remained skeptical. Food companies had to resort to selling their canned products to military forces, hospitals, hotels, and ocean-liners instead of the retail market. In response to the anxiety concerning industrially produced food, urban housewives and their servants began to preserve fruits, vegetables, mushrooms, and even meat in bottles and glass jars. Complementing older methods, like drying, pickling, and salting, home canning came to represent the healthy, do-it-yourself alternative to often unbranded industrially processed cans or glasses that anonymous producers offered on the market. Throughout Europe, home canning first caught on among the urban middle classes before workers and farmers followed suit during and after the First World War.[14]

On the basis of research done by the American-British scientist Sir Benjamin Thompson (Count Rumford) the French confectioner Nicolas Appert in the early nineteenth century pioneered the

Fig. 5.3 Do-it-yourself Food Kits:

In the 1800s, companies began to sterilize food—in cans and jars—in order to preserve it. Manufacturers also offered pricey canning kits for home consumers. Proud middle-class housewives adopted these jars, thermometers, and specialized equipment. Representatives of the modernist movement regarded such kits and other technical household equipment as symbols for the modern home. These images provide a glimpse of home economist Erna Meyer's 1926 classic, Der Neue Haushalt (The Modern Household)—*in Czech translation. The volume advocates home canning as a healthy alternative to industrially produced food, explicitly associated with the U.S.*

technology of canning. The process Appert mastered was sterilizing produce at a high temperature and preserving it under airtight conditions. The French state supported Appert's experiments, and before long, tin cans were produced in great numbers for Napoleon's army. Tin-plate canning was a comparatively complicated process that involved soldering or welding. If you wanted to make your own tin cans at home, you had to make sure that a tinsmith was on hand to seal the cans; then, as now, the soldering iron was not a household appliance. A much more practical and user-friendly method was to use glass jars or bottles—although this technique also required experience and knowledge. Users had the choice of various sealing systems. To make sure that no oxygen entered the glasses, screw caps had to be accompanied by a rubber seal or a layer of wax. Another method was to employ cork and wire, a technique used for champagne bottles. In the 1890s, the German businessman Johann Weck developed a more refined kit, a combination of glass lid, metal hinge, and rubber ring. His system included thermometers, corers, and other paraphernalia. Weck's system quickly found markets in countries like Finland, where customers eagerly adopted canning to preserve the natural abundance of berries and mushrooms, for example. By appropriating this relatively expensive apparatus, the middle-class European housewife signaled self-reliance and a certain economic status. Home canning became a way to avoid the dangers associated with industrially produced foodstuffs and a means to increase one's degree of independence from the marketplace.[15]

Appert's principles returned to France via Germany. Despite initial popularity with the middle class, Appert's home-canning methods took almost a hundred years to reach the French countryside. During the harvest season, rural women were unable to spend days on end in the kitchen; in any case, the costs of the necessary equipment were, for most French farmers, prohibitive. German middle-class housewives and better-off farm wives proved quicker in adopting this technique than their French sisters. After the German–French war of 1870–71, home canning enjoyed an upsurge in the occupied territory of Lorraine. French manufacturers offered various sealing technologies. The *Savor* system was similar to Weck's—and marketed as a French system, whereas the *Éclair* system employed a metal chain collar instead of a bale, or wire apparatus. Both were expensive. Buyers of *Savor* jars were advised to use pressure cookers for sterilization and the *Éclair* system included a specially designed kettle.[16]

The breakthrough of home canning was not a foregone conclusion but resulted from aggressive propaganda and marketing methods. After all, other preservation techniques—like drying, pickling, and salting—had been known for centuries, and it was not easy to convince users to turn to something novel. Weck's company published its own cookbook and even a monthly magazine to attract and retain customers. In other words, the company assumed the role of the expert, teaching potential users how to domesticate their canning kits in the proper way. New companies jumped onto the bandwagon, advertising preservatives and gelling agents to simplify the process. Home economics teachers who taught in schools typically covered various canning techniques.[17] Household manuals and advertising campaigns—especially during the First World War—contributed to the acceptance of this "bourgeois practice," as an anonymous villager from Northern France called it as late as the 1930s.[18] Before the war, rural women had shunned home canning. To combat resistance, one author of a 1913 French handbook claimed that the newer apparatuses "simplify the manipulations of home canning and boost its chances of success."[19] To adopt home canning was to perform a national duty: "Every good housewife, every good French woman must prepare cans of preserved food."[20]

Developments on the European Continent differed from those in the United States. Partly as a result of the greater distance between agricultural centers and large cities, canning took on industrial dimensions soon after the Civil War. The Heinz Company succeeded in convincing customers of the purity and high quality of its products like baked beans and ketchup. Early on, U.S. housewives were ready to serve their families canned corn, sweet pickles, and ready-made soups. Home canning, on the other hand, waned quickly in the United States after the First World War. For political and economic reasons, canning reappeared in Germany's *Third Reich*, where the do-it-yourself method aligned with the official autarky campaigns of the Nazi government. To guarantee successful jams, a cookbook author in 1933 advised housewives not to buy produce in the store but to rise early and pick berries in her own garden. Supported by later war-time shortages, this policy contributed to 90 percent of all German households having home-canned produce in their pantries and basements. During the war, the Nazi regime exported its policy of food autarky and propaganda for do-it-yourself techniques to the occupied countries as well.[21]

In Britain, toward the end of the nineteenth century, the use of Kilner and so-called Mason jars spread. During the First World War, home economics institutes offered courses on the proper use of jars. During the interwar years, the Scottish Women's Rural Institutes organized demonstrations for schools, stores, and even theaters on how to make one's own marmalade, jam, and chutney; the organization published a cookbook featuring various recipes. Locally organized, and supported by the Board of Agriculture for Scotland, the institutes realized how important farmers' wives were to the national economy. Lectures on how to make your own cheese and honey, and how to grow particular vegetables and fruits, were not established for romantic reasons, but strategically aimed to strengthen the country's economic base. Reformers, believing that women were easier to reach than their husbands, focused on women as agents of modernization. The Scottish Rural Institutes' cookbook was reprinted several times. Its origins in the mid-1920s illustrate the important role imputed to individual experience and skill. The editors had invited the institutes' members—nearly 10,000 women—to submit their favorite recipes; a selection of these was included in the published volume. Indeed, the Scottish example was typical of many. From the United States to the Netherlands, home economists, supported by national governments, targeted rural and working-class women through lectures on nutrition and courses on cooking in an effort to bring modernization to the countryside and low-tech, yet efficient cooking techniques like hay boxes to the city.[22]

At Home with Colonial Foods

The Scottish Women's Rural Institutes brought out a new edition of their cookbook to mark the opening of the 1938 Empire Exhibition, organized in Glasgow. By no means did all dishes belong to traditional Scottish cuisine; the editors also included foreign and so-called colonial recipes. Since the first World's Fair at the Crystal Palace in the mid-nineteenth century, exhibitions of this kind had showcased the latest achievements in industry and craftsmanship. From the start, they had included nationalist components. One year before U.S. corporations would display their grandiose visions of the future consumer society at the New York World's Fair, Scottish women explained to visitors how to make clootie dumpling and skirlie. The Empire Exhibition highlighted commodities from Scotland and various parts of the Commonwealth. Indeed, the interwar years bracketed a period when British authorities made a concerted effort to convince their subjects to buy products that hailed from the Empire. A Conservative government established the Empire Marketing Board in 1925 to support the selling of such goods, primarily foodstuffs. The Board organized public campaigns and supported propaganda movies on topics such as Jamaican bananas and Sri Lankan tea. Setting a historical precedent in the commercial film industry, Sri Lankan tea itself starred in a movie with the romantic title, *Song of Ceylon*. The Board even persuaded the British Prime Minister to accept the lead role in one of its films.[23]

 Conservative women's organizations supported the activities of the Empire Marketing Board. Earlier, the British Women's Patriotic League—known for its clear anti-U.S. stand—had organized

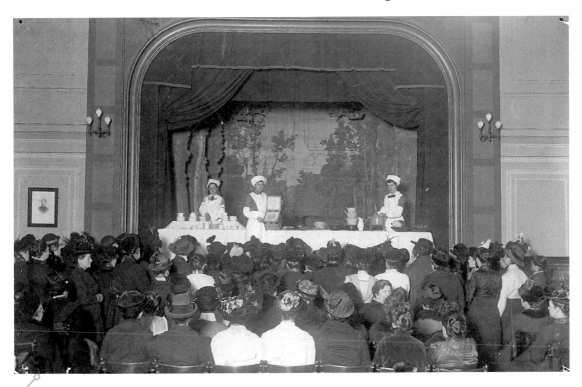

Fig. 5.4 The Home Economist as Science Teacher: *The profession of home economics was dedicated to improving social conditions. By popularizing science and technological solutions, home economists hoped to better daily life. In this photo, taken during the First World War, home economists demonstrate to an audience of Dutch working-class women how to cook using the so-called hay box, seen at center. This wooden cooking device, when filled with hay for insulation, cooked food over a 24-hour period—with a minimum of petroleum as fuel.*

so-called Empire Shopping Weeks. Given the choice between Canadian and U.S. apples, for example, the patriotic British house-wife was coached to buy Canadian—and avoid U.S. products. The middle-class Women's Unionist Organisation also took up the cause. In the second half of the 1920s, it arranged local cooking competitions to support the consumption of products from the British dominions, and it tried to convince storeowners to deal in colonial goods.[24] In 1925, its journal with the telling title *Home and Politics* reported that housewives at the British Empire Exhibition at Wembley received instruction on "the right methods of thawing frozen meat from New Zealand, [and] of soaking Australian dried fruits to make delicious summer dishes."[25] By the time the Empire Marketing Board closed down in 1933, the "Empire" had lost some of its political and commercial pull. This did not mean that the Brits gave up drinking tea and eating bananas. Nor did it imply that they refrained from putting West-Indian sugar in their tea. It indicated that British politicians found it more of a priority to support "Buy British" campaigns than to to promote advertising that featured various dominions.[26]

Other European countries also relied on distant worlds for various foods. Europe's foodways had depended for centuries on both Arab countries and far-away colonial lands. From the early days of world explorations, imperial and commercial interests had brought exotic fruits, plants, and drinks to Europe. Like tea, sugar, coffee, and chocolate were traditional colonial commodities. European consumers domesticated them into their daily eating and drinking habits in various ways. Sugar was more commonly used in Northern Europe; chocolate remained a luxury good for the upper classes well into the twentieth century.

The global history of sugar is one of conflict. It involves trade routes and slave labor, industrial developments, protectionist politics, and social differences. It is a story that highlights Europe's dependency on—and ties with—countries overseas. Sugar cane came to Europe from the Arab world in the Middle Ages. In the fifteenth century, it was grown in Spain and Portugal, in Madeira, and on other islands in the eastern part of the Atlantic Ocean. Once the Europeans began to colonize the Americas it did they soon established sugar plantations staffed with slaves brought over from Africa. Living and working conditions for the slaves were dreadful. To satisfy the sweet tooth of Europeans and North Americans, African laborers paid with their lives. The sugar cane was either refined in the colonies or shipped to Europe, where mills had been constructed to produce pure sugar for various uses. Although consumer prices for sugar dropped marginally as

Fig. 5.5   Forced Labor beyond Europe: *Despite human-rights and labor activists' campaigns, forced labor took place well into the nineteenth century—and indeed continues to this day. To satisfy Europe and America's cravings for cocoa, coffee, and sugar, laborers endured brutal working conditions at plantations in the Americas, Africa, and Asia. From the French and the Belgians to the Dutch and the Portuguese, farming these cash crops was a European imperial practice here illustrated in a 1910 postcard depicting sugarcane workers in Portuguese Madeira. The introduction of beet sugar in Europe led to lower commodities prices— which made the plight of laborers even more precarious.*

the cultivated area expanded, its use was largely restricted to the affluent. Worldwide, the great majority continued to use honey as a sweetener. Confectioners, whose handicraft knowledge spread throughout Europe from Venice, innovatively adopted sugar to create new types of delicacies. Members of the nobility with cash to spare were able to indulge in cakes and candy, liquors, and lemonades. In the eighteenth century, the use of sugar was especially pronounced in Great Britain, where it found its way not only into teacups but into innumerable desserts. In the Edwardian era, puddings became a standard item on the supper menus of the middle and upper classes.[27]

The large-scale planting of beet sugar in the mid-nineteenth century was the first step toward what may be called the democratization of sugar consumption. Sugar beets were well suited to the soils and climate of the European Continent and lent themselves to refinement on an industrial scale. Along with chocolate producers, sugar refineries belonged to those companies in the nutrition industry with the highest number of employees. Consumer prices dropped steadily in the second half of the century, and sugar could soon be found on the tables of workers and farmers. On the eve of the First World War, one observer reported that even the poorest Londoners were able to afford "tiny amounts of tea, dripping [melted fat], butter, jam, sugar and greens."[28] As in so many other areas, British consumers led the way. Statistics from the turn of the twentieth century show that Brits consumed six or seven times as much sugar as the inhabitants of Austria-Hungary and Russia. So did the Scandinavians. Southern Europeans, however, took in less sugar because it featured less prominently in their cuisines.[29]

The declining price of sugar does not account entirely for its growing popularity in The Long Twentieth Century. The social aspect also figures prominently: by adopting sweet dishes and fancy foods from the tables the well-heeled, members of the aspiring middle classes tried to improve their social standing—a process skilled workers copied in due course. The widespread practice of home canning—in which sugar also played a central role—also enhanced one's social status in middle-class circles. Having sweet snacks for tea and pudding to round off dinner were well-established practices in the British upper and middle classes by the mid-nineteenth century.[30]

A substantial amount of the sugar consumed was used to sweeten beverages such as tea, coffee, and hot chocolate. The historical connections and parallels between sugar and coffee are especially pronounced. Originally, coffee had come to Europe as an expensive trade good from the Arab world and the Ottoman Empire. Consumption was marginal. Documents from the seventeenth century show that people living in those parts of Hungary occupied by Turkish troops avoided black coffee, denouncing it as the drink of the enemy. Coffee's popularity increased during the following century, as European trading enterprises like the Dutch and the French West-Indian companies set up coffee plantations in Latin America. As with sugar, the cultivation of coffee was based on slave labor. A new kind of coffee-drinking culture emerged in cities such as Amsterdam, Paris, and London, and soon spread all the way to Vienna. Coffee houses emerged, in many countries attracting politically liberal groups. In Europe, the social ritual of drinking at a coffee house facilitated an emerging political culture. One could say the coffee drinkers of the time—the 1700s and 1800s—were pioneers of a users' group. For members of the Hungarian urban upper class, primarily in Buda, "Dutch" coffee had none of the negative connotations that "Arabic" coffee evoked. In order to distinguish themselves from their Turkish occupiers, Hungarians chose to drink "Dutch coffee" with milk—a custom unknown to the Turks. As prices dropped over the course of the nineteenth century, Europe's lower classes adopted the habit of drinking coffee and tea. By 1900, most social classes included these beverages in their daily menus; still there were marked differences between national practices. In Germany, coffee had established itself as a daily drink in all segments of society; in Britain, however, coffee retained its position as an upper-class beverage for several decades longer.[31]

Whereas sugar and coffee had become economically accessible to the majority of the European population by the end of the nineteenth century, chocolate remained largely a luxury. Preparing and drinking hot chocolate (based on cheap cocoa powder) was a poplar practice around the turn of the twentieth century, but chocolate bars remained fairly expensive until the interwar period. Records from Belgian working-class families show that they ate almost no chocolate whatsoever in the 1890s; even at the end of the 1920s, each worker's household still consumed less than 1.5 pounds of chocolate each year.

An indigenous American plant, the cocoa tree was unknown to Europeans before Columbus. In the late sixteenth century inhabitants of the Old World began to adopt cocoa beans. Cocoa first arrived in Europe by way of the courts and noble circles of Spain and Italy. From there, it spread to Paris and London in the seventeenth century. The first British recipes for a cocoa drink are recorded in the early eighteenth century. As might be expected today, one early recipe prescribed mixing the cocoa with water, milk, and sugar; the unexpected elements were egg yolks and rosewater. This brew was a transitional beverage. As Europeans domesticated native-American chocolate, they transformed it from a spicy and bitter drink that was an acquired taste, to a sweet, smooth, hot beverage to which even the youngest of children could easily become habituated. Paramount in this transformation of food habits and tastes was the addition of milk and sugar.[32]

It would take some time for chocolate to move from the courts of Madrid and Paris via the clubs of London to the nurseries of the Continent. By and large, cocoa followed the same trajectory as tea and coffee. In fact, like coffee, chocolate circulated across the globe, throughout Europe, and among classes. In the process, chocolate took on new meanings and adapted to new contexts. In part, its growing acceptance had to do with lower prices enabled by the mechanization of production. And, in part, its popularization was the outcome of new and often aggressively advertised products entering the market. In the mid-nineteenth century, companies in a number of countries introduced cocoa powder and molded chocolate. The powder could be mixed much more easily with hot water and sugar compared to the earlier cocoa paste, which was heavy and fatty. The products made with the new molds became known as chocolate bars. One historian wrote about what one could call the international division of labor in this area: "Still in the 1890s chocolate could be regarded as French, cocoa powder as Dutch, the artisan specialties were Spanish, Belgian or at most Italian, whereas Great Britain, or later America and Germany, brought to mind large industrial productions."[33] Craftsmen from the Lombardy region of Italy, particularly Milan, manufactured exclusive chocolate delights that were exported to wealthy consumers north of the Alps.

Switzerland is often associated with chocolate. Chocolate-making had a long tradition in the (Italian-speaking) Swiss district of Ticino. Here, artisans benefitted from close contact with colleagues in Lombardy. In the last decades of the nineteenth century, some of the leading players in the industrialization of Swiss chocolate manufacturing emerged from Ticino and from other areas in southern Switzerland. The rapid success of Swiss companies like Suchard, Lindt, and Tobler illustrates the importance of product novelty and the power of active marketing. Of immense significance for the growing popularity of cocoa on an international scale was the development of palatable milk chocolate by the Swiss candle maker Daniel Peter. Other confectioners had experimented with this; Peter's success was based on his use of condensed milk—a product invented by Henri Nestlé in the 1860s. The fact that Peter and Nestlé were neighbors in the small town of Vevey on the Swiss side of Lake Geneva spotlights the role of direct information exchange and close cooperation in the history of technology. Supported by extensive marketing campaigns, sweet milk chocolate—either as a hot beverage or in the form of bars or drops—became *the* lever to press in order to reach a mass-scale market. Within a few decades, cocoa products could be found on breakfast tables in most European countries as well as in schoolbags of middle-class children. Milk-chocolate powder, sometimes of dubious quality, was offered at prices that working-class families could afford as well.[34]

The cocoa–milk–sugar mixture became a commercial success. Children liked its flavor, as did grown-ups. Dairy farmers were happy to find new outlets for their milk, and nutrition scientists hurried to support the chocolate producers. Criticizing tea and coffee for lacking any nutritional value, a Swiss scientist claimed that chocolate had an important role to play in the "national diet."[35] The combination of cocoa with milk suited perfectly the obsession of the day with proteins in general and milk in particular. Nutritionists also praised chocolate as an important energy source for workers; dentists were more critical, but their voice was utterly drowned out by advertising campaigns. Within a few decades, companies and medical doctors deliberately transformed the image of chocolate from a stimulant and luxury product to a healthy food. Savvy marketing people in France consequently launched a new product, *chocolat de santé*. In Britain, Cadbury became associated with milk chocolate of all kinds; in Germany, the company Stollwerk acquired

Fig. 5.6 Chocolate & the Food Chain: *Throughout its history, chocolate has been marketed as either an exotic foreign product or a locally produced one. In the case of nineteenth-century Europe, chocolate production was colonial, while consumption was European. In this advertisement from the end of the century, a commercial poster produced by the Swiss Lindt Company, chocolate has, in a sense, conquered the world. Consumers and producers from around the world are an imagined global group celebrating chocolate. The poster upends the notion of colonial producers and European consumers.*

a similar reputation. In the Netherlands, Verkade, Van Houten, and Droste dominated the market.[36]

Swiss manufacturers were especially successful in exploiting the connections between cocoa and milk. They did so by national-izing the image of milk chocolate in several ways. Various actors perpetuated the notion of Switzerland as the promised land of milk and cheese that Swiss nutritionists and medical doctors had helped to invent. In 1888 the Swiss Charitable Society (*Schweizerische Gemeinnützige Gesellschaft*) defined milk products as "national nutrients" (*Volksnahrungsmittel*) and argued that they ought to replace coffee and potatoes.[37] Milk-chocolate producers like Suchard happily exploited the nationalist idea further, branding its products as particularly clean and original. This was based on the proposition that milk came from healthy cows, grazing the unpolluted slopes of the Swiss Alps. Advertisements typically depicted Alpine settings with calm and contented cattle alongside traditionally dressed milkmaids, hearty and vigorous. In a similar vein, a manufacturer like Tobler created a direct link between chocolate and sublime Swiss mountain peaks. In an impressively innovative manner, its triangular-shaped chocolate bar, *Toblerone*, embodied the Alps. Although the recipe had Italian origins, Tobler successfully marketed its chocolate–honey–almond blend as an edible, miniature version of the Matterhorn, the Swiss summit. This connection was of particular importance to the British and German markets: in these countries, Switzerland had a powerful allure for mountaineers and other tourists. In this connection, the high caloric value of chocolate came in handy. Ads pictured mountaineers eating chocolate to gather strength for their athletic feats. Some of these ads became more or less interchangeable with posters from Swiss tourist offices.[38]

The Swiss attempt to, as it were, nationalize cocoa was one marketing strategy. More generally, the public image of chocolate slowly but surely transformed from a colonial to a domestic product. During the late nineteenth century, bluntly colonialist and even outright racist motifs were still commonplace. Advertisements often featured either happy, dark-skinned cocoa pod pickers in a tropical setting, or smiling black servants in an upper-class, European home. The fact that most plantation workers were forced by poverty to work under slave-like conditions—had not yet entered the consumer's consciousness. The association of cocoa with Africa strengthened during these years. To meet the growing demand for chocolate in the rich countries—and as a reaction to uprisings and strikes in Latin America—globally-acting companies had begun to set up plantations in Africa and in Dutch Indonesia. Economically speaking, cocoa was a global product; culturally speaking, it had sometimes been defined as a colonial good, at other times as a national food.[39]

By the first decades of the twentieth century, the domestication of cocoa meant that chocolate had lost some of its exotic flair. It had become a product associated with harmonious, middle-class life, and with individual European countries. Caring mothers provided their children with this supposedly healthy beverage. Posters featured well-dressed children of both sexes, chocolate bars in hand. Increasingly, chocolate became defined as a children's product. Arguably, this kind of class-based advertising alienated working-class customers. Molded chocolate remained a privilege of the upper and middle classes. Shortly before the First World War, even working-class families in Switzerland only occasionally consumed chocolate products, usually on important holidays and for celebrations. Exploring the connections between healthy Alpine landscapes and chocolate, Swiss companies were to some extent more successful abroad than at home.[40]

Feeding the Common Man in Poland

Europeans were indeed the first to combine chocolate with sugar and milk. Throughout the Continent, the milk of cows, sheep, and goats has long been employed to make various kinds of cheese, curd milk, buttermilk, and cream. Given that it was expensive, butter had initially been a medicine or ointment; only the wealthy included it in their diets. However, drinking pure, non-processed milk was not a widespread habit; only people in dairy-farming regions—primarily in Northern Europe and the Alpine regions—included it in their diet. Even there, consumption was limited to infants and children. In the city, working-class families seldom had milk on their kitchen tables. The situation changed when social reformers championed milk as a better alternative to the working-class habit of drinking large amounts of alcohol. The milk bar would become an alternative place to the (liquor) bar where the working class and teenagers met.[41]

The newly developed taste for milk in the last third of the nineteenth century was an outcome of promotional campaigns orchestrated by national governments. The state had won the support of middle-class

reformers and scientists. All agreed that milk was a healthy alternative to beer and heavy alcohol. Charities began to serve milk to the
poor on a non-profit basis; city authorities in some cases subsidized
milk for school children. In the 1870s, the German Emperor personally supported the push to increase milk consumption. The state's
support of milk reached its peak in the interwar period; numerous
European countries took initiatives to convince citizens to drink
more milk in an effort to beat the economic crisis: farmers faced
declining international cheese prices and overproduction of milk.
Governments, dairies, and farmers everywhere joined forces. In the
Netherlands, the government and large dairies in the 1930s founded
a Dairy Crisis Agency (*Crisiszuivelbureau*); the crisis in question was
not consumers' health but the industry's overproduction.[42]

One state-subsidized measure to help dairy farmers to capitalize
on their surplus was the free delivery of milk to all primary schools.
In Scotland, school children and parents were mobilized to solve
the problems of the milk producers: pupils received school milk
for 50 percent of the standard retail price. The 1933 Milk Marketing
Board worked to keep producer prices up—and even raise prices.
State organizations also supported nutritional research and organized promotional events. In Norway, one measure was to force

**Fig. 5.7   Milk to
the Rescue:** *With the
upsurge of dairy farming
after the First World
War, many countries
"suffered" from a milk
surplus. With the
support of nutritionists,
home economists, and
relief workers, the dairy
industry persuaded
people to consume
more milk products
for their nutritional
value. National and
local governments
also promoted milk
products—by providing
school children with free
or subsidized milk, for
example. Throughout the
1900s, relief organizations
set up milk stations to aid
the undernourished. This
1948 photo shows Jewish
orphans eating a snack
with milk at a children's
home in Hungary.*

manufacturers of margarine to mix a specified amount of butter into their products; another tactic was to mobilize scientific expertise.[43] Behind these measures lurked a nationalistic view. Edited by the professor of Hygiene at the University of Oslo, Carl Schiøtz, the magazine *Life and Health* reminded its readers in 1936 that "the three main demands of the national dietary plan are the daily consumption of 1 liter of milk, 30 grams of butter, and 30 grams of cheese."[44] Claiming that the question of nutrition was "a national problem," the magazine regarded milk and milk products as the most important "protective foods." Schiøtz adopted this concept from the influential U.S. chemist Elmer McCollum. The League of Nations had also used the idea in various campaigns.[45] From his position as a scientist and promoter, Schiøtz supported the Oslo authorities' decision in 1932 to introduce compulsory breakfasts in all schools in the Norwegian capital. The "Oslo-breakfast" also included "Whole-wheat bread"; "Norwegian Knecke-bread"; and vegetables or fruit.[46]

The increased consumption of dairy products represented a key market development: European farmers shifted to cattle breeding when they faced a sharp decline in grain profits because of the cheap imports from North America. In response, farmers organized production cooperatives; this created demand for their dairy products. Both on an international and a national level, farmer cooperatives articulated an answer to large-scale capitalist production and commerce. As self-help associations, cooperatives enabled farmers, workers, and consumers to make larger investments; spread economic risks; and deal with market fluctuations. Fostering a combination of individual responsibility and collective solidarity, the cooperatives often went hand-in-hand with grassroots democracy. The Danish farmers' cooperatives in particular served as an international role model in successfully mastering the transition to capital-intensive dairy and cattle farming. From Russia, farmers bought cheap grain in bulk to feed their livestock; they created cooperative dairies and slaughterhouses to keep the processing of their products in their own hands; and they worked together to sell their milk, butter, eggs, and meat to urban customers in order to circumvent middlemen. Imbued with political activism and ideals, the movement spread from Denmark to the United States as well as to Russia and Poland. Opening milk bars was another way in which cooperatives tried to capture a share in the distribution of their products.[47]

In Poland, in particular, customers came to embrace the milk bar as a unique national institution. Serving milk products and inexpensive meals for the urban working class as an alternative to bars serving alcohol, milk bars emerged as an international trend. In today's Poland, the milk bar is considered to be part of the national food heritage and has recently been the subject of many heated discussions, given the proliferation of foreign fast-food restaurants.[48]

The first Polish milk bars (*mleczarnia*) were private initiatives. In 1896, landowner and entrepreneur Stanisław Dłużewski opened a "milk drinking room" (*pijalnia mleka*) in Warsaw—a standing only place where customers were served fresh milk, tea, coffee with cream, and buttermilk. Later on, he added affordable meals to the menu. The idea spread like wildfire; by 1909, nearly ninety small milk bars dotted Warsaw. Most dairy-shop owners were farmers who used them as outlets to sell to urban consumers, thus avoiding middlemen and increase their profit margins. Dłużewski had milk delivered fresh every morning from his farming estate located in the village of Dłużew, about twenty-five miles outside the capital. Evoking the lush meadows along the Świder River, where his cattle grazed, he named his milk bar *Mleczarnia Nadświdrzańska* (The Świder River Creamery). Success was such that Dłużewski soon started four more bars in central Warsaw. Traditionally, blue-collar workers made up the largest group of patrons, but artists, journalists, and writers of limited means were also sighted here. The most famous *Mleczarnia* in the Nowy Świat Street garnered a special reputation as a place where bohemian women gathered.[49]

The idea of the milk bar received more systemic support. As part of the political excitement of Polish independence in 1918, the Congress of Polish Cooperatives met in Warsaw and soon established the Union of Consumer Cooperatives (*Związek Spółdzielnia Spożywców*). The Union brought together local food cooperatives, appropriately changing its name to *Społem* (Together). As in other parts of the world, the 1930s were the formative years in the history of milk promotion and supply. To propagate the idea of milk bars (now called *bar mleczny*), the Polish Dairy League (*Polska Liga Nabiałowa*) used the Warsaw Diary Exhibition in October 1934 as a platform for launching the rallying cry "milk for all." The league sought to promote the consumption of milk through cooperative movement channels. To do this, the league counted on government assistance. The exhibition displayed model milk bars and a dairy shop in separate, well-lit rooms as part of the campaign.[50]

Fig. 5.8 Poland Eats Healthy—The Milk Bar: *It started as international social reformers' alternative to the bar that served alcohol to the working class. And for more than a century, the milk bar has been a cornerstone of urban Polish culture. In this 1954 oil painting by Józefa Wnukowa, a sympathetic-looking milk bar employee serves nutritious food. To this day, the milk bar's wholesome, affordable fare attracts everyone from workers and students to tourists and intellectuals. As a Polish phenomenon, the milk bar has been the subject of everything from satire to national nostalgia.*

After the Second World War, the socialist regime in Poland embarked on a policy of providing all workers with cheap meals at their workplace through canteen facilities. At the same time, the regime either closed down or nationalized privately owned restaurants. For pensioners, students, and workers without access to state-subsidized canteens, milk bars offered a welcome alternative. Rooted in the late nineteenth century food cooperative traditions, these small-scale eateries were dairy-based—and sometimes all-vegetarian cafeterias. Milk bars served egg dishes, cereal, and flour-based meals like the typical Polish cheese dumpling *pierogi ruskie*. The first postwar milk bar, appropriately called *Pionier*, opened in 1948 in Krakow, traditionally a leading center in Polish academic, cultural, and artistic life. A year later, there were eighty such milk bars all over the country, including seven in Warsaw. They quickly gained a reputation as the cheapest places to eat. Today, milk bars offer low-income citizens much-needed affordable fare—traditional meals touted as Poland's most authentic food.[51]

Manufacturing Balkan Authenticity

In its efforts to encourage milk consumption, the dairy industry—often supported by science—launched new products that were thought to be healthy and to foster longevity. Yoghurt is a case in point. Strictly speaking, yoghurt was not novel, but when the Paris-based Russian scientist Élie Metchnikoff linked yoghurt to longevity, he singlehandedly managed to popularize the Ottoman and Balkan product that had been unknown in Western and Central Europe.

A microbiologist, Metchnikoff was especially concerned with the process of human digestion. In 1903 and 1907, he published two popular-science books that became international bestsellers. Metchnikoff used unpublished data from a demographer M. Chemin, concerning numerous old people in Greece, Serbia, Bulgaria, and Romania; the writer claimed that "there were more than 5,000 centenarians (5,545) living in 1896."[52] Based on rather thin scientific evidence, Metchnikoff concluded that "the pure and keen air of the Balkans, and the pastoral or agricultural life of the natives, predisposes [people] to old age."[53] More specifically, he connected longevity to Bulgarian peasants and their consumption of "substantial amounts of 'soured milks.'"[54] For this insight he credited a young Bulgarian doctoral student, Stamen Grigorov, who in 1905 had identified a strain of bacillus in a yoghurt sample brought from Trun, Bulgaria to his supervisor's lab in Geneva, Switzerland.[55]

This story might have ended as a parenthesis in the history of science had it not been for Metchnikoff's almost messianic drive to advance his cause. As an employee of the prestigious *Institut Pasteur*, he was able to promote the cultures as pure and authentic samples from the Balkans. Metchnikoff's public relations campaign worked: by 1907, numerous companies had already begun to sell yoghurt's bacterial culture as a medication on the commercial market. After newspapers in Europe and the United States in 1910 had reported on the oldest living Bulgarian peasant woman, the English dairy and nutrition specialist Loudon Douglas began to popularize the Balkan legend of longevity and yoghurt further. Dairies in France and abroad eagerly began to produce yoghurts based on fermentation. The *Maya Bulgare* firm cultivated yoghurt, sold starter cultures, and published material to advertise the product for the Western market, playing on the connection that Metchnikoff had made between yoghurt, Bulgarian centenarians, and the promise of longevity.[56]

To capitalize on the consumer craze for the new health food, a number of Balkan emigrants (living in France, Spain, and elsewhere) further commercialized yoghurt's newfound reputation in Western

Fig. 5.9 Selling Bulgarian Robustness across Europe: *In 1903, Doctor Elie Metchnikoff claimed that Bulgarians lived longer than any other Europeans—because they ate Bulgarian yoghurt. To capitalize on this claim, several European companies began to market yoghurt as a healthful food in the 1920s and 1930s. This 1982 advertisement by the Finnish dairy cooperative* Valio *exemplifies this marketing strategy: Bulgaria's pristine natural environment becomes the backdrop for yoghurt as health food.* Valio's *products, while licensed from Bulgaria's state-owned dairy company, were made and sold successfully in Finland. Bulgaria, too, profited from exporting its wholesome image and yoghurt-making technology.*

Europe. The Constantinople-born Armenian Aram Deukmedjian, studying at the Sorbonne on a French scholarship, established the yoghurt brand Aram in 1912 in his creamery-restaurant named *Cure de Yogourt*, with the assistance and approval of Metchnikoff. Similarly exploiting his ethnic background and the industrial world's health fad, the Jewish Isaac Carasso emigrated with his family from the multi-ethnic Thessaloniki, situated in the border zone between the Ottoman Empire, Greece, and the Balkans, to the Catalan port city Barcelona. There he started the yoghurt company Danone in 1919. Familiar with Metchnikoff's work, he promoted his yoghurt first as a medication, distributing the product through local pharmacies and doctors' offices. But when his son, Daniel, after studying at the *École Supérieure de Commerce* in Marseille and taking a course in bacteriology at the *Institut Pasteur*, opened a family business in Paris in 1929, he decided to differentiate the yoghurt from that of his competitors. He marketed yoghurt as a healthy dessert rather than a medication. During the interwar period, many companies from Paris to Munich followed suit.[57]

Entrepreneurs sought to capture consumers' imagination by evoking the healthy air of the mountainous Balkans. The marketers mined the product's locale and authenticity rather than science and the laboratory. The immigrant shops in France borrowed the names from the geography of Bulgaria to market their products: Vardar (Mavale), Rila (Iamkoff), Sofia (Nestoroff), and Balkans (Bonzoni and Speransky). A Munich-based company used oriental images of Turks to advertise its yoghurt, Paskal. So did the Dutch company HET. More well-established food companies like the Anglo-Swiss Nestlé and German Maggi entered the booming new health-food market. In 1915, Nestlé opened a small plant, *Fabrique de Joghurt*, in Constantinople to acquire local knowledge and develop its own brand before establishing an experimental laboratory for dairy products in Paris in the 1920s. Do-it-yourself kits offered by other entrepreneurs promised that the homemade fare would protect users against the unhygienic conditions of the dairies. Yalacta instructed consumers how to make yoghurt cheaply and safely at home. The kits also suggested how users could experiment with new flavors to suit their taste. The yoghurt-making device came with a booklet from recipes of the "Orient," using coffee, chocolate, and fruit. The device, while emphasizing yoghurt's medicinal characteristics, helped consumers to produce, consume, and imagine yoghurt in new ways.[58]

The creation of Bulgaria as the land of yoghurt took place primarily in Western Europe. After the Second World War, the Bulgarian state laid claim to the local version of yoghurt as the most authentic product in Europe—if not in the world. For export purposes, the government invested in research and development. Preserving the taste of the Balkans proved difficult. For decades, Bulgaria, like other European countries, had sought to modernize its small-scale dairies. The communists only intensified the modernization and collectivization of production. The twin forces also stamped out the local knowledge and resources of peasant women who had preserved the technique of making yoghurt from mother to daughter for generations. Just as elsewhere in Europe, Bulgaria's premier dairy plant and the largest dairy producer of Eastern Europe, Serdica, used the milk of cows instead of the sheep's milk traditionally required in home production: cow's milk was cheaper to produce and satisfied the demands of mechanization and continuous production. Scientists at Serdica discovered that the industrially processed milk treated with antibiotics changed the cells of *Lactobacillus bulgaricus* and decreased the *Streptococcus* that had so impressed Mentchnikoff.[59]

Bulgarians disliked standardized yoghurt. In the early 1960s, scientists at the Higher Institute for Food and Flavor Industries in Plovdiv—and at the Serdica dairy plant—invested heavily in trying to reproduce the flavor and consistency of yoghurt made by Bulgaria's women peasants. Scientists collected yoghurt samples from farmers around the country. In the end, they managed to isolate and develop the strains that made Bulgarian yoghurt distinctive. They patented the product and methods in the early 1970s. At first, Bulgaria sought to export Bulgarian yoghurt directly, but transportation costs for such a delicate product were prohibitive. Next, Serdica sought to sell its patented and cow-based starter culture—and the company's know-how—to foreign companies.[60]

The Bulgarian state discovered that Western consumers had developed a sweet tooth. Moreover, Bulgarian yoghurt was neither an internationally recognized trademark nor patent protected. Few European companies recognized Bulgarian yoghurt as authentic. In 1967, not long after a tasting session at the consulate in Paris, Bulgarian diplomats were ecstatic when the company Yoplait concluded Europe's first license agreement to use Bulgarian yoghurt technology. While Chambourcy did promote the Balkan nation, claiming the company sold "*Le* vrai *yoghurt goût bulgare*," it did so without a license from the Bulgarian state. Another French company had trademarked "Balkan"; and to the dismay of Bulgarian officials, the French company Danone, the Portuguese producer of *Farinha lactobulgara*, and the Dutch firm Mencken (which produced *Bulgaarse yoghurt*) all claimed Bulgarian authenticity and methods. Finland was Europe's most successful nation in appropriating Bulgarian yoghurt's taste, technology, and trademark. The leading Finnish Cooperative Dairy Association, Valio, concluded a licensing agreement for their *Bulgarianjogurtti*. For the Bulgarians, however, the first truly successful export of their yoghurt innovation—and promotion of the Bulgarian mountains as a brand—happened via the Meiji Dairy Company in Japan in 1972.[61]

The Bulgarians were less successful with the Swiss cooperative, Migros. In June 1976, Bulgarian representatives held yoghurt-tasting tests in Migros' Zurich office, presenting a variety of plain yoghurts from both cow and sheep milk. Realizing that Western Europeans liked flavored rather than plain yoghurts, they also presented yoghurt samples and drinks with raspberry syrup and one with strawberry jam. The Bulgarians discovered that Migros was already successfully producing a range of flavored yoghurt based on Danish starter cultures; Migros' product line used technologies that were similar to those of the Bulgarians. Negotiations broke down. As Danone had already found out in the 1930s—and the Bulgarians discovered in Zurich in 1970s—Americans and Western Europeans had developed a sweet tooth for flavored yoghurts. These milk-based products had little to do with the yoghurt strains that Menchikoff had promoted as so beneficial. But, like the ideology of the American Kitchen, the legend of Bulgarian yoghurt has been a powerful one: to this day, many buyers of Bulgarian yoghurt believe that they are purchasing an intrinsically healthful product.[62]

Supermarket U.S.A. Meets Europe

By the time Bulgarian representatives had gotten around to negotiating with Migros in the 1970s, the Swiss supermarket company had developed beyond recognition. The pioneering store in the Seidengasse had been 120 square meters (1,300 square feet) large—in hindsight, an incomprehensibly small size. As far as size goes, European supermarkets did become increasingly similar to their big brothers in the United States. As far as content and organization were concerned, however, the European stores developed their own profile.

In the beginning, the United States had definitely served as an explicit role model. Internationally operating U.S. players—most notably journalist and businessman Max Zimmerman and his Super Market Institute—actively promoted the self-service idea. Zimmerman received direct support from both corporate America and the U.S. government. He was instrumental in organizing the First International Congress of Food Distribution, held in Paris in 1950—an event that attracted 1,000 participants from more than twenty countries. Five years later, Zimmerman published a much-cited book that summarized the principles of self-service and supermarket retailing; a Spanish translation appeared in 1959. U.S. companies also got involved. During the Third Congress, held in Rome, the National Cash Register Company (NCR) designed a complete U.S.-style supermarket featuring 2,500 different products. By this time, American efforts to roll out supermarkets throughout Europe had taken on an explicitly anticommunist approach. Supported by two U.S. ministries, the store in the Italian capital attracted nearly half-a-million visitors—in less than a single month![63]

U.S. influence went even further. One year after Zimmerman published *The Super Market*, the Rockefeller International Basic Economy Corporation (IBEC) decided to try to gain a foothold in

Fig. 5.10 The Supermarket as Cold War Tool: *For the United States, the supermarket was a capitalist tool. Not so for Marshal Tito, who tried to appropriate the supermarket as a Socialist strategy. In defining his own version of Socialism, the Yugoslavian President used the supermarket model to promote large-scale, collective farming over small-scale, private farms. In this 1957 photograph of an agricultural exhibition in Zagreb, "Supermarket U.S.A." appears to display U.S. consumerism. In reality, it was exploited by Tito, who used the American model—and Yugoslavian women consumers—to impose collective farming on his people.*

the European market. Given the success of NCR's Rome experiment—and Italy's position as an American ally against communism,
IBEC chose Milan as its first European target market in 1957. With
a number of Italian investors, IBEC founded *Supermarkets Italiani
S.p.A.*—known today as *Esselunga S.p.A.* Richard W. Boogaart, an
American businessman responsible for the undertaking, was mildly
surprised to find that the domestic partners "wanted the stores
to be run and look exactly like they do in the United States."[64] To
meet this goal, employees were even offered English lessons. And
to bridge financial shortfalls the company received advantageous
credits from the Export-Import Bank in Washington.[65]

During the same year, a similar experiment took place with a
different set of players. The U.S. government, American business,
and the Yugoslavian state tried to airlift "Supermarket U.S.A."
into the Yugoslavian city of Zagreb for an exhibit. The American
government jumped on the project as an opportunity to spread
"The American Way of Life" in a socialist country. For their part,
American businesspeople hoped to use the experience as a spring-
board for opening a consumer market for U.S. products, and
Marshal Tito sought to embarrass the Soviets in his geopolitical
game. Tito also saw the supermarket as a tool for breaking the resistance of small-scale peasants who refused to cooperate with the
state's programs for abolishing the private ownership of farmland.
Outright collectivization had failed; the American supermarket
now served as a seductive tool for inducing farmers to turn to
large-scale, industrial agriculture.[66]

As history would have it, the first supermarkets in Milan and,
later, Belgrade, would differ substantially from those that followed
the U.S. paradigm. Urban and economic structures, as well as
customers' expectations, forced U.S. organizers to modify their
original store concepts in Italy and Yugoslavia. Given that the stores
in Milan were built in urban settings with few parking lots, there
was no direct access to the supermarket system by car—unlike
in the United States. Finding food producers with the capacity to
deliver large quantities of highly standardized goods on a regular
basis proved to be extremely difficult. The American organizers of
Supermarket U.S.A. faced the same problems. For logistical help
in solving these problems, the Americans turned to U.S. Marines
stationed in the Mediterranean: the military had to airlift fruits and
vegetables every five days in a refrigerated airplane and transport

the goods in a refrigerated truck to the new supermarket; the costs were staggering. And when the supermarket exhibit was over, and Supermarket U.S.A. closed down, the Yugoslavian supermarkets bought locally produced and packed foods—not American brands as the U.S. corporations had hoped. Moreover, shoppers continued to frequent Belgrade's urban farmers markets (*pijacas*) to buy seasonal fruits and vegetables instead of oranges year-round, as American shoppers had come to expect.[67]

Many Italian customers were critical toward pre-packaged, canned, and pre-cooked foods, and demanded fresh, non-packaged bread, vegetables, and fruits. Given that the great majority of customers did not own freezers, frozen products did not sell well. To Belgrade shoppers, frozen TV dinners also made no sense: American-style self-service meat departments aroused suspicion. Yugoslavian shoppers demanded that they witness the butcher cut, weigh, price, and wrap the meat; the consumer insisted on monitoring quality and guarding against misconduct. The concept of self-service had been so foreign that the organizers of Supermarket USA had hired women students from Belgrade to act as shills: they modeled the behavior of American housewives, pushing grocery carts, and often borrowing babies from mothers in the crowd. Serbo-Croatian speaking students who had learned English acted as cashiers, explaining the benefits of high-volume, low-margin food retailing. To attract low-income groups, Italian managers decided to provide customers with hand-baskets and not to make the supermarkets look too modern. In the end, the Italian self-service store developed into a hybrid of the traditional, small grocery store and the large U.S. supermarket.[68]

The "American Way of Life," then, remained a model that European businesses, states, and consumer groups sought to emulate. But what emerged were not facsimiles of supermarkets, but stores with a highly nuanced mix of foreign influence and domestic tradition. Migros' manager Elsa Gasser was wrong in claiming that women shoppers around the globe were, in general, similar. But she was right about women shoppers when it came to one aspect of their character: in their approach to food, women were active and critical consumers. And, as an integral part of the food chain, women were also social reformers, home economists, relief workers, and experts. In all of these roles, women helped to create, tinker with, and critique the emerging food system.

6
Living in
State-Sponsored Europe

In charting the history of modern Europe, the kitchen appears an unlikely area of state control. Nor is the kitchen an intuitive focal point of Europe's engineering race to produce superior cars, computers, and nuclear missiles. The concept of the kitchen, in fact, conjures images of consumers unpacking groceries, preparing dinner, and eating home-cooked food. In the decades before and after the Second World War, however, modernist kitchens were considered technological marvels—and were indeed politically controversial. In technological terms, modernist kitchens are places filled with appliances, and they are connected to the large technological systems that came to define the twentieth century: electrical grids, gas networks, water systems, and food chains. Kitchens were also as profoundly political. Cold War statesmen Richard Nixon and Nikita Khrushchev, together with Winston Churchill (United Kingdom), Ludwig Erhard (West Germany), and Walter Ulbricht (East Germany)—all considered kitchen appliances to be the building blocks of the social contract between citizens and the state: to consume was to be a true citizen. A political focus on kitchens, moreover, helped anchor a traditional gender hierarchy at the very moment when the feminist movement, socialist ideology, and war emergencies had challenged fundamentally conventional roles for women.[1]

Fig. 6.1 The Cold War Enters the Kitchen: *In the 1950s, America chose the kitchen as a Cold War battleground, using the home to show how consumerism—rather than heavy industry—could boost Europe's postwar economy. Pictured here is a typical kitchen exhibit on display at "We're Building a Better Life" in Stuttgart, Germany, in 1952. Note how visitors to the exhibit literally look down on the American consumer model for stimulating Europe's economy. Most European welfare-state and socialist politicians favored collective provisioning—like public housing and transportation—over individual consumerism.*

As a radical innovation of the twentieth century, the modernist urban kitchen was a separate space. Consider the kitchen's evolution. In the nineteenth century, only upper-class families had separate basement kitchens complete with tables, ovens, and servant-operated pumps. In contrast, most working-class and farming families cooked on a coal or petroleum stove with a side table in the same space where they worked, socialized, and slept. By contrast, the modern kitchen was built as a separate area and was equipped with modular rectilinear appliances with a unified look: continuous, unbroken countertops; sleek cabinet covers over most appliances; and standard measurements. These electrical and mechanical units were set into an integrated, mass-produced ensemble with discreet controls. All component parts—from cabinetry to plumbing—matched to create a unified, modernist experience. It was precisely this version of the kitchen that became the flashpoint in the struggle to define how Europeans were to prepare and eat their meals.[2]

On July 24, 1959, a diplomatic drama thrust household machines onto the Cold War's center stage. At the Moscow trade and cultural fair, General Electric's lemon-yellow kitchen served as the unlikely backdrop for the now-famous exchange between Soviet Premier Nikita S. Khrushchev and U.S. Vice President Richard M. Nixon. In front of Lois Epstein, a Harvard Medical School student hired to play the housewife in the model kitchen, Nixon lectured the communist leader on the advantages of living in the United States and, more important, of American capitalist consumption. The two heads of state conducted the exchange, later dubbed the "Kitchen Debate," on behalf of women consumers on both sides of the Iron Curtain. Although U.S. officials proclaimed their 1959 Moscow exhibition to be "the most productive single psychological effort ever launched by the U.S. in any communist country," the Kitchen Debate was merely the culmination of a propaganda campaign begun more than a decade earlier. The city of Berlin was the prime battleground. There, the "first" and "second" worlds met at a still-permeable border, over the course of the fifteen years before the Berlin Wall was built in 1961. After the Berlin airlift in 1948, the German city had served as America's crucial testing ground—and kitchens served as ideal visual aids in the strategy to foster consumption.[3]

At every stop where the touring kitchens were displayed, women students and actresses played the glamorous housewife. Fashionably dressed women posing in appliance-packed kitchens figured prominently in exporting American mass culture to Europe. Photographs, movie newsreels, newspaper stories, advertisements, testimonials, and radio segments portraying the miraculous advances of the American kitchens-of-tomorrow flooded European news outlets between 1945 and 1960. Refrigerators in particular became the standard icon of American modernity. Images of refrigerators overflowing with food circulated widely, suggesting that superior American kitchens were either currently being built in Europe or were soon to arrive on the Continent's shores—and would be accepted uncritically by enthusiastic multitudes.[4]

The American kitchen was both fact and fiction. We now know that the "official" story of the American kitchen was an idealized one. In this fantasy narrative, the modernist kitchen—and other U.S.-made machinery—enchanted Europeans who eagerly embraced the United States' liberal consumption ideology as well as its consumer goods. In fact, the American kitchen was, at times, ignored or rejected by European consumers; or it was at other times appropriated as fodder for experimentation by local consumers. Social democratic and socialist regimes sought to introduce price controls, support collective provision, finance governmental research and development, and institute the consumer voice as part of a social contract of modest spending and redistribution of wealth. The U.S. technology-rich, middle-class suburban kitchen was imaginary, created as a corporate technological promise. Its gadgets sought to seduce prospective users into buying into abundance paired with planned obsolescence—not to help them cook meals. As Annie M.G. Schmidt, a celebrated Dutch writer shrewdly remarked in 1955: "The French kitchen one associates with dining well; with the American kitchen, on the other hand, one thinks of big things with push buttons... and of gigantic white fridges filled with deep-frozen lettuce and big peas. American food always seems to stay in the fridge but never to come out of it." Indeed, modern American kitchens rarely celebrated the taste of food. Kitchens were endpoints of large technological systems (food processing, energy systems, and building projects) and ideologies (consumerism). And large refrigerators, connected to the cold chain of industrialized processed food, represented the "American way of life."[5]

In one prototypical photograph from the Marshall Plan, two dowdy European housewives on a study tour in the United States happily inspected an American refrigerator crammed with food. Juxtaposing glamorous Hollywood actresses with old-fashioned European housewives was as deliberate as it was effective. This pairing visualized the promise of abundance and high-tech marvels that America would bring to war-torn Europe. The powerful image of the U.S. kitchen also ignored the design traditions of European building societies, cooperatives, and local and state governments. In reality, during the postwar period many European welfare states (i.e. British, Swedish, French, West German, and Dutch) and socialist governments (i.e. Finnish, Soviet, East German, and

Bulgarian) built housing on a mass scale. And that housing featured standardized, modern urban kitchens.[6]

Despite Europe's successful building efforts after the war, *ideologically* the "European kitchen" was simply no match for its American counterpart. In 1961, when an American model kitchen was exhibited in Helsinki, it came with the usual well-publicized story of high-tech marvels. The U.S.-sponsored narrative was so effective—and expectations so high—that, upon seeing the exhibit, Finnish visitors were prompted to question what, really, was so innovative about the American kitchen. Consumers questioned whether American kitchens were indeed more modern, more suburban, and technically superior to those of their European counterparts. Consumers wondered whether the prototypes of these techno-kitchens so successfully staged at international fairs ever made it to production. And the question remained whether or not users liked U.S. modernist inventions.[7]

Kitchen displays camouflaged deep internal divisions in Europe and on the U.S. home front. Professional women, social reformers, and housing officials confronted government agencies, corporations, and professional organizations; they contested, among other questions, what a kitchen should look like and how it should function. Reformers were more interested in efficient housework methods than in expensive kitchens packed with gadgets. In fact, real consumers and their organizations were functionally banned from these Cold War struggles; it was a battle conducted largely among corporations, experts, and states.[8]

Cooling the Cold War

Nixon and Khrushchev disagreed on many issues, but they happened on common diplomatic ground with the belief that science and technology are the true yardsticks of a society's progress. Exhibitions represented the perfect arena for the two nations to compete in this regard, comparing their scientific and technological performance. The 1959 international trade shows in Moscow and New York were key occasions. Sandwiched between the *Sputnik* satellite launch of 1957, the building of the Berlin Wall in 1961, and the Cuban missile crisis of 1962, the Kitchen Debate of

1959 marked a momentary thaw in the Cold War. In fact, shortly before the Moscow event the Americans had hosted a Soviet exhibit in New York, where the USSR showcased its most advanced and prestigious technologies, including satellites, space capsules, heavy machinery, and a model nuclear ice-breaker. Also exhibited were fashions, furs, dishes, televisions, and row after row of appliances such as washing machines and refrigerators. This demonstrated the Soviets' readiness to boost individual consumption, in addition to their commitment to collective provisioning measures like afford-able homes, vacations, and public transportation. Khrushchev had promised that the Soviet Union would match—or even surpass—the United States in consumer-durable production by 1965, the end of the Soviet leader's just-announced seven-year plan. Khrushchev's confidence in meeting this ambitious goal rested on the Soviets' spectacular successes in space and military technologies; a nation that could build atomic bombs and launch satellites into Earth's orbit would surely have no problem producing washing machines and TV sets for its citizens.[9]

Next, the United States upped the ante: a few weeks later, the American exhibit in Moscow showcased a cornucopia of consumer capitalism. Technological achievements in space research, nuclear research, chemistry, medicine, agriculture, education, and labor productivity shared the limelight with material goods for home and leisure. At this exhibit, consumerism was no longer a side-show to production and military technologies. American automo-biles, Pepsi carbonated beverages, and the latest voting machines complemented three fully equipped kitchens. The show was hastily assembled in anxious response to the Soviets' popular appeal for all social classes to have access to technological advancements. The American model house was displayed to prove that an "average" home was already available to all Americans—and was not a promise in some distant future. To the exasperation of the Soviets, U.S. officials had changed the rule of the superpower game that defined what "real" technology meant. According to U.S. boosters, from that moment forward technology was to be measured in terms of consumer goods rather than space and nuclear technologies.[10]

At this turn in the public-relations game, the Americans caught the communist regime off guard. But on the eve of the trade show, Khrushchev had good reason to show ebullient confidence in his nation's technological prowess: a mere two years earlier, in 1957,

Fig. 6.2 The Kitchen Debate: *The 1959 American Kitchen exhibit in Moscow has been inscribed in history as the "Kitchen Debate." In this photo, Soviet Premier Nikita Khrushchev and American Vice President Richard Nixon spar with one another about how people should live: as capitalistic consumers versus communists. The woman [Right] in the kitchen is Harvard Medical School student Lois Epstein, who was hired to play the housewife. And although she is included in the original picture, her image—and indeed the consumer she represented—was cut out of the photograph later circulated by the Americans as propaganda.*

the Soviets had shattered America's self-confidence with the launch of the space satellite *Sputnik*. This prompted the U.S. government to create NASA (the National Aeronautics and Space Administration) and further to increase government spending on scientific research and technical education. It is no wonder that the American public-relations-motivated redefinition of technological advancement in consumer terms piqued the Soviets. For them, the emphasis on individual consumer goods was a moot rhetorical point: Soviet leaders of the era were dedicated to making technological systems accessible to all citizens. For example, they invested in buses, trains, and taxis instead of privately owned cars. Their construction programs aimed to solve housing and labor shortages by providing apartment blocks for nuclear families along with collective facilities like child-care centers and public laundries.[11]

The policy of rebuilding production capacity rather than encouraging individual consumption was not limited to the communist countries. Most politicians in Western European welfare states focused on reigniting heavy industry rather than on stoking the fires of consumption. Indeed, all postwar societies in Europe faced massive housing shortages well into the 1960s. In the Soviet bloc, as well as in social-democratically governed states, reconstruction planning therefore favored apartment buildings—which were built with prefabricated concrete slabs in standardized modules to resemble socialized forms of housing—rather than the detached homes that symbolized individual consumption.[12]

Facing similar problems, European governments in both East and West chose technical solutions yielding housing and kitchens that were strikingly similar on both sides of the Iron Curtain. Initially, the Kitchen Debate appeared to be a fundamental controversy between the Cold War's two superpowers. On closer inspection, though, the debate looks more like a transatlantic clash between American corporations and European welfare-state/socialist visions of technological development. Still deeper analysis also reveals a brewing conflict between two groups: the alliance of policymakers, architects, reformers, and the building industry on the one hand; and middle-class housewives and working-class women with their practical experience on the other. Many women's organizations and female professionals were either caught in the middle or sought to negotiate the ensuing tensions on behalf of users when it came to new technologies like kitchens.[13]

Users United

Nixon was rather late in discovering household technology as a political tool. Almost a century before he traveled to Moscow, the design of modern kitchens had become the focal point of utopian socialists, social reformers, and women activists. The nineteenth-century middle-class ideology of the home was founded on the concept of the parlor; socialists and modernists politicized the kitchen. On both sides of the Atlantic, a rich palette of ideas about modern kitchens developed. Based on the international ideas of Charles Fourier, Robert Owen, and August Bebel, socialists envisioned "kitchen-less" residences paired with in-house, communal food preparation and laundry spaces. These thinkers focused on outsourcing women's housework to the services of cooperative food delivery and washing facilities outside the household living space.[14]

Theoretical European discussions were transformed into practical and detailed American solutions on how to save, share, and redistribute household labor. Nineteenth-century American utopian communities like the Shakers were noted for their radical designs with communal kitchens. Layouts were spacious; kitchens sported large stoves and were equipped with the most advanced appliances to handle hundreds of meals at a time. In 1876, Boston feminist Melusina Fay Peirce proposed cooperative housekeeping in which married women would carry out all domestic work collectively. These middle-class wives would save costs by purchasing goods in bulk; sharing capital-intensive equipment; and specializing through division of labor. Peirce's kitchen-less homes featured a cooperative housekeeping center; sewing or fitting rooms; as well as bakery, cooking, and laundry rooms. Every function had a room unto itself. All services were available to communal users at fair prices. The existence of an accounting room in Peirce's plan best expressed the expectation that housewives would manage purchases collectively and charge their husbands for their work. Socialist feminist Charlotte Perkins Gilman saw the professionalization of housework as a key to women's emancipation by outsourcing the work to professional nannies, housekeepers, and cooks. She recommended redesigning homes to maximize women's potential for creativity and leisure; that meant including studios and eliminating individual kitchens and dining rooms.[15]

For the utopian socialists, communal organization was the essence of the kitchen, but they also exhibited an interest in the latest available technologies. Their focus on professional household equipment made sense: before the First World War, most household appliances were oversized and user-*un*friendly. The early models of vacuum cleaners, for example, looked more like bulky steam engines than handy labor-saving devices. Similarly, early refrigerators with their exposed coils resembled cooling machines on a factory floor rather than user-friendly food storage spaces. At that time, manufacturers had in mind industrial users like hospitals, restaurants, hotels, convents, and army camps. Consumers living in individual households were not yet the target market.[16]

With his utopian novel *Looking Backward* (1888), Edward Bellamy sowed the seeds of a user movement demanding collective solutions to household chores. His book popularized the idea of cooperative housekeeping for an egalitarian society without the evils of capitalism. Women would initiate cooperative laundries and kitchens by organizing the collective acquisition of capital-intensive equipment like gas and electric stoves as well as heating, lighting, and laundry systems.[17] For at least two decades, cooperative kitchens

Fig. 6.3 The Cooperative Kitchen: *From the 1880s until the 1950s, cooperatives thrived—farming co-ops, housing co-ops, and food co-ops included. The cooperative kitchen was among the most radical initiatives. Cooperative kitchens like the one in this 1920s photo cooked for their members. Healthy food was served in the co-op's restaurant; hundreds of meals per day were home-delivered. The co-ops menus were planned with home economists according to the latest insights of nutritional science. The cooperative kitchen eased the homemaker's burden—and enabled middle-class women to work outside the home. Note in this 1920 photo the towers of nesting metal containers used for transporting cooked meals.*

or "dining clubs" servicing existing homes were a successful solution to the challenges of women's labor-intensive housework. Such dining clubs outsourced the work of preparing meals. Around the turn of the century, at least a dozen cooperative dining clubs operated in the United States for some time; many had been inspired by Bellamy's novel, but they remained limited in scope.[18]

In Scotland, Austria, Belgium, the Netherlands, Norway, Denmark, France, and Italy, dining clubs were founded by associations advocating cooperatives; home-economics teachers promoting nutrition science; and housewives' organizations interested in alternative ways of arranging housework. The Amsterdam cooperative kitchen established in 1903 provided its members with 40,000 meals a year. Based on progressive ideas, it used the latest cooking technologies and insights from nutrition science. A dozen Dutch cities followed suit, serving on average 300 meals a day. The food was packaged in an ingenious nesting tin box and home delivered by horse-drawn carriages. Between 1900 and the 1920s, cooperative kitchens served middle-class families and single urban professionals living alone on a small budget. Members could benefit from healthy, inexpensive meals in the intimacy of their homes or in the cooperative's attractive dining hall. The promise of

Fig. 6.4 Sharing the Cooking: *This 1920s poster advertises the Amsterdam Cooperative Kitchen in the Netherlands. In his right hand, the chef displays a miniature version of the pushcart used for delivering food to co-op members. The text boasts of 1,000 dinner deliveries made daily. In his right hand, the chef holds up a tableau of the co-op members dining in the collective's restaurant, which served from noon till 8 p.m. Artist Charles Verschuuren created the Art Deco-inflected poster.*

all cooperative kitchens was that their members would save time by pooling resources and by dividing labor.[19]

Inspired in part by American examples and Fabian ideas, German socialist Lily Braun proposed the *Einküchenhaus*, or apartment hotel, in 1901. Braun's so-called one-kitchen-house sought to strike a socialist balance between privacy and collectivism by "insourcing" collective services under one roof. These buildings would offer individual apartments with one communal kitchen for all. Other facilities would also be shared, including central heating, a central vacuum cleaning system, a laundry, library, kindergarten, shops, and roof gardens.[20]

Braun's ideas rippled through the transnational socialist network of architects, activists, and middle-class women's organizations. The small wave of initiatives that resulted helped to shape a Europe that set itself apart from the United States. Two years after Braun's proposal was published, the first collective project was realized in Copenhagen, followed by other European cities connected by the network: Stockholm (1906 Hemgården Centralkök and 1932 Kollektivhuset), Berlin (1908 Einküchenhaus Charlottenburg, 1909 Einküchenhäuser Lichterfeld, 1909 Einküchenhäuser Friedenau), Letchworth (1909 Solershot House), Zurich (1916 Amerikanerhaus), Hamburg (1921 Ledigenheim Dulsberg, 1930 Boardinghouse des Westens), Vienna (1922 Heimhof), Amsterdam (1927 Het Nieuwe Huis), Moscow (1928 Narkomfin), Breslau (1929 Ledigenheim Werkbundsiedlung), and London (1933 Isokon Building). Some were private initiatives, others were cooperatives; still others were supported by local social-democratic and communist governments.[21]

Orthodox socialists such as the German Clara Zetkin dismissed as hopelessly bourgeois such feminist-socialist attempts to lift the household burden by delegating the work to low-paid employees. In 1919, Bolshevik Vladimir Lenin phrased his objections somewhat differently. A woman, he said, "continues to be a domestic slave, because petty housework crushes, strangles, stultifies and degrades her, chains her to the kitchen and the nursery, and she wastes her labor on barbarously unproductive, petty, nerve-racking, stultifying and crushing drudgery." Organizing kitchen-less homes was merely a stopgap measure. Cooperative initiatives would distract the working-class from the revolution. Only communism would bring true structural change. "The real emancipation of women, real communism, will begin only where and when an all-out struggle begins (led by the proletariat wielding state power) against this petty housekeeping, or rather when its wholesale transformation into a large-scale socialist economy begins."[22] In the midst of the Russian Revolution, socialist thinker and activist Alexandra Kollontai argued that, to be economically independent and freed from domestic work, women had to be integrated as equals into all sectors of society and domestic work outsourced to state institutions like communal kitchens, laundries, and nurseries. Most communists, though considering women as workers for the revolution, primarily expected them to fulfill their traditional roles. Women's liberation would come once communism had arrived; no extra revolutionary attention to women's housework was required.[23]

In the wake of the Revolution, socialists won political majorities in a host of cities throughout Europe. Both in competition with and inspired by the successful cooperative movement, many socialist-governed cities sought to externalize household services. These new city governments devised municipal policies to collectivize household services that had been private: cooking, bathing, washing, and child-rearing. In Vienna, Amsterdam, Stockholm, Hamburg, Cologne, and Frankfurt, social-democratic majorities ran public services to lighten the burden on working-class women. These institutions were public facilities with subsidized municipal public kitchens, baths, laundries, and child-care centers as well as libraries, school lunches, and vacation clubs for their working-class constituencies.[24]

Designers & Imagined Users

Middle-class reformers sought individual answers rather than collective solutions to decrease household labor. In their search, reformers fostered and appropriated new technical solutions. The vision for the middle-class kitchen that was most grounded in practicality came from Catharine Esther Beecher, founding mother of Christian feminism and daughter of America's famous Calvinist minister Lyman Beecher. Domesticity was women's highest moral calling and the necessary counterpoint to the evils of industrialization, they believed. The servant-less woman worked individually as a full-time middle-class housewife and mother. Her 1869 best-selling book, *The American Woman's Home*, projected an individual user without a servant in a family house in which the kitchen occupied center stage. A streamlined, single-surface workspace had a mechanical core of water pipes, heating systems, and ventilation equipment. The layout wedded organization with mechanical tools.[25]

Beecher created her design from the homemaker's viewpoint, seeing the perspective of the residents who labored in—rather than visited—the home. That perspective went against the contemporaneous convention that homes be laid out to please visitors, first and foremost. Beecher's ideas undermined the notion that display—rather than utility—was the best yardstick of domesticity. Building on this new foundation, middle-class women's groups throughout Europe took the radical stance that kitchens should be moved out of damp basements into the center of the home. Houses that were easy to clean and to maintain should be the true measure of good design. Servant-less women, neither aristocrats nor working-class, began to judge house design through the prism of middle-class household work rather than (aristocratic) leisure. The kitchen, not the Victorian parlor, was to be the heart of the ideal modern home.[26]

In the nineteenth century, middle-class women empowered themselves as a user group, insisting on tinkering with—if not designing—the kitchens in which they worked. The kitchen was becoming these women's domain: they espoused the labor-intensive ideals of domesticity—without the servants of their aristocratic counterparts to help them. Over time, they would lose their pioneering position to new professionals. Architects, urban planners, and policy makers developed solutions on behalf of users they imagined at their drawing tables. They designed kitchens, homes, and cities according to the projected needs of imagined future users. The interwar period witnessed an enormous expansion of experts speaking on behalf of users they otherwise ignored. Improved housing, they thought, offered the best point of intervention in creating a better society. For their part, female home-economics professionals claimed housing and the kitchen as the woman's legitimate and professional domain of operation. These women experts emphasized the scientific principles guiding a modern kitchen. Several decades later, modern architects seized on the kitchen in order to further their own professional ambitions. Utilities agents, food marketers, and chemical corporations got involved to satisfy their commercial ends. Last to discover the kitchen as a policy instrument were politicians like Nixon. In the end, then, all of these groups came to share a keen interest in the kitchen.[27]

Some groups mobilized the kitchen as a locus for reform by stressing hygiene; others claimed it as a place for professional identity formation by calling attention to science; others mobilized it for governmental goals by emphasizing governance intervention and education; and still others exploited it for marketing purposes by accentuating comfort. Each social group projected another type of user in their kitchen design.

Of all the new professional groups, home economists were the most systematic in claiming the kitchen as their area of expertise. As early as the 1870s, these experts built an international and transatlantic network. Model homes, kitchens, and cooking demonstrations became their favorite visual, tactile, and educational instruments. Promoting this new knowledge domain, they even aimed to found an academic discipline that included science, nutrition, budgeting, and pedagogy. Their concept of "domestic science" tied the kitchen to the chemical laboratory. On the political level, the term "home economics" inserted the home into the larger polity and reform movement. Since the mid-nineteenth century, home-economics professionals claimed the testing and teaching kitchens as their knowledge domain, using them as labs for demonstration, instruction, and experimentation. By the 1920s, the engineering visions of Christine Frederick (*Household Engineering: Scientific Management in the Home*) and Lillian Gilbreth (*The Home-Maker and her Job*) had grown popular, particularly in Germany, where the discourse of rationalization and efficiency struck a cultural chord. While socialist feminists, utopians, and home economists sought to advance women's interests, Frederick and Gilbreth were the key American ideologues of the antifeminist, pro-consumption, suburban home. They advanced the kitchen as a business opportunity to sell products. Although their ideas circulated widely in the Netherlands, Finland, Sweden, Turkey, and beyond, they were often translated into more austere visions of modernity. Erna Meyer, an advisor to German government and industry, and the French Paulette Bernège represented their generation of women who sought inspiration in the American debate—but appropriated the transatlantic currents in their own ways.[28]

Utility firms, food corporations, and chemical companies comprised the second group to discover home economists and their test kitchens as research and development tools in the search for new consumers. In the 1890s, gas was still a novel, expensive,

unreliable, and sometimes lethal technology. Hence, gas companies launched fierce public relations campaigns to counter reports
about exploding gas stoves, nauseating fumes, resistance to new
cooking requirements, and high prices. Between 1892 and 1910,
to protest expensive gas, French consumers organized strikes and
founded associations in cities from Marseilles and Toulon to Lyon
and Bordeaux. Reports of similar strikes emerged from Spain and
Britain. Courts dealt with gas explosion victims. Many customers
preferred the slow cooking of coal stoves and their dual function as
heaters; loyalists to coal disliked the intense heat of gas that required
constant attention and vastly different cooking techniques.[29]

Taking their cue from the Singer Company's use of women as
instructors and salespeople, the British gas industry hired women
cookery teachers to boost sales. Initially, stove manufacturers
complained about purchasers' improper use of the stoves. But the
manufacturers came to realize that their post-sales practice—that
is, neglecting to educate customers about how to operate and
maintain the novel technology—cost them dearly in returned
goods. Through public demonstrations and home visits, women
home economists were to fill the gap in convincing customers to
adopt the proper cooking techniques for gas. These female British
gas instructors, the so-called "lady demons," also carried out small
repairs during home visits when necessary. Such repairs were not
included in their job description, however.[30]

Next, it was the electricity industry's turn to copy and improve
upon the gas sector's tactics. This time, professional cooks were
hired to convince customers to adopt the innovation. Manufacturers
and power companies assigned women to write cookbooks to introduce users to electricity, and to what they deemed proper practices.
Representing the newer technology, the electrical industry competed
with gas and coal through campaigns that tied their system, which
was more expensive, to modernity and to modern middle-class
womanhood. Associated with an older technology, the gas and coal
companies found their most loyal clientele in working-class homes;
electricity companies at first targeted upper-class homes. Gas and
electricity providers, in aggressive competition hired well-known
home economists to argue why their approach to cooking was the
better choice. In Norway, Henriette Schønberg Erken, founder of a
cooking school on her farm in 1908 and author of the country's best-
known cookbook, *Stor Kokebok* (1895), advocated electric cooking.

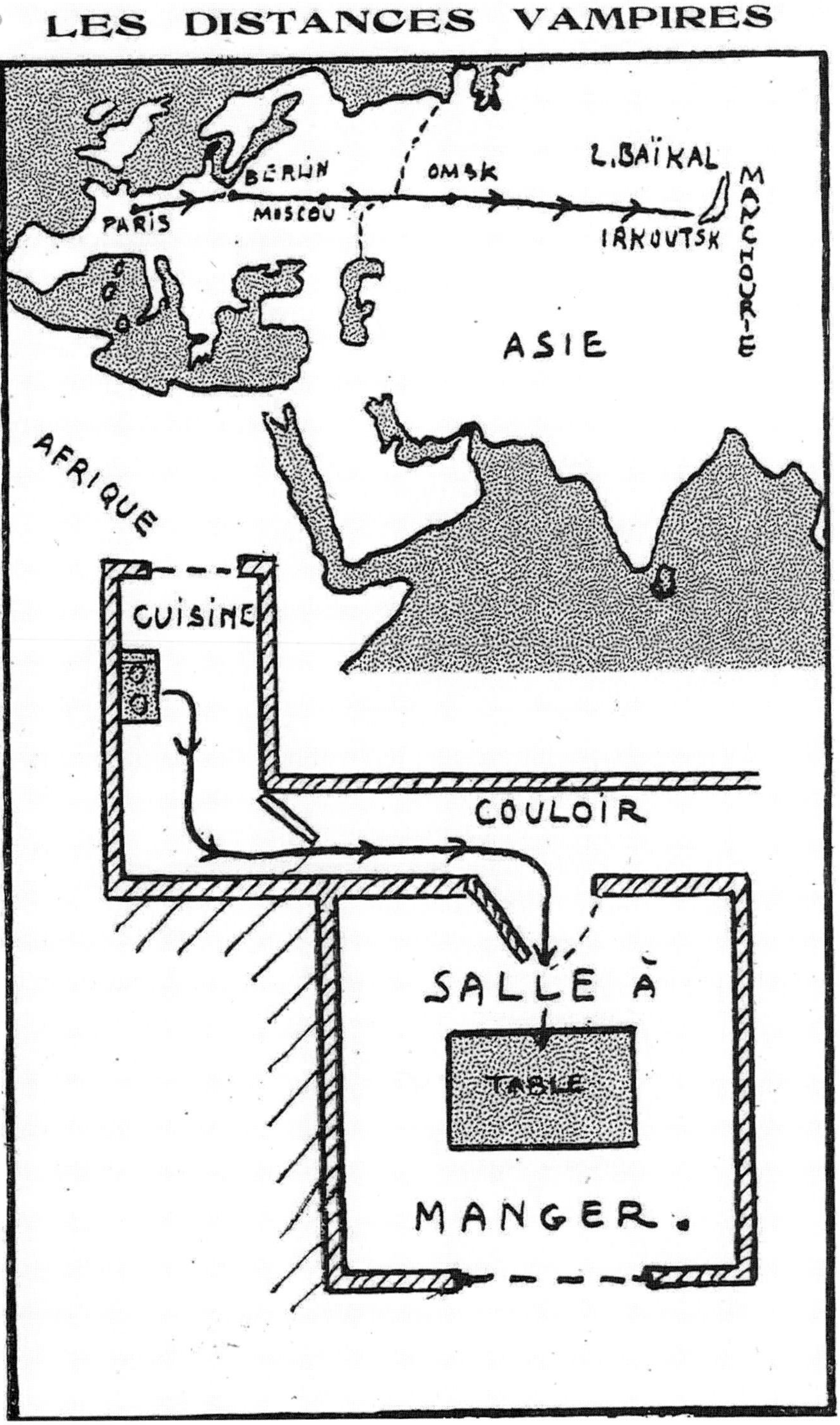

Fig. 6.5 Hazards of the Ill-Planned Home: *In 1928, French home economist and writer Paulette Bernège published* Si les femmes faisaient les maisons *(If Women Built Houses). This book fostered the efficiency movement's ideas about the "rationalization" of household engineering. This page, titled "The Vampire Distances," diagrams how a woman living in a poorly designed home could unwittingly walk the distance from Paris to Irkutsk (Siberia) in unnecessary steps. Bernège based this calculation on the space between the kitchen (cuisine) and the dining-room (salle à manger) via the hallway (couloir). Bernège helped to pioneer housework as a profession.*

LES DISTANCES VAMPIRES

PARIS BERLIN MOSCOU OMSK L. BAÏKAL IRKOUTSK MANDCHOURIE ASIE AFRIQUE

CUISINE

COULOIR

SALLE À MANGER.

TABLE

In Britain, Marguerite Patten was employed as a home economist first by an electrical utility company and then by Frigidaire. During the First World War, she was hired by the Ministry of Food to help women deal with wartime-rationed food through the radio program *Kitchen Front*. These home economics professionals had many colleagues in other European countries.[31]

Women home economists were not just corporate sales tools but took on the role of negotiating between manufacturers and users. Prior to the First World War, women's organizations and home economists believed passionately that gas and electricity would relieve the burden of women's household labor. Supported by electricity companies, middle-class women's organizations and home economists established women's electricity associations in Britain (The Electrical Association for Women, 1924) and the Netherlands (*Nederlandsche Vrouwen Electriciteits Vereniging*, 1932). Home economists also worked in the research and development departments of the food and chemical industries. Multinational food corporation Unilever hired the celebrated Dutch home economist Riek Lotgering-Hillebrand to run its flavor, taste, smell, consistency, and pricing department. The company acknowledged the importance of women's perspectives for product innovations like margarine. As a result of such corporate interest, academically trained women found employment with utility firms: marketing departments began to see home economists' professional services as the missing link to a vast market peopled with women consumers. Home economists found a professional niche as mediators between producers and consumers in research and development of kitchens' testing programs. Often, professional women collaborated with housewives' organizations. The demonstration kitchens became science instructors, knowledge authenticators, and visual aids. Home economists believed they could give consumers a necessary voice in the negotiations with manufacturers, and help manufacturers shape the innovations of their time.[32]

A third group, architects, also lay special claim to the kitchen. Seeking a professional identity between pure art and the marketplace, modernist architects discovered the kitchen as the prime vehicle for their own expertise and ideology. A whole generation of women architects found a professional niche in kitchen design. In Finland, women architects such as Elsi Borg and Salma Setälä drew kitchens that took women's household work as the point of departure. German theorist Erna Meyer collaborated with De Stijl architect J.J.P. Oud on the Weissenhof Settlement kitchen for the 1927 Werkbund exhibit in Stuttgart. The socialist architect Margarete Schütte-Lihotzky from Vienna became famous for her 1926 "Frankfurt Kitchen." In close collaboration with the German Housewives' Association and the manufacturer Georg Grumbach, she developed the Frankfurt Kitchen for the modernist showcase suburb of Römerstadt. Still, the design reinforced the framework of the Weimar Republic's state regulations, demanding individual houses rather than communal living.[33]

Lihotzky's influence was wide and deep. Her model kitchen, for example, installed in more than 10,000 homes, banished the act of eating from the kitchen to the parlor. Soviet architect Mosei Ginzberg turned the Frankfurt Kitchen into a workstation cabinet that minimized housework altogether. After visiting the Frankfurt display, a Dutch middle-class women's organization adapted Lihotzky's design to fit traditional Dutch buildings. Lihotzky's template, representing the pinnacle of modernism in the young Kemalist state, also circulated widely in Turkish magazines during the 1920s and 1930s. And when the Nazis came to power in Germany and Austria, Lihotzky and many fellow socialists accepted exile in Istanbul at the invitation of the Turkish government to work on designing modern housing. Her influence consolidated after the Second World War, when Europe's damaged countries undertook large-scale housing projects. Her kitchen also became a contested terrain among state officials, architects, and women users. Home economics experts and women architects, caught in the middle, sought to mediate the tensions.[34]

Home Economists Speak for Users

Before the First World War, the state showed scant interest in taking the lead with either building projects or food provisioning. The war, however, served as the dress rehearsal for kitchen design: in most industrializing countries, efficient housework and kitchens became the business of the fatherland. It was during the war that governments discovered the value of the kitchen as a site of resource processing and women's utility as experts. In the

subsequent decades of economic crisis, war, and reconstruction—
from 1914 until circa 1957—policy makers found home economists
extremely useful in devising programs to teach housewives to
feed their families with fewer calories and healthier ingredients;
to save energy; to mend and remake old clothing; and to design
efficient houses for the benefit of the national economy and the
military. In short, women were appointed the task of shaping the
national economy "from below"—from their position as mediators on the home front. The German, French, Finnish, Dutch,
and American home economics associations collaborated with
their governments in targeting agrarian women and housewives'
organizations as a way of improving the rural economy. For their
part, home economists acted as mediators between governments
and their citizens, believing times of crises offered them unique
opportunities to advance their expertise. Having worked first on
emergency food programs during the war, many home economists forged inroads into more institutional settings when the
hostilities ended.[35]

In the German Weimar Republic, the Home Economics
Group (1921) of the National Advisory Board of Productivity
(*Reichskuratorium für Wirtschaftlichkeit*) was an initiative of Charlotte
Mühsam-Werther, the president of the housewives' organization,
and Marie-Elisabeth Lüders, an MP for the Democratic Party.
Inspired by the American example of the U.S. Bureau of Home
Economics, the group brought together women's organizations
and home economics expertise with engineers and representatives
of government and business; their task was to reform housework
in order to boost the nation's productivity. In the most systematic
approach of all, the Nazi regime mobilized women both at home
and in the occupied countries for its autarky policy to support the
war economy. Home economists opened offices in other countries as
well. In 1943, the Finnish government established its Reconstruction
Office, a Department of Home Economics at the Work Efficiency
Institute, and the Swedish Government followed in 1944 with its
Home Research Institute. After the war, the socialist East German
Ministry for General Engineering instituted a working group to
research, develop, and produce household technologies to make
women's housework easier and increase the nation's production in
a coordinated way. Similar initiatives took place in the British and
Dutch welfare states.[36]

Fig. 6.6 The Home Economist at Work:

According to home economists, homemaking was a serious profession, and the home a place of work—hard work, at that. Science and technology informed home economists' teaching. They hoped to shape homemakers into professionals skilled enough to solve any modern household problem. In this 1940s photo, British home economist and electrical engineer Caroline Haslett instructs housewives in home wiring. Haslett edited the 1934 Electrical Handbook for Women—*a housewife's user manual for electrical technology. Britain's Electrical Association for Women sponsored the manual.*

After the Second World War, an alliance of experts, women's organizations, industry, and state agencies coalesced around kitchen plans that were particularly European. The collaboration, however, was haphazard; often competitive and fraught with tensions. Members sometimes worked at cross-purposes. Welfare-state and socialist governments, ranging from Great Britain, France, Finland, and the Netherlands to East Germany and Turkey, engaged modernist architects, women home-economics professionals, and manufacturing firms in the building of modern houses with separate kitchens. Sensing a unique opportunity to advance their cause, women's organizations lobbied for a voice in setting national housing policies. Assisted by empirical sociologists, home economists and women architects carried out surveys with residents, asking them to express their needs and desires for future housing designs. These women experts sought to mediate between the modernist architects' tenet of separating spaces according to their function and women's demand for more centrally placed, user-friendly kitchens. They also sought to harmonize the modernist belief in a spare aesthetic and the user-defined meaning of efficiency.[37]

Definitions of efficiency and logic were politically contested. Modernists claimed that, as trained architects, their values were professional, correct, and compelling. These values included rational-functionalist principles; a masculine, machine aesthetics; clean lines; and hygiene; all expressed most famously in Le Corbusier's principle, *La maison est une machine à habiter.* Functionalists celebrated simplicity as efficiency and authenticity as truth. They abhorred knickknacks and detested bourgeois homes. In one typical expression of the international modernist ethos, a Belgian commentator found stuffing one's house with useless things "at once extremely illogical, ridiculous and depressing."[38] A French architect echoed the same sentiment: "We condemn…slipcovers and house slippers, parlors, corridors, decorative facades and shutters."[39] The modernist principle of separating spaces by function not only influenced their aesthetic, but also dictated their blueprints. Modernist architects deemed combining the kitchen with living and dining functions to be particularly misplaced and old-fashioned. One French architect summed up the ideologues' antipathy for what he called the peasant kitchen by describing it as the "hearth around which the family gathers, where you have a radio, and the laundry drying…a place where you shouldn't be."[40]

The modernist principle of separating spaces according to their functions brought architects in direct conflict with rural and working-class users. Their families preferred kitchens in which they could eat together and do the laundry as well. This was to be a central and lively rather than an isolated space. Architects' academically-based knowledge clashed with the experience-based users' understanding that tenants brought to the table. In listening to working-class families living in modernist apartments, French ethnologist Chombart de Lauwe concluded that "Renters do not like having a bedroom right off the living room…a larger kitchen permitting family dining is desired, particularly by working-class and lower-middle-class households."[41] When asked, they criticized the lack of privacy for parents and the absence of a proper hallway;

as well as the illogical placing of radiators, light fixtures, doors, closets, and windows. Working-class women found the separation of the kitchen from the dining area (*séjour*) inefficient because it meant more back-and-forth walking between rooms, and extra cleaning work for the dining room. Further, the division made the task of supervising small children more difficult for mothers while they were preparing meals for their families.

Everywhere, the transnational community of modernist architects came in conflict with kitchen users, women in particular. In France, the conflict over the absence of the eat-in kitchen was framed in the starkest terms in 1958. On behalf of the French Ministry of Housing, Jeanne Picard, a woman reformer with extensive experience of working-class families, interviewed tenants and consulted with three home economics organizations and another thirty women's groups to come up with appropriate designs. Through her mediation, residents offered plans that better suited their lives, including, for example, a bathtub instead of a shower, because mothers found the tub more child-friendly. The so-called "referendum apartment," featuring a hallway and extra electrical outlets, greatly offended the architecture and design community.[42]

French architect Veillon-Duverneuil denounced the attempt by Picard and the Ministry to involve "amateur" women in the planning process: "a charming learned assembly of matrons, all housewives, right-thinking and well intentioned," motivated by "ridiculous vague impulses...presented a 'referendum' apartment, which seemed at the same time—we can say it now—both a practical joke and a do-it-yourself home!" He wrote, "architects, who, when these ladies permit them, also deal with questions of housing...reacted discreetly, as well-bred men...careful not to upset this witty bevy of delightful little faces (delightful, but sure of themselves!) They...thought...that perhaps they, too, by joining forces, who knows? [*sic*] could present the apartment of their dreams."[43] Another architect, André Wogenscky, threw up his hands in exasperation: "Women, in particular, are astonishing; they describe exactly what they want," before claiming architects' absolute expertise in the matter of design. "Naturally, the architect should take people's desires into account, to the extent that their opinions seem justified to him," Wogenscky wrote.[44] Another architect defended his profession's expertise in a similar way: "The public cannot do this work and neither can three million women,

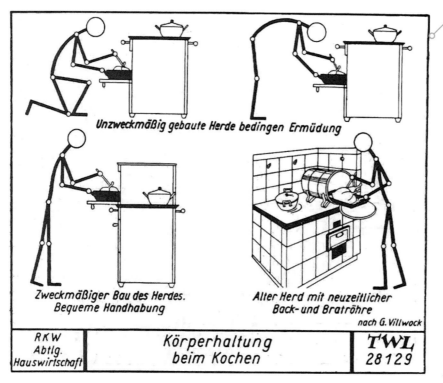

Unzweckmäßig gebaute Herde bedingen Ermüdung

Zweckmäßiger Bau des Herdes.
Bequeme Handhabung

Alter Herd mit neuzeitlicher
Back- und Bratröhre

nach G. Villwock

| RKW Abtlg. Hauswirtschaft | Körperhaltung beim Kochen | TWL 28129 |

Fig. 6.7 Housework Made Efficient: *In the 1920s, engineers and home economists mobilized for the rationalization of housework. This diagram, produced by the German National Advisory Board for Productivity (Reichskuratorium für Wirtschaftlichkeit, RKW) exemplifies their work. "Ergonomics of Cooking," reads the title [Bottom Center]. "Inappropriately constructed ovens cause fatigue," reads the top caption. The upright figure at the oven [Bottom Left] is captioned, "Appropriately constructed ovens make for easy handling." And the genderless stick-figure with the cylindrical oven [Bottom Right] is described using "A modern oven for baking and roasting." Note how closely this 1920 diagram resembles an engineering blueprint.*

even though they are users."[45] Dismissing architects' protests, the Ministry of Housing heeded Picard's mediation and Chombart de Lauwe's research. In response to tenants' concerns and preferences, functionalist architecture tenets were modified and new standards issued. The new housing norms rejected a strict interpretation of the model kitchen; made the apartments larger; required better quality materials; increased storage space; and mandated higher minimum temperatures for central heating.[46]

Home-economics experts embraced simplicity and practicality without sharing architects' visceral disgust for the middle- and working-class fondness of mementos and furnishing for comfort and intimacy. Home-economics professionals were more accepting of user practices than the modernist architects. They sought both to present their own expertise and to give voice to the experience-based knowledge of users. Women's organizations offered solutions based on ideas from the efficiency movement; functionalist architecture; and socialist utopians. In the Netherlands, most towns involved women as spokespeople on behalf of users. The advisory committees wanted "the voices of future residents to speak during the process of building, especially those of housewives, starting with the experience of practicing housekeeping."[47]

International modernism failed to gain wide popularity with the general public and clashed with working-class and rural practices. European modernist architects were able to realize their ideas only when national governments hired them for large-scale public-housing projects. Home-economics professionals were caught in the middle. Carving out spaces of their own, working-class residents steeped in agrarian traditions undermined modernist attempts to discipline their families by separating the functions of eating, sleeping, and working. Tenants rejected functionalist spatial separations between kitchen and living rooms. Some tenants could not get used to eating in the dining room and took their meals while standing in the small kitchen. To the horror of modernist architects and social reformers, walls separating kitchens from living rooms were demolished, tables were pushed back in, and beds were installed for the occasional nap. This restored the kitchen to its former role in rural and working-class homes: the place that integrated the activities of preparing meals, eating, working, and sleeping.[48]

In Hungary, modernists and the state clashed with citizens in even more fundamental ways. To their dismay, architects faced a grassroots movement of people building their own houses. They saw their profession as engaged in an existential "fight against 'botch-architecture', 'kitsch-design', 'bad taste' and the 'unhealthy vogue of the family house'."[49] Their outrage focused on the "pyramid-roof house" (*sátortet ős ház*), a single-family home that non-urban Hungarians often built in villages and small towns. According to the modernist architects, the pyramid-roof house was "illogically" built; they derisively referred to the self-builders of spontaneous architecture, as "botchers" (*kontár*), who mistakenly defied architects' expertise. For their part, the enterprising Hungarian self-builders viewed architects as mere agents of the state, useful only in helping to navigate government bureaucracy to get the necessary permits. They saw themselves as "self-help builders"—tinkerers who, forced by the housing shortage and a failing state, established informal networks of family, friends, and neighbors. No money changed hands, only skills. As dissident writer János Kenedi explained this informal system in his book,

Do It Yourself!: "Your sister-in-law's uncle's dentist treating the daughter of a factory supervisor can be the key to your electricity supply."[50] The U.N. Housing Committee in 1965 worried that such self-building burdened ordinary Hungarians—rather than the state—disproportionally in coping with the acute shortage of building materials and other resources. By 1975, the houses that Hungarians built themselves accounted for 60 percent of all new construction—and the proportion was still growing.

The Contested Kitchen

In the United States, modern housing was mainly a commercial enterprise. Accordingly, American architects often surrendered to residents' rejection of strict modernist blueprints, offering prefabricated neoclassical designs and open kitchens as part of the living space. (Georges-Henri Pingusson, a typical French architect, characterized the open kitchen disapprovingly as "both an American-bourgeois and a French-peasant trend."[51]) America's Levittown on Long Island, New York, came to symbolize suburban living and open kitchens most famously. The Levitt Company turned out thirty prefab houses per day by slashing production and labor costs despite strong union opposition. The company's greatest innovation took the form of neither miracle materials, nor gadgets, nor production techniques. It was a sales innovation that the Levitt Company perfected: the installment plan. Kitchens played a key role in the company's marketing strategies: targeting working-class couples, the company lured prospective buyers with a Bendix washing machine in the built-in kitchen, a GE Hotpoint electric stove and refrigerator, a stainless steel sink, white-enameled steel cabinets, and a York oil burning furnace. American intellectuals worried that these prefab houses, as one song phrased it, "all look[ed] just the same," and that the new middle-class Organization Man and his suburban housewife lacked any individuality; they bought these gadgets in a futile attempt to counter the dullness of their predictable lives and assert individual taste in their mass-produced boxes.[52]

Once they had moved in, Levittown homeowners and residents proved to be much more resilient and creative than predicted, tinkering with these prefab houses, altering them beyond recognition. Residents added everything from white-columned porches to garages and patios, bay windows and shutters to stone facades and extra stories. To the consternation of architectural preservationists, not a single ranch house of the 18,000 once built by the Levitt Company has been left in its original state. Working-class families contested the alliance of cost-cutting builders with modernist architects, who subscribed to an aesthetic of simplicity. As they moved up the economic ladder, homeowners and residents nurtured a domestic culture of ornamentation, knick-knacks, flashiness, and color—what one scholar calls an aesthetic of more-is-better.[53]

Levittown holds an iconic place in America's suburbia, and post-Second World War European planning circles often cited the United States as an important role model. Despite this, government building programs in Finland, Sweden, and Belgian Flanders developed their own versions of suburbia. The Finnish government embarked on an intensive modernist housing program that was rooted in agrarian traditions. Implemented in the countryside, the program aimed to promote home ownership for the country's refugees and veterans. Here too, kitchens were centrally located and more spacious than in Lihotzky's urban prototype. Pingusson's focus on the open design of the American kitchen as hopelessly anti-modern might have made ideological sense, but it was not entirely accurate.[54]

The gadget-packed American kitchen came to symbolize a different economic theory of consumerism—and a deepening transatlantic divide. During the Second World War, U.S. corporations had already begun to criticize Roosevelt's New-Deal politics of universal provisioning. Through highly effective advertising campaigns, American corporations promoted the U.S. economy as a uniquely productive system of free enterprise that was at once dynamic, classless, and benign. The Second World War and the Cold War provided the political context for American home economists and architects to carry their agendas overseas. The U.S. federal government claimed modern architecture as a typically American form of expressing individual freedom and democracy, ignoring the fact that modernism had been an international movement prior to the war. American architects and designers sought to disassociate modernism from communist ideas by linking it to corporate industrialism, consumerism, and individualism.[55]

Next, the ideological contest between austerity and abundance moved to the international and transnational stage: the United Nations and the U.S. Office of Foreign Agricultural Relations enlisted home economists to help spread American democracy, market capitalism, and consumption around the globe. This ideological shift resulted in dethroning the European-based Frankfurt Kitchen as an important role model. After all, the Frankfurt Kitchen was socialist-inspired and internationally-embedded. Worse—and perhaps most menacing to the Americans—it had been created by Grete Lihotzky, who had remained a committed socialist, if not a communist, during the Cold War.[56]

The American kitchen became synonymous with its gadgets, just as it was synonymous with the open design. It became the most beloved visual aid of the U.S. corporations' and U.S. government's campaigns in Europe. The food-filled fridge served as the best evidence that a bright future for Europe was right around the corner—if only it would adopt the American way of life. These kitchens, and their oversized refrigerators, were meant to stimulate consumption rather than to solve the postwar housing crisis.[57] Both American and European citizens had to be convinced that consumption was not a personal indulgence but a civic duty that would lift the nation out of its economic rut in the transition from war to peacetime. To keep consumers' desires vital during the shortages of war, U.S. corporations had flooded women's magazines and general-circulation media with futuristic fantasy kitchens.[58] Critics believed that corporate America was "super-dupering the war"—more interested in profits than in patriotism.[59] In 1943, the famous French-American industrial designer Raymond Loewy blasted these fantasies claiming "the public is being misinformed systematically about the wonders that await them."[60] A former federal housing administrator found the vision misguided: "Let's examine this super-electronic, radio-activated, solar-energized miracle house of tomorrow. Nobody works here...not even the servants. All that stuff is done by electric eyes and levers and things.... Meals cooked by polarized atoms roll right out to you in a mobile kitchen. Could you stand it?"[61]

The American promotion of abundance often clashed with the policies of war-damaged European countries. Like their fascist predecessors, welfare and socialist governments of Europe redirected state and market resources to heavy industry and reconstruction instead of encouraging consumption. They elected to underwrite mass-scale urban housing programs to deal with the pressing shortages. To facilitate austere consumption regimes, their propaganda campaigns asked the private sector and citizens—women in particular—to tighten their belts; to mobilize their ingenuity in conserving resources; to invest in repair; and to help invent surrogate products. Meanwhile, the U.S. government sought to instill values of consumerism in Western Europe. American Marshall Plan officials urged Western Europeans to rebuild their economies by shifting their focus from heavy industry and to deciding to "Think Consumer." Marshall Plan organizers were ecstatic when, in 1955, representatives from Dutch women's organizations and home-economics professionals embarked on a study tour of the United States; theirs was the first European group to show interest in consumer issues. The officials hoped that this group would serve as an example to other Europeans. The study tours made stopovers at women's organizations, home-economics schools, and the test kitchens of utilities companies. And there was the obligatory touchstone: the visit to a model kitchen, this one at the New York Good Housekeeping Institute.[62]

U.S. corporations including General Motors, General Electric, RCA/Whirlpool, and General Mills exhibited gadget-filled kitchens that toured Europe's capitals. Sponsored by the U.S. government, they came to symbolize America's newly acquired global power. Ironically, the models were never produced in series, but remained mockups and prototypes. Most required props to make a techno-logical promise for the distant future and used paid actresses "to hover among the futuristic appliances, not really doing nothing but not doing anything either: They opened refrigerator doors, poured beverages into glasses, pushed buttons."[63] Indeed, the American corporate displays did not seek to sell designs and appliances, but an aspiration and an idea. Khrushchev's irritation with the Miracle Kitchen was thus quite understandable. After all, Whirlpool had originally conceived the model only as a test laboratory for new products, never as a prototype to be readied for production. The Company's Mechanical Maid promised to scrub the floor before putting itself away, prompting Khrushchev's jibe: "Don't you have a machine that puts food into the mouth and pushes it down?" At one point in the Kitchen Debate, a guide pushed a button on a dishwasher to demonstrate another Miracle Kitchen marvel to

Khrushchev and Nixon. The dishwasher started to career towards them. Both politicians shook their heads. Khrushchev said, "This is probably always out of order," which Nixon affirmed with a succinct, "Da."[64] Even the American Vice President had to admit that most of the kitchens at the Moscow fair often proffered far-fetched promises.[65]

Khrushchev's criticism echoed that of the Catholic Church, as well as many female professionals, women's organizations, and housewives on both sides of the Iron Curtain. While some journalists faithfully copied General Motors' press releases promulgating the life of leisure that awaited housewives ("Robots will literally do your laundry"), other members of the media were more critical. The only thing missing from the Frigidaire kitchen-of-tomorrow, one Catholic daily believed, was a food-swallowing machine. Even more skeptical, a popular women's magazine proclaimed: "An automated household will be a housewife's nightmare. She refuses to believe that robots can do her job."[66] Commentators in Soviet bloc countries, as well, questioned the merits of the American-style kitchen. After comparing Soviet bloc, Scandinavian, and West European models, East German architects and engineers came away most impressed with the American kitchen, brimming as it did with refrigerators, freezers, food processors, toasters, dishwashers, and the like. But one female party official, Hilde Krasnogolowy, begged to differ with the endorsement. What mattered most for East Germany, she said, were designs that helped to reduce women's household labor. "It is a joke when the shops offer a machine... whose sole purpose is to crush sugar cubes and spices," she said and went on to criticize engineers who designed gadgets like "an automatic machine for pouring liqueur from bottles into jiggers."[67] Finnish visitors, too, were equally unimpressed when, in 1961, they toured the kitchen where the famous Nixon–Khrushchev debate had raged two years earlier. Americans, the Finns felt, could not lay sole claim to modernity.[68]

The rejection of gadgets in favor of practical solutions was articulated even in the United States. In 1943, the popular women's magazine *McCall's* invited readers to write an essay about whether they preferred a kitchen with appliances that could be bought in stores or a Day-after-Tomorrow-Dream-Kitchen equipped with pedal-operated faucets, glass cabinet fronts, oven, and refrigerator. Two-thirds of the 12,000 contestants—many of whom still cooked with wood, coal, or kerosene (23 percent) and had no access to hot water (25 percent)—passed over the dream version for the present-day, tested kitchen.[69]

Female professionals—from architects to household scientists and teachers—believed that modern women could be active housewives and economically independent women if kitchens were transformed to save labor. Their efforts centered primarily on low-tech solutions. Finnish, German, and Dutch housewives' organizations, home economists, and women architects exhibited little interest in expensive and energy-demanding dishwashers, for example. Instead, they designed more practical innovations. Indeed, time and again, women's organizations focused on low-tech solutions like materials to make cleaning easier; they designed and tested cost-effective dish-drying racks and devised handy solutions for organizing cabinets.[70]

Nixon's Cold War narrative implied that European homes, practices, and desires lagged behind those of Americans. Yet not a single U.S. corporation built the iconic American kitchen—the honor went to a European firm. On its own initiative, the Dutch office-furniture company Ahrend in 1953 produced an "American Kitchen" for the high-end market. Ahrend licensed the kitchen's streamlined design from America's most famous designer, the French-born Loewy. To boost sales in economic downturns and shift from wartime to peacetime production, corporations sought industrial designers to help improve the look and function of their products. The Aviation Corporation, for example, diversified from building aircraft to manufacturing kitchens. This strategy aimed to cushion the transition to peacetime conditions. The company produced a steel-pressed Loewy kitchen in the designer's signature streamlined style: smooth metal surfaces and rounded corners. Loewy's high-end "American Kitchen" became the standard kitchen installed in public housing in the Netherlands. This happened only after Eurowoningen N.V., the Dutch property developer, secured a better price for the kitchen from Ahrend than from the incumbent leading supplier of built-in kitchens for public housing. These American Kitchen cabinets featured aluminum handles and cream-white surfaces—reminiscent of General Motors' famously-tall Frigidaire refrigerator/freezers. The chrome handles became the most tactile and visual representatives of American cars and fridges in the European context.[71]

Yet the adoption of the American Kitchen in the Netherlands did not signal the triumph of American modernity, technology, and consumerism in the European context. In practice, none of the American Kitchen models came with a refrigerator or any other

Fig. 6.8 Users as Advisers: *Since 1945, the Dutch government has mandated that Women's Advisory Committees (VACs) promote user-friendly solutions in the home. In this 1970 photo taken in Utrecht, the Netherlands, a woman provides consulting advice to local social housing authorities. A network of more than 600 such independent local organizations still operates across the country. These groups provide user-based knowledge on facets of domestic life from optimal building materials to preferred home design. Note the blueprints in the background and the skeptical expressions on the men's faces.*

high-tech gadgets. Initially, the company offered an option for an American GE Hotpoint electric cooking range, soon replaced by a German AEG gas oven because Dutch housewives disliked cooking with electricity. By the 1990s, household design dealers were buying up Loewy's now-vintage American Kitchen for a new generation of consumers. These dealers retrofitted, rearranged, and reassembled the kitchen for clients who were no longer seduced by America's promises of tomorrow. These users signaled an ironic relationship with the United States by kitting out their retro kitchens with the "America" of yesterday. In material terms, the American *corporate* kitchen remained a curiosity in Europe.[72]

Indeed, Nixon's public-relations campaign around the Kitchen Debate ignored the very existence of the millions of modern homes that European welfare and socialist governments were building. Hundreds of thousands of kitchens—integrated, ensemble kitchens, no less—were constructed in Europe as a part of massive state-sponsored housing programs. The kitchens were erected by powerful national coalitions of European socialist and welfare states, architects, manufacturers, and women professionals. Between 1944 and 1951, the British government built more than 150,000 temporary apartments with state-designed kitchens inspired, in part, by Lihotzky's Frankfurt prototype. Ironically, British residents came to believe that their kitchens were based on U.S. standards rather than European modernist designs, proudly calling them American kitchens because they included electrical cookers and fridges. Modernist French urban planners in the 1950s also incorporated the German example of a separate kitchen space into their housing projects, thus challenging French women's preference for an eat-in kitchen. In the Netherlands, in almost all new housing projects by door manufacturer Bruynzeel, a stripped-down version of Lihotzky's original was installed. It combined the women's organizations' interwar experimentation with Lihotzky's kitchen; the Belgian Cubex model; and the demands of mass production. The austere version, featuring standardized, wood-paneled cabinets with chrome handles, appeared in more than 90 percent of all newly built, government-sponsored apartments. This was neither a General Motors techno-kitchen displayed as a mock-up at international fairs, nor Loewy's high-end, streamlined kitchen cabinets, reflecting the wartime aeronautics industry's search for new peacetime markets. Indeed, this was a minimalist interpretation of the modernist kitchen prototype, and few, if any, of the kitchens built according to this blueprint featured the iconic American refrigerators.[73]

Despite intense publicity campaigns, General Motors' Frigidaire kitchen never gained a foothold in the European market, where refrigerators remained quite small. European advertising emphasized the fridge's clever "fit" within the kitchen. And instead of stressing size and "technological" features, designers in much of Europe often tried to hide or disguise refrigerators as another form of kitchen cupboard. Indeed, many kitchens in the late 1950s had no fridges, let alone freezers.[74] Moreover, cooperative practices persisted. In several rural areas, alternative methods developed. Norwegian and Dutch farm women, for example, organized communal facilities with large freezers in which families could rent space. In urban settings, the oversized fridges—with designs based on market research and input from American housewives— were too tall for standard European subsidized housing. Kitchens provided by the European welfare state and the socialist state were sober and standard products supported by mass-scale government rebuilding programs in coalition with modernist architects.[75]

The modernist kitchen was practically everywhere in urban Europe. Despite its omnipresence, though, the kitchen never acquired symbolic status. Most European governments restricted their public-relations fanfare to press conferences announcing the milestone of the millionth house. These events lacked the Hollywood glamour that American corporations so successfully mobilized in collaboration with the U.S. State Department. During the Cold War, anything German or socialist in origin became suspect in Western Europe and was suppressed.

Most telling about corporate America's true intention behind its investment in the kitchen, General Motors showed little interest in marketing its model kitchens. In fact, the corporation never sold a single kitchen—including the model that toured internationally. The company focused more on stoking consumer desire by selling technological promises for some time in the indefinite future. Such promises could be applied to even broader public-relations purposes. Kitchen displays helped domesticate innovations like nuclear reactors that threatened to disrupt daily routines; were politically questionable; and were detrimental to public health. American corporate kitchens evolved into a means of encouraging market capitalism, an instrument of U.S. foreign policy, a carrier of the politics of the domestication of consumer regimes, and a vehicle for technological systems. As visual and tactile aids, kitchens were instrumental for integrating technological innovations into familiar routines. Nixon's kitchen epitomized the masterful story of appropriating modernism for America and its corporate consumer regime.[76]

In contrast, no such bold attempt was made on the part of "corporate Europe" to appropriate modernism for a "European kitchen." The European Economic Community (EEC—precursor of the European Union) did make one feckless effort to sponsor a European kitchen. At an exhibit, in 1961, the EEC assembled its own propagandistic model kitchen that represented industry rather than the users. The slapdash display combined German techniques and taste, French craftsmanship, Italian style and design, Swiss precision tools, and Dutch gadgets, "that would please even the American woman."[77]

As a tool for industrial growth—to showcase the best of the EEC nations—the European corporate kitchen also came to naught. Ultimately, Lihotzky's vision was more enduring for European socialist and welfare states through two postwar generations. The socialist-inflected kitchen combined individual consumption with collective provisioning. It gave many of Europe's urban citizens equal access to affordable housing and to water, electricity, and gas networks. The Marshall planners depicted European kitchens as lagging behind American examples. In appropriating modernism, they also disregarded the rich European and international building traditions. In terms of kitchens actually constructed, European national governments were far more successful. Well into the 1980s, when the welfare state and the socialist visions of collective provisioning came under fire, millions of urban European homemakers shaped a continent with surprisingly uniform kitchen structures and similar ways of life.

As heads of state, both Nixon and Khrushchev claimed to be speaking on behalf of consumers. Despite their rhetoric, they were remarkably uninterested in the consumer as a living subject, a real user, an active agent in shaping new consumer goods. This disregard for the user is evident in the public-relations campaign for the America kitchen: Nixon's team cut out Lois Epstein entirely from press photos. Epstein, the medical-student-as-American-housewife at the Kitchen Debate, was on her way to a professional career, and hardly represented consumers at large.

Despite the dominance of Cold War politics, kitchen consumers managed to create their own organizations, many of which acted on the transnational stage. For example, in 1964, five non-governmental consumer-testing organizations established the International Organization of Consumer Unions (IOCU) to demand tough protection. In 1985, in the fight over the U.N. Guidelines on Consumer Protection, one side advocated an unfettered global marketplace, the other argued for consumer protection. The jury is still out concerning who won the battle, much less the war. Throughout The Long Twentieth Century, consumers sought to shape the kitchen, its appliances, and the infrastructures linked to them. The century saw a steady progression of empowered users in Europe, from the gas and electricity consumers of the 1890s to the kitchen communards in the 1910s, the housewives organizations of the 1920s to the consumer testing institutes of the 1930s. Through it all, home economics professionals mediated between states, the market, and non-governmental organizations—on behalf of women users.[78]

III

BEYOND THE 1960S: USERS EMPOWERED?

7
Saving the Nation, Saving the Earth

The time was the early 1970s, the conflict, deep concern for "Spaceship Earth."[1] Prompted by this environmental issue, worried citizens in Western Europe and Scandinavia mobilized. Users demanded that glass bottles be returned to manufacturers. They appropriated cargo bikes as an alternative to automobiles. They invested windmills with the hope that they would one day replace nuclear power plants. Old technologies were revived and reassigned new uses; old artifacts assumed new meanings. Concern for Spaceship Earth was at once transatlantic and transnational. Many environmentalists in the European context found inspiration with their U.S. counterparts. Yet the actions of Europe's citizens were decidedly distinct from those of Americans. Europe's recent past, encompassing war and economic depression, shortages and autarky, informed its frugality. Responding to the burgeoning activism, European Union officials in turn viewed the environment as an excellent arena in which to demonstrate the relevance of their economic and political mission to skeptical citizens.

In January of 1972, an outraged citizen wrote a letter to the West German Ministry of Interior: "When the first company started to dump waste into the river Rhine, when the first plastic bag hung on the market stall, someone should have said STOP." Seven months earlier, high-school student Volker B. had submitted a petition with 3,600 signatures attached: "We protest against the ever increasing pollution of the environment and the deterioration of the living conditions of men, animals, and plants through emissions by cars, industrial waste, non-degradable waste of affluence (plastic packages, throwaway bottles etc.)." And architect Werner V. had, similarly, articulated his critique in the most politically explicit terms of all: "I must bitterly reproach the governments we have had in place since 1945 for the fact that they have heavily favored a materialistic thinking in terms of prosperity. I need only remind you of the disastrous promises of property made by Prof. Erhard."[2] The German architect had zeroed in on Ludwig Erhard, the politician most directly associated with Germany's "Economic Miracle"—and by association, the development of that nation's consumer democracy. This consumer democracy was the centerpiece of the U.S.-sponsored policy for Western Europe.

By 1950, the consumerist foundation for West Germany's economic power was firmly in place. This was symbolized by the reopening of the once Jewish-owned—and later bombed-out—*Kaufhaus des Westens* (Department of the West, also known locally as the *KaDeWe*) in West Berlin. During the years of reconstruction, the U.S. government promoted a democracy based on consumption in its Cold War battle with Soviet communism. Citizens were to be consumers, first and foremost. And West Germany became America's showpiece of abundance as a defense against the communist threat. In stimulating individual shopping instead of collective provisioning, West Europeans were encouraged to consider garbage as a nuisance, to be disposed of rather than minimized, salvaged, and recycled. The politics of individual consumption deliberately and radically sought to break with Europe's past of economic crisis, war shortages and Nazi autarkic policy, presenting a challenge to the communist world. U.S. policy, as articulated by Erhard, had promised his citizens prosperity. In fact, economic growth and individual consumption would be Western Europe's escape route from decades of economic crisis, political instability, and military destruction.

Slowly but surely, though, a growing number of citizens and organizations began to voice their unease with the consequences of a lifestyle based on high turnover of energy, raw materials, and water. Rachel Carson's 1962 international bestseller, *Silent Spring*, and the Club of Rome's official report, *The Limits to Growth*, published ten years later, introduced a sustained critique of the devil-may-care mentality. In 1972, the United Nations held its first Conference on the Human Environment in Stockholm. Academics analyzed the impact of industrialization; tourists were upset with road-side debris; housewives were confronted with new sets of resource norms; and municipal waste officials faced unmanageable landfills. Citizens and policy makers alike began to realize that endless economic growth came at a price: the Earth is not a cornucopia that can replenish itself magically; nor is it a waste dump that can be used forever.[3]

Indeed, the "one-way" economy, with its disposables and non-returnables, is a parenthesis in history, an anomaly made possible by cheap fossil fuels. During the First World War and the Great Depression, families were forced to "make do." And during the

Fig. 7.1 Taking to the Streets:
Starting in the 1960s, environmentalists were rebels with a cause. Waste was one concern. Pollution of all kinds was another. In taking protest to the streets, environmentalists aligned with fellow rebels, who fought for peace, equality, and other human rights. Activists identified toxic waste, nuclear power, smog, and acid rain as threats to Spaceship Earth. Environmentalists and other protesters drew inspiration from U.S. scientists and French philosophers, among others. At stake was a radically new, improved—if not utopian—society. In living their commitment to "Think globally, act locally," these groups formed well-connected, transnational networks.

Second World War, government-sponsored measures compelled all Europeans, regardless of class, to save resources for the sake of the nation and the war economy. When, in the 1960s and 1970s, they called for the recycling of bottles and paper, activists reinvigorated the salvage and recovery practices that had been commonplace in various European countries before and during the war. Their activism was rooted in old practices, but their discourse shifted radically: reusing old materials no longer was to benefit the fatherland, but to save planet Earth. To declare oneself an environmentalist was to become a citizen of the Earth—and as such, to rebel against the state and big business to care for the planet.[4]

Salvage & Recycle

In the spring of 1972, two Dutch women, concerned about the alarming increase in waste that they observed, positioned a large container for empty bottles in the center of Zeist, their hometown. Soon, action committees in many cities followed, turning the Netherlands into the first West European country in which glass was collected separately and systematically on a wide scale. Within six years, glass recycling bins dotted the entire country, making the Netherlands, along with Austria, Belgium, Sweden, and Switzerland, Europe's leading examples, with a glass recycling rate of 90 percent or higher.[5] The women's direct action was an extraordinary and an enduring success.[6]

In many countries, the packaging and distribution sectors' pressure to bury the bottle-deposit system they had once established prompted responses from consumers. The critique of the non-returnable bottle and the call for the recycling deposit bin came to symbolize the environmental movement's success. A year prior to the Dutch women's direct action, media-savvy environmentalists in London had leaped onto the stage. As their founding act, in May 1971, the British branch of the California-based Friends of the Earth deposited more than 1,500 non-returnable bottles at the doorstep of the U.K. Cadbury-Schweppes headquarters. The Friends of the Earth demanded that industry and the government set up a nationwide recycling program. The activists pursued their campaign on the street, in the press, and through novel advertising ("DON'T LET THEM SCHHH...ON BRITAIN") for over a year. Their political strategy combined street theater, consumer boycotts, marches, and rallies with science-based analysis and political campaigning. When the company responded that manufacturers don't litter,

Fig. 7.2   Media-Friendly Protest: *In 1971, the Cadbury-Schweppes company ended their policy of using returnable bottles only—and politically organized users rebelled. In this photo, a member of Friends of the Earth in Britain deposits 1,500 non-returnable bottles on the doorstep of the Cadbury-Schweppes' London headquarters. Leveraging the media, the activist group raised great public awareness of recycling and anti-nuclear issues, among others. As with the like-minded Greenpeace, Friends of the Earth was a transnational pioneer of direct activism.*

litterbugs do, the nascent environmental movement struck back: the Friends of the Earth bottle protest advanced the more fundamental critique of squandering resources. The protest turned into a frontal attack on the consumer throwaway society. To put pollution on the agenda, this new generation of environmental activists mobilized the well-established idea of austerity through reclaiming resources. For economic reasons, activists chose to promote recycling over the tradition of conserving nature and landscapes.[7]

Financial reward had been part of the bottle-deposit system. Yet, the willingness of ordinary users to make the effort of returning glass was not a given. The Norwegian brothers Petter and Tore Planke discovered this when they sought to export their innovative electronic bottle-deposit machine to the United States. The success of the brothers' recycling machine depended on subtle negotiations they conducted with users willing to carry empty bottles; grocers searching for a solution to the labor-intensive bottle logistics; a beverage industry looking for new markets; a glass industry facing materials shortages; and law makers responding to the outcry over littering. Starting in 1972, the Planke brothers had developed their high-tech machine, Tomra, in a Scandinavian context of thrift, and within the legal framework of a deposit system. The entrepreneurs succeeded in exporting their system to all of Europe. They failed, however, in New York State, where recycling was associated with poverty rather than with good housekeeping and responsible citizenship. The Tomra company ascribed their failure to what they perceived as the dearth of thrifty housewives willing to return bottles— combined with the hordes of homeless and the Afro-Americans who, they believed, recycled bottles to earn money. The company had assumed that the thriftiness of Scandinavian women was a given all over the world.[8]

The Norwegian entrepreneurs had underestimated the extent to which their sophisticated deposit-return machine depended on state laws that rewarded industry for high recycling rates. Also underestimated was the readiness of average citizens to carry empty bottles to a collection point. Even more fundamental to this issue of recycling in the 1970s was the deposit bottle; the deposit system was on its way out, eliminating an important financial incentive for consumers to return bottles to the store. The Norwegian brothers realized that the device they had designed to help grocers solve a logistical problem was more successful when marketed as an innovation to help the environment. They quickly jumped on the environmental bandwagon and promoted their deposit-return apparatus as a green machine.[9]

Yet returning bottles as part of a daily routine, rather than as an incidental act of environmentalism on Earth Day, had a deeper history. The habit of cleaning and neatly sorting bottles by color in countries like Norway, Germany, and the Netherlands before depositing them into special containers was the "kind of discipline [that could not] be expected in the United States," as one American environmental engineer later concluded.[10] Why European but not American households—women, in fact—were willing to take on the responsibility of returning glass without any financial gain has continued to puzzle the environmental movement, the glass industry, and government officials. The answer derives in part from the divergent war experiences that had marked a generation—in America and in Europe.[11]

A 1985 nationwide survey of 20,000 households in the Netherlands showed that the generation most disciplined in recycling had been born before the war and had come of age during the German occupation. The motivation for citizens to resist the emerging throw-away consumerism stemmed not only from environmental concerns, but from a tradition rooted in the culture of frugality. Indeed, this attitude of frugality contrasted with that of the younger generation of environmentalists. One survey of West European baby-boomers offered insight into the tensions of their generation. Having grown up under materially improved conditions, baby-boomers now claimed non-materialist values; they embraced autonomy, self-expression, and quality of life; and they cared about the environment. Although they eagerly protested nuclear power and voted green, the younger generation did not necessarily adhere to traditional values such as thrift. Quite ironically, in their much-celebrated "post-material" behavior, baby-boomers actually colluded with wasteful practices in environmental terms. Their new self-indulgence thus opposed the norms of the older generation, to which the two Dutch women activists belonged.[12]

In the 1970s, many West European citizens who had been born before the war experienced the era's abundance and the sudden, pervasive rise in waste as an assault on their core values. For the Dutch women activists, the first plan of action against this assault was an attempt to reinstate the deposit-refund system of recycling. This entailed the cleaning and the reuse of each bottle roughly twenty times. Of the two bottle recycling methods available, the deposit-bottle system was the more environmentally sustainable. This was also the system that industry had come to oppose. For this reason, the women's activism focused on the politically more feasible goal of recycling disposable bottles by crushing them into ready-to-be-melted glass cullet. The women advocated this form of glass recycling as a compromise.[13]

The two women, Babs Riemens-Jagerman and Miep Kuiper-Verkuyl, came to their activism after completing a five-day women's leadership course on environmental issues in 1970. The course's organizer, Elisabeth Aiking-van Wageningen, had been profoundly impressed by *Silent Spring*—and shocked by the critical report published by President Kennedy's committee investigating the impact of pesticides on health and the environment. Aiking-van Wageningen found her moral anchor in the Bible. In accord with the theological doctrine of humans' responsible stewardship of the Earth, she shared the notion that Christians have an obligation to maintain rather than exploit the gifts that God has bestowed on mankind. She felt that academic ethical concerns needed translation into direct action and daily practice; middle-class women, in their role as household managers, were best positioned to turn the environmental tide. Such a belief went back to the nineteenth-century idea of municipal housekeeping, which taught middle-class women that they had a special moral role to play in stemming the excesses of industrialization.[14]

Organizations like Protestant church groups and the Young European Federalists seized the new issue, integrating the cause into their own agendas.[15] So did the conservative middle-class women's movement. In the early 1970s, for example, many Dutch women's organizations sponsored subcommittees on environmental pollution in which they kept abreast of the latest chemistry journals; discussed academic calculations on the relationship between demographic data and the country's total amount of waste; and shared the evidence-based information with their rank-and-file members. Aiking-van Wageningen's course targeted middle-class women because they "viscerally know how 'precious' life is…[and] through them, one reaches the entire family."[16] The 1970 leadership course attracted more than three hundred mid-level representatives of women's organizations and featured concerned academics as teachers. Participants were asked to sign a form confirming their willingness to "make an effort to the best of their abilities to disseminate the knowledge obtained during this course" and were urged to lobby officials in their local communities rather than to take to the street in protest.[17] This non-confrontational tactic contrasted with those of activists entering the political arena, who sought to provoke the authorities through "direct action" against what they called "the system." This was a political strategy that had been especially successful in the American civil rights movement.[18]

Riemens and Kuiper explained that the growing garbage levels contradicted everything they stood for. "Having been through the war, we just couldn't stand this kind of waste," they said, articulating a feeling broadly shared by their generation.[19] Throwing away potentially useful products constituted a moral affront to the experiences of poverty, shortages, and hunger. One lecturer at the leadership course in 1972 referred to the Netherlands' growing U.S.-style consumerism as an external attack on national values, comparing it with the German invasion of the Netherlands in 1940. The lecturer characterized the urge to consume and the German invasion as a foreign assault on "feelings of safety and security in honorable traditions."[20] Ironically, these honorable traditions had been reinforced, if not introduced, by the Nazi regime in the early 1940s. The government and the glass industry had a large stake in the bottle issue. Yet recycling's success ultimately rested on women's willingness and eagerness to collect and separate empty bottles on behalf of the environment rather than on behalf of the government. Women were willing to clean and sort the bottles by color and carry them when shopping—without the financial reward the older deposit system had provided. The economic incentive for saving materials had disappeared. In its place, the motivation to save derived from moral injunctions that women had internalized during decades of government sponsored thrift and frugality campaigns. The first wave of campaigns occurred during the First World War and the economic crisis of the 1930s. A more systematic approach to thrift and frugality was enacted through autarkic policies during the German occupation. Thrift and frugality were then

Fig. 7.3 **Be Thrifty, Be Frugal**—*Right Now*: *This was the implicit campaign message. In 1972, Dutch activists Wilhelmina Kuiper-Verkuyl and Babs Riemens-Jagerman kick-started a movement with the first glass recycling bins in Western Europe, pictured here, in the town of Zeist, the Netherlands. Beyond anti-pollution and anti-littering, the women's motivation was thrift and frugality. For these activists and their generation, recycling was a religious and moral act. Salvaging was a state-sponsored practice ingrained during the Depression and German occupation. Thanks to these activists and their sister organizations, a network of recycling bins across the country emerged.*

reinforced by the public policies for reconstruction, which encouraged a culture of saving well into the 1960s.[21]

Industry and policy makers either took for granted or overlooked the importance of consumers' willingness to return bottles. While no longer an economic necessity, the practice of reusing glass persisted; it was so deeply engrained in women's daily routines that new arguments were hardly needed to reconstitute old habits. There are additional explanations for the quick acceptance of glass recycling. While the image of the environmental movement has been dominated by the anti-establishment young baby-boomers, the initiators of the movement were much more diverse and had a longer history. For example, bird protection activists, some of whom were quite radical, based their critique on new ecological paradigms and reproduced doomsday visions of the generation born and raised before the Second World War. A long-term view shows how the ideology of environmental engagement was built on a European tradition of thrift and reuse. This tradition clashed with the culture of abundance—the culture that affronted the German architect; spurred the activism of the two Dutch women; and foiled the Norwegian bottle-machine entrepreneurs.[22]

The Fatherland Deploys Waste

During the First World War, European states were already experimenting with collective saving measures, in fact. After the Depression, governments began to encourage such activities in a more systematic fashion. Rather than raise wages and consumer spending, the government sponsored home economists to teach working-class and farmers' wives to produce good meals with little money; to preserve the summer's harvest and winter's slaughter; to refashion old clothes into new; and to make their own mattresses filled with straw. Toward the end of the 1930s, governmental organizations and private companies needed to amass raw materials for salvage purposes. During the Second World War, these national practices intensified in countries including France, Britain, and the Netherlands. They became more systematic when, to support the war economy, the German regime exported its autarky policy to the occupied countries.[23]

As the liberal feminist and politician Marie-Elisabeth Lüders recalled, the First World War had turned German women into "rag pickers, junk dealers, bookkeepers, and employers" overnight, forced to conserve the nation's resources. Housewives "were not allowed to throw anything away and had to collect everything from hair and nails, to string, paper, cans, and kitchen slops."[24] At the outset of the war, the waste politics of the *Kaiserreich* had been rather uncoordinated and decentralized. After all, the country had been cut off from imports and lacked raw materials. This situation left a great deal of room for local initiatives from "below"—that is, from users themselves. In fact, urban German housewives had been active in food-waste collections in patriotic support of the war long before the state first launched salvage drives in 1916.[25]

Nearly twenty years later, in preparation for the next war, national-socialist waste collections, by contrast, were imposed from above. Drawing on the earlier experiments, the *Reich* Commissioner of Scrap Utilization called on "the wisdom and insight of the German woman." It was her "responsibility" to make sure that no valuable resources were wasted, and her "duties" included handing over any useful materials to the state: "Do not forget that 80 % of the national income passes through the hands of the housewife!"[26] Be it old newspapers or magazines, food leftovers or bones, rags or discarded clothes, scraps of rubber or metal—everything that could be recycled had to be saved and passed on to an authorized collector or an official organization. During special campaigns, parents were asked to have their children bring their "important raw materials" to school.[27]

Hermann Göring, the infamous Supreme Commander of the German Air Force, appointed Wilhelm Ziegler from the Nazi party's elite Storm Troopers (*Sturmabteilung*, SA) as *Reich* Commissioner in 1937 to optimize the collection of waste material. The Commissioner's task was to make sure that as many resources as possible—including material that was formerly considered garbage—were recovered and made available to industry. The housewife's practices of saving and separating represented only the first steps in a longer salvage process. The waste that her children could not bring to school was picked up by scavengers, who in turn were forced to deliver the refuse to local dealers.[28]

The recycling of waste was part of a much larger undertaking to ready Germany for the future war. Ziegler was responsible for a section within the Department of Trade and Industry; its activities were dictated by the Nazi Four-year Plan. Göring had announced the first of the four-year plans in 1936. Like Walther Rathenau during the First World War, Göring tried to regulate and optimize

Fig. 7.4 The History of Salvage: *Until the 1930s, waste collection in Europe was an independent trade, a livelihood for Jews, in particular. In this 1925 photo, so-called rag-and-bone men in the city of Amsterdam collect worn-out clothes and other textiles. With the rise of anti-Semitism and the looming war, the Nazi government took over the salvage-and-recovery trade in Germany and the occupied countries. Jews were banished from the trade and later deported. Facing raw-materials shortages, women and children were enlisted to collect waste "for the nation." This practice of citizens salvaging and re-using inflected the environmental movement of the 1970s.*

the provision of raw materials within the national economy. The plan needed to guarantee industry continuous access to crucial commodities such as iron and wood, mineral oil and other chemical products. The plan also needed to ensure the *Reich*'s ability to feed its population under all circumstances. Albeit on a small scale, the recovery of scrap metal, textiles, and organic waste from households would contribute to a higher degree of self-sufficiency.[29]

As with many other Nazi policies, the approach to salvage and recycling had contradictory elements. On the one hand, Göring and his companions hoped that the recovery of household and industrial waste would make Germany less dependent on imports; Nazi policy makers also believed that recycling would support the future war effort. Recycling was a key ingredient of their so-called autarky policy. On the other hand, the Nazis' anti-Semitic laws and regulations drove Jewish scavengers and waste mongers out of business, thus challenging existing, well-functioning waste-recovery structures. First, "Aryan" waste collectors were forbidden to deliver their material to Jewish middlemen, and, after 1938, no Jews remained in the trade. The government tried to rationalize the business by eliminating the local rag dealers altogether and "Aryanizing" the remaining mid-sized and larger companies. In

Fig. 7.5 The State Recycles: *Recycling was an important part of the Nazi government's future-war preparations. This policy of autarky (self-sufficiency) aimed to make Germany as independent as possible, free from importing raw materials. Housewives, school children, and Nazi youth organizations were required to save everything from rags and glass to hair and love letters. Trucks like this one, from Cologne and Aachen, Germany (1937), collected organic waste from cities and suburbs— and delivered it to farmers to use as pig fodder. The slogan on the pig reads: "We collect for me! I eat: potato peels, onions, vegetable leftovers!"*

the first six years of Nazi rule, more than two-thirds of all firms in this sector disappeared. Consequently, Ziegler had to organize the recovery and recycling process anew. Traditionally, scavengers and waste mongers had collected, separated, and processed waste from households; the task was now delegated to school children and housewives.[30]

Given that waste collection was not very profitable and rag gathering was not a particularly attractive profession, Ziegler had to find ways to market waste-processing that went beyond traditional business strategies. Here, the Nazi regime mobilized its many organizations. Merely a year after the Nazis had been voted into power, the so-called National-Socialist People's Welfare (*Nationalsozialistische Volkswohlfahrt*, NSV) had already initiated the collection of rags. Using a language of solidarity with the poor and the unemployed, the organization asked housewives to save old clothes and worn-out fabrics: "For once, please refrain from the small sums that you ordinarily receive by selling rags."[31] Then, in 1937, the *Reich* Commissioner assigned the NSV the task of collecting organic waste in cities to help feed pigs in the countryside; in earlier times, this had been a welcome source of income for urban families. To this end, the NSV founded a special Nutrition Relief Organization (*Ernährungshilfswerk*).[32]

The Nazi party mobilized additional groups to amass various types of waste on an honorary basis. The official Nazi union—the German Labor Front (*Deutsche Arbeiterfront*, DAF)—was responsible for the collection of valuable leftover materials at industrial plants, and the Hitler Youth (*Hitlerjugend*, HJ) was entrusted with this task in households. Even the redoubtable Protection Squadron (*Schutzstaffel*, SS) engaged in this national mission. During special initiatives, the first of which date to 1935, 10–18-year-old *Hitlerjugend* boys and girls went from house to house to secure such things as bottle tops and corks, metal foils and tubes. And in 1937, the Storm Troopers carried out a "clearing-out campaign," during which they rummaged even attics and cellars in the search for usable objects and recyclable materials. District leaders (*Gauleiter*) were responsible for these household collections, and they were supported by women's national-socialist organizations (NS *Frauenschaft*) as well as by local salvage traders.[33]

Nevertheless, conflicts with the threated members of the recovery and recycling trade were bound to develop. The socially disadvantaged but still proud rag men found themselves being both harassed and economically threatened by aggressive *Hitlerjugend* members, whom they derided as "amateurs" and "innocent angels."[34] By appealing to the benevolence of housewives, Nazi organizations transformed recovery and salvage from an economic necessity into a political and moral requirement. Waste could be recovered without any financial compensation. By collecting products that were not commercially profitable, the Nazi groups tried to stigmatize the ragmen as greedy egotists who did not care about the national economy. Squeezed between political propaganda and business interests, the trade tried to motivate the rag men: "If you do not only accept high-quality woolen textiles but also stuff of lower value, then you do not only assist the housewife but also yourself and the German economy."[35]

The government's recovery program intensified as war drew closer. In 1938, the *Reich* Commissioner turned all remaining ragmen into "duty collectors" (*Pflichtsammler*) and assigned each of them a certain area of town. To emphasize the importance of this (by now) ethnically-cleansed profession, its members were requested to dress properly—incredibly, a collar and tie were recommended. Instead of getting people's attention by shouting in the courtyards, rag men were asked to knock on each and every door, thus approaching housewives and others face to face. Such direct confrontation put pressure on citizens to donate materials rather than to sell them off. To some extent, the new collection system worked. Nevertheless, the "Aryanization" and rationalization of the trade also produced serious bottlenecks on the recycling side. In 1939, the *Reich* Commissioner reported that large amounts of scrap paper became mildewed, and that bones putrefied on the premises of the recovery plants. The system suffered from its own success.[36]

To simplify the work, the authorities obliged citizens to separate various kinds of recyclable material. The police decreed that organic waste was to be collected in special containers and threatened to conduct inspections if citizens failed to follow orders. Local party officials deployed collecting bins in public places; in larger apartment buildings, janitors were required to set up separate bins for different materials. Recovery and recycling were officially sanctioned activities that tied the household directly to the national economy. Thus the Nazi government effectively instrumentalized housewives and children for the war effort. It is hard to tell whether they accepted their roles happily. Registered complaints concerned hygiene and odors. In Hamburg, for example, tenants objected to smelly organic-waste bins in their stairwells and the public-health department protested against the manual collection of bones on the part of school children.[37]

The salvage programs then took on a new character. After 1942, the Nazi fortunes of war reversed due to the German debacle on the Eastern Front and the Nazi extermination policy having caused the economy of the Jewish poor to collapse. The weapons industry demanded iron and steel in all shapes and forms, and soldiers and bombed-out citizens badly needed clothes and shoes. When the general labor shortage grew acute, the mobilization of school children became ever more important. At many German schools, bins and containers for various recyclables were put in place, and the most zealous pupils received prizes for their effort. To guarantee the regular collection of ordinary waste, some cities employed forced laborers who had to work from dawn to dusk with minimal nourishment. Newspapers in the city of Hamburg complained in 1943 that these mostly foreign prisoners-of-war failed to carry out such "hard work."[38]

Sometimes, the hunt for recyclables went overboard. Before the war, authorities in 150 German cities had already started the practice of ordering women and children to gather tin cans in the cities' waste dumps. The Storm Troopers organized a campaign to persuade homeowners and landholders to dismantle their iron fences and sell them to the local junk dealer. Even before the war started, in 1938, a dramatic poster insinuated that such fences could be turned into airplanes, submarines, and tanks in order to "protect Greater Germany." In the midst of the war, almost nothing was sacred any more. Church-bells, lightning rods, door handles, coins, and copper kettles were all turned into weapons. Women and children served as the foot soldiers on this warfront for autarky.[39]

Public engagement in waste recovery was also practiced in Fascist Italy. During the Abyssinian war, the Fascist party asked Italian wives to sacrifice their gold wedding rings for their fatherland. This drive peaked on December 18, 1935, the "day of the wedding ring," when hundreds of thousands of women throughout the country sacrificed this precious personal item for their *patria*.[40] During a well-staged ceremony on the Piazza Venezia in Rome, Queen Elena of Savoy took the first step and put her ring in a huge offering jar before the enthused masses. Next to make the sacrifice was Rachele Mussolini, the wife of *Il Duce*, and then, "the poor in threadbare clothes and often hatless; others in fur coats, officers and clerks." All expressed their solidarity by exchanging their golden rings for others made of cheaper metal.[41]

Despite denials that it took its cues from the Nazi model, Britain was also not immune to the fanatical hunt for recyclables. At the peak of the Second World War, W.A. Foyle, a British book-seller, expressed deep concerns over the "many priceless, rare, and irreplaceable books [that] are being destroyed owing to the campaign for waste."[42] His misgivings appear to have been well founded. Many old books, letters, and archival material fell prey to the initiatives of the British Directorate of Salvage and Recovery. Reportedly incunabula were also threatened. Herbert Morrison, Minister of Supply—to which the Directorate belonged—made it clear that "even old love letters can be turned into cartridge wads, meat bones into explosives, tin cans into tanks, and garden tools into guns."[43] His officials asked the population to hand in all "the old forgotten books—all the old treasured programs—the useless receipts—sentimental letters,"[44] and they did not shy away from melting down priceless historic trophies of war and church bells. Although the Brits were aware that the Germans had established a compulsory waste recovery program, for several years, recycling in Britain relied on citizens' patriotism and voluntary engagement. Apparently, though, pleas for patriotism were not enough: not even a contest promising a prize of £20,000 achieved the hoped-for effect. In March of 1942, then, Morrison resorted to issuing an edict that criminalized burning or discarding paper and other kinds of recyclable waste. As in Germany, school children took part in widespread collections. Overall, the German policy was more systematic, however. And, perhaps more significantly, the coercive state policy was exported to the occupied countries.[45]

Waste, War, & Nazi Politics for Europe

The *Wehrmacht* established a Supreme Command (*Oberkommando*) that formulated recovery and recycling strategies for the occupied territories. Archival material has survived that confirms that the military forces' efforts to organize the collection of waste—especially scrap iron—in the Soviet Union, Hungary, and Rumania. Also documented is how this material was to be transported to Germany for reuse. Under the Nazi regime, the conquered countries, especially those in Eastern Europe, were reduced to the status of colonies, and expected to supply the fatherland with agricultural products, natural resources, and energy, as well as recyclable materials. The policy of autarky was complemented by a "philosophy of prey," or exploitation. In occupied and unoccupied Western Europe, as well, the policy made deep imprint on the daily lives of ordinary citizens.[46]

During the First World War, all governments had experimented with nationalizing waste and recycling, but Imperial Germany had taken the lead; the Nazis made it a permanent drive from the very start. In the years leading up to the war, newspapers in France, Great Britain, and the Netherlands, for instance, had followed Göring's coercive recycling policy with growing amazement. Whereas the Dutch observed his measures with intense curiosity, French and British commentators at first considered his policy a sign of Germany's economic weakness and rejected it outright. As national governments faced war, all criticism vanished, however— practically overnight. Policy makers insisted that their strategy of saving essential resources had nothing to do with the German measures, but their discourse and initiatives largely mimicked the Nazi example. Thus, by the time the Germans occupied other countries and took control of their government structures, local politicians and civil servants had already adjusted mentally to the Nazi approaches. Local politicians and civil servants collaborated closely with the new government in reorganizing the waste sector according to the German model: establishing a special bureau, nationalizing the waste collecting, establishing waste collection sectors, expropriating Jewish firms, deporting Jewish peddlers, and mobilizing school children and women to do the work.[47]

The Second World War intensified and nationalized efforts of recovery and salvage, but the war did not create these practices. The main difference was that the war helped frame recycling not only in monetary terms, but in political terms as well. Waste collection became a part of the moral economy, and as such it was resurrected in the 1970s. Some of the actual war practices were carried forward into the postwar period. Well before the world wars, collection and recycling had been established commercial businesses; it was a highly international trade. Scrap metal, rags, paper, and animal

Fig. 7.6 Mobilizing for the War Effort: *Foreseeing war, the German government instituted waste-recovery programs in 1935. The Allied countries criticized these measures—but later launched their own. In 1939, Britain, for example, started persuading citizens to save resources and recover diverse materials. The government urged Brits to donate scrap paper and old books, discarded pots and pans, plus organic waste. This war-time poster urges citizens to save food leftovers. City authorities collected this material and distributed it to local pig and poultry farmers. Like in the German-occupied lands, the procedure mimicked that of Nazi Germany— right down to the pig motif.*

hides circulated between European countries and between Europe and other parts of the globe. In the interwar period, China and Japan emerged as important exporters of cotton rags—mostly used for paper and cardboard manufacturing. The Polish textile industry imported increasingly larger quantities of recycled wool—which was re-spun and applied to manufacturing new, cheaper wool products. Scrap paper also circulated in substantial quantities between various countries in Northern and Central Europe. In the case of glass, Sweden and France were exporters, and Czechoslovakia, Poland, and Austria were listed as the main importers. During the Depression, however, protectionism grew—which bolstered the argument in favor of countries' need for self-sufficiency. In 1930, for instance, the United States raised substantially its import duties on rags. And, whereas Germany had been an important exporter of scrap paper and rags, the Nazi autarky policy effectively put a halt to the outflows of such products in 1936.[48]

The outbreak of the Second World War was a further blow to the international scrap and recycling trade. Traditional trade channels largely collapsed. The Nazi-occupied territories were integrated into the *Reich*'s waste-recovery program. In the East, this policy most often boiled down to pure exploitation; in the West, waste-recovery policy took somewhat different forms. In France, Belgium, and the Netherlands, Ziegler's successor as *Reich* Commissioner of Scrap Utilization, Hans Heck, initiated recovery and recycling activities similar to those in Germany. During the same period, several of the ideas and practices reappeared in other countries, such as Britain. Strikingly similar campaigns were set up in Fascist Italy: the government had already appealed to the Italian population to hand over garden fences and pots—and special containers had already been installed for the collection of scrap paper.[49]

The situation in France is particularly noteworthy. When Heck visited Paris in early 1941, he met with representatives of both the occupied North and the Vichy section of the country. Before this meeting, the French had recently founded a Central Office for the Distribution of Industrial Products (*Office central de répartition des produits industriels*, OCRPI) to coordinate the collection and recycling of waste. Concerning technical matters, the Office drew heavily on the knowledge provided by German experts. As in Germany, the French authorities forced scrap dealers and rag men to register, and they forbade citizens from throwing away or burning paper and other recyclable products. To solve the bottleneck of collecting waste that resulted from the expulsion of Jewish scrap collectors and dealers from the trade, French authorities organized drives to recover textiles and, like their German counterparts, had school children collect waste paper. When Heck went to Paris, he happily reported he had already successfully reorganized salvage operations in occupied Belgium.[50]

Salvage drives took different forms in the warring nations. While the British Empire suffered like other countries, its maritime trade offered an escape route for raw materials. Like many other governments, the U.S. government established highly publicized wartime recycling programs. But these were largely propagandistic tools to boost morale rather than to promote substantive policy measures. In fact, even before war's end, corporate America had successfully lobbied and geared up for peacetime expansion and consumption, unlike its European counterparts.[51]

Creating, Repairing, & Crafting

The Nazi-inspired waste policies in Germany and the countries it occupied shifted the job of waste collection. The professional, predominantly Jewish proletarians who had earned income from private-sector waste-collecting were replaced by a non-professional army of users: schoolchildren, charitable organizations, youth clubs, and women who worked for free.

The household task of waste collection, separation, and delivery was an activity that did not impart personal pleasure but did increase the workload—of housewives, in particular. One part of the job, though, despite its labor-intensive nature, did stir feelings of personal pride in the women who engaged in it: the refashioning of clothes. Invoking such feelings, governments had launched *Make-Do and Mend* campaigns during the Depression; these efforts intensified during the war. The main targets were poor working-class women and upper-class women deprived of their seamstresses. The British, American, and Canadian governments, in collaboration with home-economics professionals, mobilized a culture of preservation, maintenance, and recycling of underwear and outerwear to cope with war shortages. The problems were even more acute in German-occupied countries like France, Belgium, the Netherlands, Norway, and Bohemia-Moravia, because the Nazi regime forced them to fuel the German war economy. Textiles—in addition to paper and iron—were especially sought after by scrap

Fig. 7.7　Reinventing Fuel: *To cope with fuel shortages of the Second World War, citizens tinkered with alternatives to wood and coal. The woman pictured here makes briquette-like "fireballs" out of scrap paper. To boost its energy density, paper was first soaked in water. It was then mixed with a binding agent like sawdust; shaped into balls; and set to dry. This 1940 photo shows a Dutch woman in The Hague at work on an apartment balcony.*

collectors and government officials. Clothing and cloth, however, were unique; they could be recycled at home and did not need to be given away.

Countless women spent long evenings turning worn-out textile products into "new" clothes, bed sheets, and quilts. During periods of economic crisis, such activities were undertaken by women of all classes. In wartime, governments encouraged middle- and upper-class women as well to be creative in making households self-sufficient. This helped to institutionalize the culture of crafting and tinkering. Governments established large-scale campaigns of make-do and repair to help women deal with tight budgets and textile shortages. New clothing styles, the economic crisis of the 1930s, and the Second World War—as well as the disappearance of seamstresses in the postwar era—would alter the phenomenon of home dressmaking. During times of economic downturn and war, dressmaking was a necessity for all. During peacetime, it was a pleasurable activity for the middle classes. In the 1950s and 1960s, home design, paper patterns, home sewing, and the sewing machine gained renewed popularity, particularly with the middle classes.

The rationing of fabrics demanded a great deal of knowledge, self-reliance, and inventiveness on the part of women responsible for clothing their families. Governments set up distribution systems, first in response to the 1930s economic crisis and later in answer to the needs of the war economy. During the 1930s Depression, government policy was often restricted to poor and rural families, seeking to teach them how to make do with what they had. Home-economics classes emphasized household management, thrift, home care, and home repair. Sewing lessons focused on conserving clothing and textiles, patching, mending, and other money-saving practices—rather than on designing and constructing anew. The Dutch government encouraged the poor, working-class, and rural families to consume less and repair more. At the same time, however, the state sought to increase middle-class and upper-class consumption through, for example, the gas and electrical utilities. The war changed that two-pronged approach: it had the unintended "democratizing" effect of instilling a culture of thrift for all classes.[52]

In wartime, governments needed textiles for uniforms and parachutes. Repeating the slogan "Make-Do and Mend," they began to discourage the buying of textiles, pushing women of all classes to use instead their skills to repair, refashion, remake, and recycle what clothing and linens they had. The British Board of Trade admonished women's voluntary organizations in an open letter in 1942: "If every single garment now in the homes of Britain, every pot and pan, every sheet, every towel is used and kept usable until not even a magician could hold it together any longer, the war will be won more surely and more quickly," and concluded: "we must do all in our power to make the whole nation conscious of the need to mend and make."[53] In the Allied countries, mending and repairing was a matter of patriotism. By contrast, first in the occupied countries and later under communist regime, such policies became symbols of suppression and failure, despite the fact that some of the policies had started before the war.

While the British government under the Conservative Neville Chamberlain only reluctantly intervened in the textile market, the Dutch government did not hesitate, instituting textile rationing in 1939. After occupation, the German regime—in the Netherlands and elsewhere—forced economic restructuring; implemented the "Aryanization" of Jewish textile firms and fashion houses; and confiscated local textile industries for the fabric, yarn, and wool to manufacture German war uniforms. The deportation of the Jewish-Dutch citizens who had dominated the garment trade also had a

URGENT
ALUMINIUM IS WANTED
FOR SPITFIRE & HURRICANE
PARTS.   TURN OUT ALL
YOU HAVE IN YOUR HOME
THE NEED IS GREAT &
YOU CAN DO WITHOUT
MOST THINGS.

IF YOU HAVE NOTHING
IN THIS METAL GIVE A
SMALL DONATION IN THE
BOX PROVIDED & WE
WILL BUY & PRESENT TO
THE COUNTRY.

Fig. 7.8 Campaigning for the State: *After the outbreak of the Second World War, all warring countries organized salvage and recovery campaigns according to Germany's self-sufficiency model. For example, patriotic women's groups in the United Kingdom went door-to-door collecting household items for the war effort. The placard in this 1940 photo directs housewives to hand in their aluminum pots and pans for remaking them as spare parts for Spitfires and Hurricanes—fighter aircraft. The rally was organized by Britain's Women's Volunteer Service for Civil Defence, established in 1938.*

profound impact on the textile and clothing industry. Shortages forced enterprises to become creative. The textile and clothing business devised new ways to manage the intricate and complex distribution and coupon system. In the Netherlands, rain jackets, for example, could be sold without a rationing coupon, so Dutch entrepreneurs sold waterproofed tweed jackets as a way to bypass the distribution system. When, in turn, raincoats were banned, tailors started to waterproof sheets to make rainwear out of them. And when, in 1942, cotton became scarce for the manufacturing of new products, the expropriated Jewish underwear factory Tweka began to offer a repair service for worn underwear. The footwear industry as well explored alternative materials such as wood, textiles, and woven rushes for novel uses. Skate straps and candlewicks, for example, were used for sandals. During the German occupation, the shortages for home consumption became even more pressing, spurring women to be inventive in repairing, redesigning, and remaking old clothes into new. Alongside government policy makers, women's magazine editors and home-economics teachers played an important role in helping women to develop needlework skills in clothes making for their families.[54]

The British government and women's organizations ran roughly 25,000 *Make-Do and Mend* classes as well as evening dressmaking classes during the war. The emphasis was on basic garment-making and traditional making-over techniques: turning collars and cuffs, the darning of socks, and the patching of trousers. In particular, upper-class and middle-class women who had always relied on tailors, dressmakers, and seamstresses found themselves in need of instruction on how to use scissors, needle, and thread. The cutting of fabric—the ability to gauge how to make the most of the available yardage—was a highly valued skill.[55]

Garment-making—and re-making—was also emphasized in the occupied Netherlands, where newspapers and magazines instructed housewives how to transform a man's old suit or a woman's skirt into children's clothes. Crocheted bedspreads and tea cozies were unraveled to make sweaters and jackets. A lively barter trade in clothing and shoes emerged in response to the acute shortages of consumer goods. Women also started to use available materials in alternative ways, making dresses out of non-rationed curtains, for example. In November of 1940, this creative practice prompted a new rule: curtains could only be sold with a special permit. In 1942, when women discovered printed lining fabric as an alternative fabric for dressmaking, the result was an acute shortfall that forced the temporary shutdown of lining-fabric production and the sale of coats. Women incorporated cotton gauze and bandages into clothing, and they dissected flour bags and parachutes into separate threads, to be reused as sewing material for jackets and sweaters. Hats were made out of scraps, cardboard and the material placed under carpets; shoe soles were made out of wood, tires, cork, reed, and bark; uppers were shaped from parchment lampshades, leather briefcases and chair upholstery, carpet scraps and felt hats.[56]

Government measures were strikingly similar in all the countries facing economic depression, war, and reconstruction. In the planned economies of the Soviet Union and its satellite countries, the policies of saving and repair—and the injunction not to buy consumer durables—continued well after the war, yet acquired a very different postwar meaning. In Paris, for example, there was a full-fledged fashion rebellion against the wartime constraints: Dior and his colleagues revived the luxury fashion trade with a vengeance. Their fashions conveyed a new femininity that spurned austerity in favor of flamboyance. And by expanding licensing, Paris reached new clients on the U.S. market, where the war and rising wages had made consumers wealthier. In contrast to the celebration of luxury, the legacy of saving and repair took a different shape in Western Europe and Scandinavia. Here, as disposable income rose and leisure time increased, the repair and tinkering ethos morphed into a do-it-yourself fad. With its roots in the United States, the do-it-yourself user movement bridged the wartime valuing of crafting, tinkering, repairing, and *bricolage* to an emerging postwar consumerism. In Eastern Europe and the Soviet Union, the rationing policy continued well beyond the 1950s, both out of necessity and for ideological reasons. To the communist parties, consumption continued to represent excess and luxury.[57]

Communist regimes incorporated older criticisms of the aristocratic culture of leisure, consumption, and fashion, echoing what Tolstoy had articulated in *The Kreutzer Sonata* decades earlier. Ideologues in communist regimes refashioned this existing criticism as a specific socialist critique of excess and luxury. Communist fashion was supposed to convey solidarity rather than individuality. Applying mass-production measures and simple designs was a logical step for communist regimes to take in order to clothe their citizens. The campaigns were ideologically charged, particularly in the Soviet Union and East Germany. There, in the 1950s and 1960s, women tinkered with needles and sewing machines; they participated in pattern making, cutting, and sewing classes like their sisters in Western Europe. The decision to learn these skills was not a tribute to tradition, but an outcome of government policies and citizens' response to a system in which industrially-produced products were uniform, badly cut, and failed to offer different sizes.[58]

Communism and crafting went hand in hand, as it were. In socialist Hungary as well as in the socialist Germany, home sewing; hiring a tailor; or asking relatives, friends, or acquaintances to make clothes was common practice. While in the 1960s authorities estimated that the ratio of store-bought to homemade clothing in West Germany was 9:1 and in the U.S. 13:1, in East Germany almost two-thirds of the clothing was still made or purchased in alternative ways: refashioning a garment, changing dresses by adding or removing parts, lengthening or shortening hems of old garments to fit new fashions. As in other countries, women in East Germany fashioned clothes out of military uniforms, linens, and curtains. Initially, the regimes aimed to increase the industrial production of ready-made garments; home sewing was regarded only as a temporary measure until the textile and garment sectors resumed full production. Yet, the regimes' later policies sought to encourage home sewing as a virtuous task and claimed that women took up sewing voluntarily. Consumer magazines published paper patterns for every conceivable clothing item. By the late 1960s, however, the Berlin chair of the party women's organization, the *Demokratischer Frauenbund Deutschlands,* asserted that its sewing classes were popular not because women "view it [sewing] as their hobby," but out of necessity. In fact, the organizer pointed to failed government policies as necessitating women's sewing: "We also don't see it [sewing] as the solution for trade and economic functionaries' absent feelings of responsibility."[59] A survey among readers of a women's magazine and market research report confirmed the critical sentiment.[60]

In postwar Russia, Soviet citizens also negotiated shortages. They appropriated and customized the few standardized factory-made items of clothing available for purchase by creating, altering, or embellishing garments. These methods were, however, not unique to the communist system. The working poor and middle-class women in the whole industrializing world had relied on home sewing when ready-made clothing was still expensive, of poor quality, or too standardized to fit most women. The Soviets practiced a culture of repair, darning, or "wearing out" (*donashivanie*) of clothes handed down from one generation to the next, or given to neighbors, colleagues, and friends. Social networks that extended beyond family and blood relations were shaped by the accessibility to official or unofficial distribution channels. Catering to this culture of crafting, repair, and redesign were special books and articles in women's magazines. The 1958 reference, "Three Hundred Handy Hints" (*300 poleznykh sovetov*), was one example. Even beyond the cycle of preserving, repairing, darning, and handing down, worn-out textiles always found a secondary use that went beyond repair or redesign: rags were used for dusting or cleaning floors; stockings for storing onions or "pulled over a broom to keep it from shedding or ... used to make a plaited rug."[61] These practices were survival skills known to women all over the world.[62]

In the Soviet bloc countries, the use of informal networks complemented the home-making strategies that did not necessarily match one's income, social group, profession, place of residence, or party membership. To have a family member working at a store could be a great help; working as a seamstress on the side, for pay or as a favor, could become a form of barter. On the black and gray markets, clothing also circulated through various channels. Packages were sent from the West via retirees who returned with goods—or from relatives from West Germany who brought gifts during visits. East German market researchers estimated that 20 to 30 percent of all the socialist country's clothing came from beyond the Wall. By 1988, clothing arriving from the West through packages was worth almost four billion Deutschmarks annually.[63]

Going Green—Locally & Globally

Transnational from the beginning, the environmental movement would reset these politics of limited resources for a new era in Western Europe and Scandinavia. In the environmental discourse of the 1970s, old recycling practices for the sake of the nation's (war) economy were recalibrated to suit a new moral frame of reference. The goal was no longer to save the nation but to protect "Spaceship Earth." While the culture of austerity was part of Europe's postwar fabric in both East and West, the socialist shortages in the 1970s and 1980s transformed consumers into consummate *bricoleurs*, with the artistry and ingenuity to combine, repair, recycle, remake, and trade.[64]

The Dutch anti-waste activists Riemers and Kuipers came on the scene just seven years after the nation's recent shift from a policy of saving and investment to one of spending and consumption. Housewives had been socialized to reuse and recycle for decades by both national and Nazi policies. After defeating the Nazi regime, the United States turned West Germany into its showpiece in the Cold War. The economic policy radically reversed what had gone before. Under U.S. tutelage and supported by currency reform, West Germany abandoned autarky, abolished rationing, and adopted Keynesian economics as early as 1948. For American policy makers, West Germany became the prototype for all of Western Europe. Other countries in Europe thought otherwise, however. The Spanish (1952), British (1954), Dutch (1955), and Austrian (1955) governments continued rationing until the mid-1950s. Norway and Finland allowed the free import of Western cars only in the 1960s. Despite the Cold War propaganda, many West European governments resisted the Marshall Plan's emphasis on consumption. In some cases, governments even lowered wages. The Dutch government's rebuilding efforts focused on investing in heavy industry and collective consumption, keeping wages down, and discouraging individual consumption; only in 1963 did the Dutch leaders allow incomes to rise as a means of stimulating the economy. The anti-consumerist policy of many of Europe's welfare states dismayed Marshall Plan policy makers. In the Soviet-dominated parts of Europe, governments similarly discouraged individual consumers and invested in collective provisioning of goods and services. East Germany moderated the restrictions on consumption in 1958 at the 5th Party Congress, when its communist leaders pronounced that socialism would prove its superiority over capitalism by producing more—and better—consumer durables. In short, West Germany represented the European exception rather than the rule.[65]

Decades of intensive European government-sponsored socialization yielded a deeply engrained morality of thrift. In the Netherlands, at the time when the two women activists decided to restore the production–consumption chain by recycling glass, the first signs of consumer society had just become visible. In 1970, nineteen out of twenty Dutch citizens favored a more stringent and restrictive environmental policy. In Britain of the 1970s, experts worried about the changing composition of municipal waste, as it increasingly contained plastic packaging resistant to quick decomposition. Tories and the Labour government rejected recycling on the grounds that waste was merely minor collateral damage in the context of a fundamentally sound economic policy. The British New Left rejected recycling on the argument that it was superficial. Instead, The New Left embraced the environmental movement's anti-affluence critique and reiterated the movement's call for a redistribution of wealth. Even in West Germany, where the culture of abundance had been a successful political instrument in creating an "economic miracle" after a painful past, the protest was intense—as the German architect and hundreds of his fellow letter-writers showed. The officially sanctioned, new culture of abundance made many Western Europeans feel slightly uncomfortable. Throwing away valuable materials contradicted the state policy of autarky and the morality of thrift that ordinary Europeans had internalized for at least four decades.[66]

Ironically perhaps, the environmental movement in Western Europe and Scandinavia was inspired ideologically by the examples of American countercultural resistance to the politics of abundance. Showing how pesticides affected the food chain and devastated birds, Carson's *Silent Spring* reset the environmental agenda in many countries in Western Europe. The American Earth Day in April 1970 showed the power of citizen activism, grassroots politics, and legislative action. Newspapers in Europe reported how an astonishing twenty million students, along with many housewives' organizations, celebrated Earth Day. In her book, *What Every Woman Should Know—and Do—about Pollution: A Guide to Good Global Housekeeping*, from the same year, Betty Ottinger explained the grassroots activism of middle-class women. These countercultural sources of inspiration reframed national discussions deeply rooted in habits of scarcity, thriftiness, and frugality into a global discourse about environmental protection. The dire warning about Earth's limited resources originated in transnational policy and scientific circles; the warning was articulated in the well-publicized Club of Rome report and gained further traction as the first so-called oil crisis unfolded.[67]

Environmental groups were not alone in operating across national borders; government actions were also transnational in scope. Starting in the late 1940s, policy makers within the framework of the United Nations worried about pollution, population growth, limited resources, and national security, pushing hefty policy papers and designing complicated rules on the global stage. By 1970, U.S. President Richard Nixon, shrewdly sensing the youthful vibe and protest of the street, jumped on the environmental bandwagon and took the lead. European parliamentarians also saw the environment as an ideal issue for boosting the much-needed legitimacy of the European Economic Community's institutions. In 1970, when Nixon established the Environmental Protection Agency, and the Council of Europe announced the European Conservation Year, the regulatory response signaled that government actions had become truly transatlantic and trans-European. During a 1972 visit to the United States, French Minister Robert Poujade, who managed environmentalism in France, admiringly called Washington, "*le Mecque de l'environement*."[68]

On the transnational stage, many non-governmental environmentalists mobilized, working to translate their activism into national and international laws. Media-savvy and agile internationally, they lobbied and succeeded in getting a slew of environmental laws passed worldwide. When the British branch of the Friends of the Earth deposited empty glass bottles on the doorstep of Cadbury-Schweppes in 1971, the protest generated a vast amount of free publicity and good-will. Empty bottles conveyed the notion of limited resources more concretely than any of the thick policy papers that the international community produced. Friends of the Earth founder David Brower, a Californian, sought French, Swedish, and British allies to build a broadly based transnational environmentalist movement. Although the Friends of the Earth also addressed nuclear power, for example, bottle activism proved easier to translate into direct action. From the start, Friends of the Earth operated transnationally to support grass-roots activism locally.[69]

Advancing the environmental movement took place on so many levels: the governmental and the user level; the national and transnational level; and the high-tech and low-tech level. In the world of bottles, the Norwegian Parliament was the first to compel industry to reuse its packaging through a heavy tax scheme in 1970. The new law made the deposit-return system a profitable innovation for the

Fig. 7.9 Pollution & Protest: *In the 1970s, air pollution became a transnational phenomenon; "acid rain" became the shorthand for this problem. Two decades later, environmentalists also waged actions around toxic-waste disposal. In this 1993 photo, rebels representing Greenpeace—a non-governmental organization—protest toxic "exports" of aluminum slag. At stake was the issue of toxic waste circulation in Europe. Stationed at the Swiss/French border, international activists inveigh against Swiss waste entering Portugal and France. Their priniciple: the polluter pays the price. The white and orange suits denote protective gear for toxic clean-up; gas masks punctuate the protesters' point.*

Planke brothers. In Denmark, bottle activism helped to outlaw aluminum beverage cans. But in 2002, the European Court of Justice deemed the environmentally friendly legislation a violation of bottlers' free access to the local market—and forced the Danish government to enact a container-deposit law, instead.[70]

Within the transnational world of environmentalism, the Swedish welfare state occupied a special place. Suffering from trans-border air pollution that drifted from the Ruhr area—and from mercury pollution created by the nation's paper industry—Sweden initiated the first U.N. conference on the environment in 1972. In legislative terms, the country became Europe's model for "green." Having been neutral during the war, Sweden's economy had boomed, allowing its citizens to enjoy an increase in disposable income and access to consumer goods. Rooted in a traditional affinity to nature and wilderness, conservationists of the older school suddenly found themselves in alliance with the younger, more urban environmental movement. In response, the Social-Democratic majority established the world's first Environmental Protection Agency in 1967 and the far-reaching Environmental Protection Law in 1975. The latter law instituted the polluter-pays principle that became standard for many EU and OECD countries. Next, Sweden established a system in which industry was obliged to choose a method of recycling containers—or confront a ban against throwaway containers.[71]

The old glass-deposit schemes had been voluntary systems initiated by beverage firms; now, many governments introduced legal measures that made industry responsible for recycling. In the end, European countries dealt with the international environmental concerns in national ways. Countries with centralized welfare states—Sweden and France, for example—relied on legal frameworks for social change. In federalist Germany, the Green party would go on to exercise considerable national and regional influence—greater influence, proportionally, than its electoral representation implied.[72]

In contrast to the Continental countries, both the United States and Britain's two-party system fostered an environmental movement that blossomed outside official political channels and thrived in non-governmental organizations. Based on the American example, the nascent British environmental movement published do-it-yourself books like the *Consumers' Guide to the Protection of the*

Environment, teaching consumers how to organize themselves into recycling clubs. Another publication, the movement's *Environmental Handbook*, suggested that consumers mobilize the law and rely on "maintenance and repair of existing products" rather than buy into the logic of "planned obsolescence" of the consumer society.[73] The San Francisco Bay entrepreneurs in Stewart Brand's Whole Earth network placed their greatest hope for the environment on technology users. Brand advocated a do-it-yourself culture and believed in the transformative power of relevant technology. *The Whole Earth Catalogue: Access to Tools*, the first edition of which was issued in 1968, listed products and services available on the market and offered people access to resources for a just and sustainable society.[74]

Although highly individualistic, Stewart Brand's ideology, symbolized by the famous NASA image of planet Earth on the catalogue's cover, was also infused with notions about the common good. The catalogue provided education and "access to tools" so a reader could "find his own inspiration, shape his own environ- ment, and share his adventure with whoever is interested."[75] Brand's definition of tools was broad indeed. The listed devices included informative instruments like books, maps, professional journals, courses, and classes, but also well-designed special- purpose utensils, including garden tools, carpenters' and masons' tools, welding equipment, chainsaws, fiberglass materials, tents, hiking shoes, and potters' wheels—even the very first synthesizers and personal computers. His goal was both social and ecological, and it was based on the ethic of do-it-yourself crafting, tinkering, and self-reliance. The point of the *Whole Earth Catalogue* was that low-tech and high-tech tools were all part of the same universe. The publication was a kind of shopping catalogue for the environ- mental movement.[76]

Some countercultural Danes resembled closely the self-builders that Brand envisioned. The global protest against nuclear energy was translated into local, cooperative terms. The Danish affiliate of the transnational Friends of the Earth (NOAH in Denmark) published the highly influential "Some Information about the Planet on which We Live Together" in 1970.[77] Inspired by visions of Spaceship Earth's vulnerability, engineering students and science professors established an anti-nuclear organization (OOA or The Organization for Information on Nuclear Power) to protest against the Barsebäck nuclear plant in Sweden, only 20 kilometers from the Danish capital, across the strait dividing the two countries. The OOA mobilized the public with the famous slogan and button *Atomkraft? Nej tak* ("Atomic Energy? No Thanks"), designed by activists Anne Lund and Søren Lisberg. Immediately successful, the button became the official trademark of the international movement and provoked the Danish government to "Make [energy] plans without nuclear power."[78] In gale-prone Denmark, wind power looked like a natural alternative to the nuclear plants across the water.

To practice what they preached, one Danish school decided to build the world's biggest wind turbine to produce non-nuclear energy for the school and demonstrate the power of self-reliance and sharing. In 1974, the Tvind School in Ulfborg, a progressive branch of the nineteenth-century People's High School movement, embarked on this bold engineering and social experiment. In a cooperative effort that lasted three years, the wind turbine was created by more than three hundred volunteers: anti-nuke teachers, eco activists, international exchange students, and engineering students and professors. The resulting concrete tower was 53 meters high, with three wing blades 27 meters long. Inspired, in part by the high-

Fig. 7.10 Wind Power to the People: *In the 1970s, Danes designed and built a modern, power-producing windmill. The idea originated with two "people's high schools" in Tvind. Experts were recruited from the Danish University of Technology; small businesspeople and hobbyists from around the country also shared their knowledge and skills. Within several years, the people had built "Tvindkraft," the world's largest wind power plant. Tvind became a transnational information center—and a place of pilgrimage. In this photo, volunteers work together on one of Tvindkraft's three colossal propellers. Denmark is currently the world's largest exporter of wind energy.*

tech design of German aeronautical engineering professor Ulrich Hütter, the turbine was large-scale and integrated high-tech aerodynamic principles. The towering windmill, called "Tvindkraft," was painted in red and white, the colors of the Danish flag. Ironically, the turbine's large-scale, high-tech design was inconsistent with the school's ideology of craft and self-reliance. Although mediagenic for the movement, the design could not be reproduced easily by small-scale users. The school, visited by more than 100,000 "eco pilgrims," nevertheless became a mecca for alternative energy builders. The school hosted a network of "wind energy offices" (*energikontorer*) nationally and internationally. Tvind became a place of pilgrimage for the alternative energy movement. Like the American Kitchen, the giant windmill became an icon for both an ideology and a technology. In this case, the towering turbine symbolized the anti-nuclear alternative-energy movement in general—and the genuinely high-tech, scalable windmill solution in particular.[79]

In 1975, the technically and practically oriented activists of the Danish anti-nuclear movement produced the "Organization for Sustainable Energy" (*Organisationen for vedvarende energi*, OVE). It brought together amateur self-builders of windmills and environmentally critical consumers and cooperatives. The organization's "Wind Meetings," and do-it-yourself books like Ole Terney's *Vedvarende Energikilder* (*Sustainable Energy Sources*), and Claus Nybroe and Carl Herforth's *Solenergi, vindkraft: En håndbog* (*Solar Energy, Wind Energy: A Manual*) instructed tinkerers, craftsmen, and cooperatives how to build their own backyard windmills, often with used parts. This informal user coalition, comprised of amateur self-builders, technically savvy users, and small firms, created the Danish technical and political standard for wind turbines. This standard was inspired by the craft-oriented designs of Johannes Juul and Christian Riisager. Tailored to serve cooperatives and individual households, these wind turbines were flexible, reliable, safe, and small scale. This grassroots user movement built the foundation for a new wind-power industry. It was these technically sophisticated and politically active consumers who were able to push the wind-power industry in a new direction.[80]

Instead of using the large-scale prototypes (based on the military research and development designs of NASA in California and of Nazi Germany), the Danes developed widely-scattered, small-scale wind turbines. In striking contrast to the Danes, the Dutch, with their centuries-old tradition of windmills, ultimately missed the opportunity to develop a broadly user-based wind-power industry: Dutch wind power matured under centralized control, driven by research and policy rather than by well-organized small-scale users. In Denmark, on the other hand, the users were highly informed technically, and the small manufacturers involved had government support, in order to generate close user – producer interaction. The collaboration produced a better product: the custom-made designs emphasized safety and reliability that ultimately led the Danes—not the Dutch—to be the world's leading exporter of wind-power technology.[81]

The Danish community traditions provided another model of sharing resources for the flourishing environmental movement. As "Europe's commune center," the Danish communal living experiments were decidedly *urban* experiments that practiced grassroots democracy, gender equality, and anti-materialism. Communes mushroomed from approximately 10 in 1968 to more than 15,000 communities involving an estimated 100,000 residents in 1974. Most experiments did not set out to be self-sufficient and abolish ownership, but to pool resources. Most famously, Copenhagen's commune Christiana turned the cargo bike—an old technology that postmen and milkmen had employed for decades—into a new vehicle for sustainability. Christiania's communards redesigned the cargo bike to transport children. This was to support couples who subscribed to equal sharing of the child-raising burden—and rejected the car for urban transport.[82]

Denmark's do-it-yourself windmill and cargo-bike builders were probably the most successful environmental user groups in Europe. These consumers managed to combine local craft ingenuity and tinkering skills with countercultural transnational concerns for planet Earth. They mobilized grassroots democracy; national legal frameworks; as well as political ideals of community-based self-reliance and social cooperation with global sensibilities. As a result of successfully integrating these arenas—the local, the national, and the global—concerned Danish citizens gave the world old technologies reconfigured for new, alternative, and sustainable uses: windmills, cargo bikes, and reusable glass bottles. None of these technologies were novel—but their *applications* were novel. In the hands of Danish tinkerers and users, these technologies became radically innovative and trendsetting.[83]

Indeed, countries with strong traditions of participatory democracy like Denmark and the Netherlands proved to be fertile breeding grounds for grassroots movements that yielded innovative regulatory measures. It is hardly a coincidence that the Dutch women bottle activists and the Danish wind-power advocates used the same method to achieve their goals: both combined their volunteerism with the collaboration of local and national governments. In contrast, the story of the Dutch windmill is a stark reminder that, by leaving out users out of the process, successful national outcomes could not be taken for granted. As the Danish bottle case showed in the mid-1980s, these national differences in the efficacy of user movements took on yet other meanings when the legal framework of the European Union entered the stage. In a landmark decision, the European Court left intact Denmark's strict bottle recycling legislation. The Court ruled—over the objections of a powerful international bottling industry—that the original Danish legislation indeed protected the environment rather than the Danish beverage industry. The court thus ruled that protecting the environment was, in fact, one of the European Community's essential objectives. Indeed, early on, the officials in Brussels had taken a keen interest in the environmental movement. The European Community recognized the movement's potential to close the enthusiasm gap between citizens and government by focusing on protecting the environment as the EU's key mission. In the end, the European Union sought to balance more trade and more environmental protection with rules that were weaker than those the greenest member states had wanted, but stricter than the previous standards set by global practices. The EU's rules and regulations would change the dynamic of Europe's local user movements in the decades to come. Environmentalists had successfully challenged the alliance between the state and big business; this transnational activism had transformative consequences.[84]

8

Toying with America, Toying with Europe

Play has long been a serious act. Historically, play has been part of every society's process of preparing younger generations for adult life. In the nineteenth century, middle-class parents gave their children toys expressly to accustom them to using emerging technologies: sons received tin soldiers and miniature steam engines; daughters were given small stoves for their dollhouses. Construction kits reflected the industrial world of adults; kits encouraged boys to build houses and bridges—to design, to craft, to engineer, and to tinker. For the middle class in the nineteenth century, play required adult approval and supervision. But, less than a century later, in the 1960s, play took on a new meaning entirely. For the children of the 1960s, play became a lifestyle unto itself, a form of pleasure and even protest. For this younger generation, play embodied personal freedom; play was distinct from "ordinary" life; and play defied the ideology of profit making. Some 1960s teenagers playfully challenged authorities— from their parents, to the state, to corporations—by wearing blue jeans. Other teens broke rules for the sheer pleasure of it, as with hackers, who also popularized personal computing. Practically all young people viewed play—without adult approval—as a human right.[1]

Consider a precursor to the 1960s revolution in play. In the opening scene of a twentieth-century Dutch young adult novel, *De Jongens van de Hobby Club* (The Hobby Club Boys), a physics teacher invites his students to imagine a future embracing space travel to the moon and Mars. That future, the teacher says, will come about only if scientists and engineers work together closely. Further, the novel is about an association of teenagers who share a passion for technical hobbies. The novel's author, Leonard de Vries, was a 22-year-old Jewish chemical lab trainee for British Petroleum. (De Vries wrote the novel from his hiding place in the German-occupied Netherlands.) Earlier, he had written a do-it-yourself book on radios —a manual credited with guiding many in building radios illegally during the German occupation. *De Jongens van de Hobby Club*, published only after the war, in 1947, was translated into other languages as well. Responding to readers' overwhelming enthusiasm, De Vries launched a monthly magazine with instructions on how to start unsupervised youth clubs. He believed the war-damaged Netherlands needed tech-savvy teenagers. Two months after the launch of De Vries' magazine, seventeen hobby clubs were thriving in the Netherlands, Belgium, and colonial Indonesia. Soon, that number would jump to eighty-five clubs.[2]

The hobby clubs owed their greatest appeal to their emphasis on self-governance without interference from authorities and institutions. For his club, De Vries envisioned no leaders, only "instructors." Anyone aged 23 and above need not apply. Without the supervision of adults or the authorities, boys—and girls—would govern themselves to prepare for life. De Vries framed the generational divide in starker terms: "We are totally opposed to it [the old system of law-and-order], blindly obeying orders…the Second World War has once again demonstrated what this *Befehl-ist-Befehl* ('order-is-order') stuff can cause."[3] In the south of the Netherlands, where religious institutions and the mining industry dominated, the Catholic Church tried to offer an alternative to such unsupervised play, but ultimately failed to bring youth into the fold. The U.S. State Department saw De Vries as potential ally for its cultural diplomacy efforts at to instill values of freedom and democracy throughout Western Europe. But, as a self-described socialist, De Vries could not accept American values wholesale. His hobby club was to be a community of peers based on shared interests and knowledge—without regard to class, ethnicity, or religion.[4]

Fig. 8.1 The Hobby Club Experience: *Dutch-Jewish chemist Leonard de Vries began writing* Boys of the Hobby Club *while in hiding from the Nazis. A bestseller, the novel sparked the creation of more than 70 science-themed hobby clubs after the war. This is the cover illustration of De Vries' manual on how to establish one's own hobby club. The manual enabled boys and girls to engage in play without adult supervision.*

De Vries and his hobby clubs were positioned between an older world, where serious play was seen as a middle-class path towards a successful career or a better society, and a newer world, where unsupervised play could be an end in itself. By the mid-1960s, De Vries and other commentators regretted the younger generation's waning interest in hobby clubs. And these critics bemoaned the younger generation's new toys, for example, marketed for teens, who were expected to follow the instructions outlined in manuals and drawings. The objections were that these toys failed to foster creative tinkering with limited resources in a communal setting; the new category of toys reduced the required amount of knowledge and skill needed for play; and it limited young adults in problem-solving. Such play, they thought, would turn the younger generation into individual customers and passive users rather than cooperative tinkerers. Despite these dark predictions, children and young adults did not abandon their curiosity: teenagers of the baby-boom generation sought activities ranging from fashioning clothing to scientific experimentation and computing—and without adult supervision, as DeVries had long advocated.[5]

It was also in the 1960s that young users of both sexes offered unexpected vistas on the technological landscape—highlighted by the story about how European teenagers on both sides of the Iron Curtain restyled their blue jeans from the United States. As an act of rebellion, European teenagers took their standard, "perfectly nice" blue denims and worked them over, bleaching, ripping, patching— tweaking and remodeling their pants into a fashion statement that mocked careers in corporations and the government. In the hands of these irreverent users, ordinary clothing like jeans and business machines like computers turned into innovations. The novel uses they invented gave pleasure in ways that the state, entrepreneurs, and engineers had never envisioned. The state and industry tried to catch up with these entirely new, unsupervised expressions of youth culture, of play and tinkering no matter what the consequences. The subcultures continued, and users dismissed the 1980s' conservative national regimes that took hold in response to the 1970s. The administrations of U.S. President Ronald Reagan, British Prime Minister Margaret Thatcher, German Chancellor Helmut Kohl, and the Polish Prime Minster and Party leader Wojciech Jaruzelski all built their power on a law-and-order ideology. Rather than confronting each other directly, subcultures and these official cultures co-existed as parallel universes during the 1980s, just before the wall dividing Eastern and Western Europe came down.

Corporate Toys & Middle-Class Boys

As twentieth-century historian Johan Huizinga emphasized in his famous book *Homo Ludens* ("Man the Player"), games are an essential ingredient of all cultures. In the industrializing world, however, educators believed playing with toys should not be whimsical but always useful. Manufacturers argued—and parents believed—that chemistry kits and construction sets would prepare boys for a bright future as innovators of science and industry. Girls were expected to play with miniature kitchen utensils and dolls-house paraphernalia. Toys were serious devices intended to pave the way for later roles in life. Indeed, toys closely mirrored the adult industrial world. Boys were cast as crafters and innovators, girls as their admiring playmates. While purpose-driven play helped create those roles, both boys and girls also tinkered with the toys through the art of playing.[6]

Nuremberg had a long history as the center for craftsmen and dealers in the European toy industry. Most toys were initially made of wood, but sheet metal became increasingly popular in the nineteenth century. On the eve of the twentieth century, the Liverpool-based Meccano Ltd. set new standards in the manufacture of sheet-metal toys. Its products allowed young adults to build toy houses and bridges, bicycles and locomotives, with perforated metal strips and plates, joining them with nuts and bolts. To connect with young customers, company founder Frank Hornby launched local Meccano clubs and published a youth magazine featuring new products, competitions, and awards. Unlike DeVries' vision a half century later, the Meccano Clubs were paternalist affairs, led by grown-ups, usually a teacher or a father in a church or school room.

In 1919, Hornby went one step further and founded the Meccano Guild. Every dues-paying young man would receive "an enameled badge of membership, which he must undertake to wear at all occasions."[7] The Guild would "encourage boys in the pursuit of their studies and hobbies, and especially in the development of their knowledge of mechanical and engineering principles."[8] Although Hornby and his staff hoped the company's clubs would spur sales, he also expected the organizations to instill upper middle-class values of "clean-mindedness, truthfulness, ambition, and initiative."[9] The intention was that, later in life, Meccano boys would embark on engineering careers, a path requiring that they "be determined to learn and to make progress."[10] Despite the fact that Meccano also sold its products abroad, the magazine's editors did not shun nationalist formulations. Tinkering, creative youth would "grow to be sturdy, prosperous, and useful Britons."[11] During

the First World War, the company advertised: "THE MECCANO
TANK MADE ENTIRELY OF MECCANO PARTS," with the reassurance that "ANY BOY CAN BUILD IT."[12]

Meccano was not the only manufacturer of metal construction
systems. In Czechoslovakia, the *Merkur* Company led the way; in
Germany, *Stabil* had substantial market share. The German firm
Märklin held a strong position in railroad models—an area that
soon attracted legions of keen amateurs, organized in national
and international networks. By the eve of the First World War,
Meccano was already serving markets outside Britain. After 1918,
expansion continued. One Swedish youth spent most of his spare
time tinkering with his Meccano set: "In the thirties I was a boy
in a Swedish upper-middle-class suburb west of Stockholm," he
recalled. "We boys used Meccano to build things. It was not a toy. It
was miniature engineering."[13] The boy's name was Jan Myrdal, and
he was the son of the celebrated social-reformer couple Gunnar and
Alva Myrdal. Although his progressive parents preferred that he
play with dolls and learn to sew, they finally gave in and allowed
him to experiment with Meccano and to build radio receivers. Such
kits were at least better than air rifles and tin soldiers, so his pacifist
parents relented.[14]

For Jan, Meccano construction was a serious pastime. Although
he became a journalist rather than an engineer, he continued as an
adult to build ever more complicated gadgets. In sharp opposition
to his theoretically minded, social-scientist parents, he embraced
the hands-on skills of the technician. It was, in Jan's analysis, not
social-democratic ideologues who truly built the modern welfare
state, but rationally acting engineers and scientists. As an adult,
Jan claimed that several of his childhood playmates had grown up
to become the very technicians and scientists who constructed the
technological foundation of modern Sweden.[15]

Technical and scientific toys were particularly popular in Germany
and the Anglo-Saxon countries. Especially successful were the
Kosmos chemistry sets, filled with exciting crystals and powders of
various colors. The *Franckh-Kosmos-Verlag* in Stuttgart had a strong
position in Germany through its popular-science magazine *Kosmos
Zeitung*, with a circulation of 200,000. Modeled after American
firms like A.C. Gilbert & Co.'s do-it-yourself Construction and the
Electric Engineering kits in the 1910s, *Kosmos* extended activities
into an area it called "teaching toys" (*Lehrspielzeuge*) in the 1920s.

Fig. 8.2 Engineers in the Making: *Meccano toy construction kits were marketed to middle-class parents all over the world. Most tinkerers were middle-class boys, who enjoyed building with the colorful metal strips, nuts, and bolts. Instructions showed how to build planes, cranes, buildings, bridges, and more. Some children, like the boys in this mid-1930s photo, played with Meccano on their own, while others joined so-called Meccano Clubs, as well. For many parents with sons, construction kits were the introduction to a projected future in engineering.*

Over time, manuals became increasingly like textbooks. In Britain, the largest supplier of toys was Lines, which continued to expand in the 1950s. When, in 1957, the Russians sent the *Sputnik* into space and instilled the Western world with the fear of losing the arms race, demand for such sets increased. School curricula were modified to include more mathematics and science, while popular-science writers like the physicist Heinz Haber and Leonard de Vries fed the growing interest in the sciences.[16]

Kosmos met serious competition from other producers, including the Scottish company Thomas Salter Ltd., which exported to more than ten European countries. But its greatest rival was Philips. Between 1955 and 1985, the Dutch electronics company developed electronics, mechanics, chemistry, physics, computer, and video kits. Initially, it sold the construction kits, experiment boxes, and component-parts packages for promotional purposes. Soon, new products in Germany, France, Italy, and Norway helped open up markets in these countries. Philips' electrical engineering series became particularly popular, requiring teenagers to tinker with screwdrivers and wrenches. The company's Radio, Gramophone, and Television division designed the Mechanical Engineer and *Philiform* kits; another branch offered hi-fi amplifiers, tuners, speakers, and components packages; Philips also published elaborate manuals for do-it-yourself customers. The firm's German division developed electronics kits with extension capabilities, adding Chemistry, Physics, and Mineralogy boxes; the Belgian division produced another range of boxes. After conducting an international research survey of 10–16-year-old boys' spending, Philips ended such decentralized activities and went global in 1965. In the years that followed it sold more than a million electronics kits, 340,000 radio kits, and 130,000 mechanical construction sets worldwide—including countries like Brazil, Australia, Argentina, and Mexico. Philips' experimental sets were cutting edge. The electronics firm developed into a sophisticated supplier of electronic experiment kits based on standard industry component parts: in-house engineers incorporated the latest technological developments in their designs. For example, for vocational-school teachers, Philips Research in Hamburg developed a computer series that incorporated the latest processors. By 1983, Philips offered young men a genuine, do-it-yourself microcomputer construction set, the MC 6400. Although meant for young teenagers, the sets evidently attracted an adult hobbyist clientele.[17]

Investments in science and math education also boomed in Europe's Soviet bloc. The Soviet manufacturer *Reakhim* exported chemistry kits to various socialist countries. In East Germany, such sets emphasized the collaborative aspects of play and learning. East German manufacturers explicitly suggested that children and teenagers work together to perform experiments. The publicly owned enterprise *Laborchemie Apolda* asked its young customers in 1960 to "join up with other friends and found a 'Young Chemists' working group within the framework of the 'Young Pioneers' or the FDJ [*Freie Deutsche Jugend*]."[18] Such collaboration reinforced the socialist belief in collective activities under official supervision. Encouraging kids to play collaboratively went a step further. In the mid-1980s, the Publicly Owned Works for Technical Glass (*VEB Kombinat Technisches Glas*) in Ilmenau, in addition to producing test tubes and other laboratory equipment, designed a chemistry set in partnership with two educators from the Halle-Wittenberg University. The project also sought to involve students in the production process itself. Under the slogan "Pupils produce for pupils" (*Schüler produzieren für Schüler*), the state-owned factory approached a number of polytechnic secondary schools to persuade them to manufacture parts for the equipment needed. More than 10,000 sets were completed in the few remaining years before the fall of the Berlin Wall.[19]

In short, on both sides of the European political divide, toys for boys meant serious business. Reflecting the adult world of industry, toys helped the young to establish careers later in life and play their proper roles in society. The Meccano Guild, the *Kosmos* boxes, and the *Reakhim* and Philips sets were boys' worlds. After 1960, girls occasionally appeared in advertisements for *Kosmos* chemistry sets. Earlier, some of the Walther Company kits featured girls, but only as admiring spectators who did not participate in the construction process. Later, while Walther marketed *Stabil* boxes as a "learning aid for small engineers, metal technicians, mechanical engineers, [and] mechanics," for girls, it designed *Stabila* boxes for building little lamps, sofas, and swings out of metal strips and wool, as well as miniature strollers and doll houses. Essentially, building sets remained an exclusively-male domain. Jan Myrdal claimed that neither his mother, Alva, nor his wives ever really understood the charm and beauty of a Meccano crane or automobile. The technical toys prepared young men "[to] create a thoroughly rational, really beautiful and new technical world," as Jan put it. Construction sets were also instrumental in encouraging the bond between fathers and sons.[20]

Believing in a just world of social and gender equality, Leonard de Vries envisioned girls as members of his hobby clubs, and recommended technical activities like photography to attract young women. In proposing technical toys for boys *and* girls, De Vries was in the minority. His thinking was displaced by a far more powerful cultural current: from the 1950s, shortly before the next wave of feminism reached the shore, the German *Bild-Lilli* and American Barbie dolls came on the scene. These dolls came to represent a young, sexually active, modern woman.

Crafting, Designing, & Girls at Play

In her own peculiar way, Barbie was subtly subversive. Unusually for a doll, she had prominent and pointed breasts. Secretly—beyond the prying eyes of the state and the family—men liked Barbie, of course: she was the ideal woman, pocket-sized. Girls loved Barbie. And mothers and grandmothers, as it turned out, also loved Barbie—secretly. Even teen-aged girls, famously judgmental consumers, loved Barbie: although the dolls projected conventional consumer roles for girls, young women enjoyed crafting this sensual new world with Barbie at its center. But there was conflict: culture critics and parents, in their roles as educators, were shocked by Barbie.

In 1966, a Catholic weekly, in trying to understand the newest American Barbie doll rage engulfing Western Europe, offered this explanation: "it seems clear that Barbie in her 'maturity' is most attractive to teenagers and teens-to-be…'Why do you like her?' a mother asked her eight-year-old daughter … Because she has a bosom,' came the prompt reply, to which the mother quickly added: 'and therefore so mature that, in her eyes, anything is allowed.'" Equally perturbing was the doll's commercialism: "Soon these little girls are grown-ups. By then she has diligently practiced for years how to buy and buy, and think of more needs, and, so helped by the manufacturer, buy each new Barbie accessory on the market." The writer concludes: "I can well imagine that the clothes-making, dressing and fantasizing for some children is a wonderful game. Yet Barbie is an exponent of an unhealthy emphasis on sex….It is really no sour brainchild of strict teachers to see that in America, the child is pushed in the direction of a sex-conscious teenager as early as possible to incorporate it in the army of consumers."[21]

The doll, as an artifact, has a long history; girls of all social classes played with dolls. But the expectation that girls would fashion their dolls' clothing *did* depend on class. In eighteenth-century Europe, fashion dolls taught aristocratic young women the dress codes of the court. Daughters of the aristocracy did not have to sew their dolls' outfits: later in life, these aristocrats would employ their own seamstresses and order dresses and gowns at a fashion house or a professional dressmaker. In contrast, the bourgeois cousins of aristocrats, were expected to know how to sew dolls' clothes as part of a proper preparation for womanhood. The *haute bourgeoisie* created a new generation of dolls, *Parisiennes*, outfitted by leading fashion designers. *Lilas*, produced in atelier *Maison Rohmer*, came richly adorned with fifteen elaborate outfits, thirty-two hats and bonnets, six pairs of shoes or boots, two sets of jewelry, a parasol or umbrella, a fan, underwear, and handkerchiefs. In protest against industrialization and the cosmopolitan Parisian fashion system, regional figures wearing local costumes from Alsace, Brittany, the Basque provinces, and Tyrol gained popularity. By the 1920s, dolls—either fully assembled or in parts—had spread to all classes. After the 1950s, the biggest change was the use of plastics: in Italy, *Ottolini* produced *Sonia* and *Bonomi*, as well as the *Jenny* doll. In Germany, *O&M Hausser* created *Lilli*, and in the United States, Mattel cloned the German *Lilli* into the American Barbie.[22]

The American Barbie doll taught girls, like their aristocratic forebears, how to consume goods rather than how to design and make clothing. The 1959 Barbie prototype showcased a ready-made wedding dress and gown, two sets of underwear, two fashionable bedroom outfits, and sixteen other Parisian ensembles. The miniature wardrobes demanded a great deal of skill and dexterity to produce—all of them were hand-sewn by low paid, highly skilled women workers in Japan. Mattel designers took their inspiration from Parisian *haute couture* shown by Givenchy, Dior, and YSL. When the American Mattel Company launched the Barbie doll, and sales took off, the company's Japanese sub-contractor was forced into overdrive to produce 20,000 dolls a week. With a blond ponytail, zebra-striped swimsuit, open-toed shoes, sunglasses, and earrings, Barbie was 11.5 inches tall. Her prominent breasts and hips were controversial from the start. Barbie became the icon of white, middle-class American girls' culture. Over time, the company responded to the demand of critics and consumers for diversity, introducing the black Barbie in the 1980s.[23]

Fig. 8.3 Beyond the Original Barbie: *She was white, middle-class, American—and she had the perfect figure. Barbie's admirers, on the other hand, were far more diverse. To enhance Barbie sales by creating a common identity with consumers, the Mattel company encouraged national and local Barbie clubs, which sponsored design contests. This photo from 1989 shows one contest winner, Syreeta Trustfull, a six-year-old Dutch-Surinamese girl. Girls worldwide appropriated Barbie, adapting her outfits and lifestyle to their own.*

But, crucial to the question of U.S. versus European cultural values around play, Barbie was not native to America; she was copied, if not plagiarized, from a German plastic doll. This German older sister, *Lilli*, had a ponytail, red lips, and red-painted fingernails; she held in her hand a miniature *BILD* newspaper, and she had an elaborate wardrobe. Between 1955 and 1964, the German doll was marketed as a male gadget, an adult toy appealing to men, sold in greasy bars and tobacco shops, as she was often given to guys as a joke or gag gift. She was modeled after the *Lilli* character in a popular 1952 comic strip. The original *Lilli* was created by the commercial artist and sculptor Reinhard Beuthien for the *BILD Zeitung*. His *Lilli* was a young, economically independent secretary who gossiped with friends; had a sharp tongue; frankly boasted about her sexual pursuits; and talked about her penchant for rich men. In the cartoon, *Lilli* made off-color quips like "I could do without balding old men but my budget couldn't!" In another frame, she challenged a policeman who told her that two-piece-swimsuits were banned: "Which piece do you want me to take off?"[24] Beuthien found his inspiration by visiting the *Reeperbahn*, the entertainment and prostitution district of Germany's port city Hamburg, where young, working-class women, including sex workers, took their cues from Hollywood movies. The fictional *Lilli*, who vaguely resembled the actress Marlene Dietrich, reflected the world of young German women, eager to get away from the war-torn past and the designation of "women of the ruins" (*Trümmerfrauen*). This new generation shunned the iconic image, in books and films, of women clearing rubble in postwar Germany's heroic but gruelingly unglamorous rebuilding efforts.[25]

In 1955, Germany's largest toy manufacturer, *O&M Hausser*, was asked to design a doll version of *Lilli* the cartoon character. For *Hausser*, this was of great political importance. The company was still infamous for its series of "personality figures" created in the *Third Reich* (Hitler, Hess, and Goebbels) and Fascist (Mussolini, Franco) eras, made of its trademarked *Elastolin* plastic. After the war, *Hausser* was forbidden to continue producing German military figures. Supervised by the Allies' postwar denazification program (1946–51), the company was authorized to produce only a new line of pre-approved U.S. army figures. It was not until after the superpowers allowed Germany and Austria in 1955 to once again commission their armies that *Hausser* was permitted to resume producing German and Austrian toy soldiers. The company later contested in court Mattel's rights to the Barbie doll. Despite a clear case of unauthorized copying, the German company lost the suit.[26]

The German and U.S. dolls may have looked quite similar, but on a commercial level, their differences were quite pronounced. *Lilli* was a high-end toy for an adult market, and she was extraordinarily expensive at 12 German Marks. Barbie, on the other hand, targeted young girls from the start and cost only $3 (USD $125 and $23, respectively, in today's terms). Mattel's strategy was to sell the high-quality doll cheaply; penetrate the market; and capitalize on sales of ready-made Barbie clothes and other accessories. The difference between the dolls becomes even more pronounced, if we consider that U.S. average family incomes in the 1950s were considerably higher than those in West Germany. Thus buying the German original would have taken a bigger chunk from a paycheck in Bonn than purchasing a Barbie would in Des Moines.

Girls were to learn how to consume rather than to craft and tinker. The American company employed its largest workforce in Japan, a country that was firmly stationed militarily and economically in the U.S. orbit during the Cold War. At the Kokusai Boeki factory, more than five thousand women sewed the labor-intensive, intricate trousseaus for Barbie and her circle of friends. Mattel combined the craft knowledge required to sew Barbie's miniature *haute couture* style with the demands of mass-production for mass-scale consumption in a global market.[27]

Like Meccano, Mattel tried to increase its branding power by founding a network of local Barbie associations—adjusted to match national sentiments. For example, to balance the international world with national sensibilities, the first Dutch fan-club newsletter told girls: "Barbie is an international girl who feels at home everywhere in the world." Barbie went to visit Amsterdam, celebrating its flower market, the countryside's windmills, meadows rife with roaming cows, and clean rows of flower bulbs: "Imagine Barbie skating, with an elegant, red shawl trailing her, hand in hand with Ken and Skipper." In concert with its international strategy, Mattel launched a Travel Costume Series in 1964—a year in which the company claimed one million fan-club members in the United States and about 100,000 in Western and Southern Europe. The

Anzug für Marina

Material: Je 50 g Türkis (Fb 7), Violett (Fb 19), Pink (Fb 23) und Rosa (Fb 8) Qual „Taipeh" (100 % Maulbeerseide, LL = 125 m/50 g) von AUSTERMANN. 1 Knopf. 1 Paar INOX-Schnellstrick-N Nr 3 und 1 Häkel-N Nr 3.

Rippenmuster: 1 M re, 1 M li im Wechsel.

Gl re: Hinr re M, Rückr li M. Beim Farbwechsel der Fäden auf der Arbeitsrückseite miteinander verkreuzen, damit keine Löcher entstehen.

Kr re: Hin- und Rückr re M.

HOSE

Ausführung: Am Taillenrand beginnen. Je 19 M in Violett und Türkis anschl (= 38 M insges.) und im Rippenmuster str. Nach 16 R abk die Arbeit in der Mitte teilen und beids für den Schritt je 1 M zun = 21 M je Hosenbein. Im Rippenmuster weiterstr, dabei nach 20 R und nach weiteren 15 R beids je 1 M abk. Die restl je 17 M nach 66 R ab Anschlag abk.

Fertigstellung: Hintere Hosennaht und Beinnähte schließen.

PULLI

Vorder- und Rückenteil: 44 M in Pink anschl und für den Bund 6 R im Rippenmuster str. Dann gl re wie folgt str: 4 M Rosa, 16 M Türkis, 24 M Rosa. Nach 20 R ab Bund: 4 M Rosa, 16 M Türkis, 24 M Violett. Nach 26 R ab Bund: 8 M Rosa, 6 M abk (= Armausschnitt), 8 M Rosa, 8 M Pink, 6 M abk (= Armausschnitt), 8 M Pink. Zunächst über die je 8 M die Rückenteile weiterstr. Nach 13 R ab Teilung die M abk. Dann über die 16 M des Vorderteils weiterstr. Nach 8 R ab Teilung für den Halsausschnitt die mittl 6 M abk und beide Seiten getrennt weiterstr. Nach 13 R ab Teilung die restl je 5 M abk.

Li Ärmel: 14 M in Violett anschl und für den Bund 6 R im Rippenmuster str, dabei in der letzten R gleichmäßig verteilt 6 M zun = 20 M. Dann gl re in Pink weiterstr. Für die Ärmelschrägungen beids in jeder 4. R 4 x je 1 M zun = 28 M. Nach 18 R ab Bund in Rosa weiterstr und gleichzeitig für die Armkugel in jeder 2. R 4 x je 1 M abk. Nach 12 R Armkugelhöhe 4 x je 5 M zus abk.

Re Ärmel: Genauso str, jedoch nach dem Bund 18 R Türkis und 12 R Pink arb.

Fertigstellung: Schulter- und Ärmelnähte schließen. Aus dem Halsausschnitt in Türkis ca 34 M auffassen und 14 R im Rippenmuster str, dann alle M abk. Rückennaht bis auf die oberen 5 cm schließen. Ärmel einsetzen. Am oberen Rand des rückwärtigen Halsrandes eine Knopflochschlinge arb. Knopf annähen.

Tasche: Je 8 M in Pink und Rosa anschl (= 16 M insges.) und kr re str. Für die Schrägungen beids in jeder 2. R 3 x je 1 M zun = 22 M. Nach 20 R ab Anschlag je 11 M in Türkis und Violett str und die Tasche gegenpl beenden.

Fertigstellung: Seitennähte schließen. Für den Henkel mit je 1 Faden Türkis und Pink eine Lftm-Kette häkeln und an den oberen Rand der Tasche nähen.

Mütze: 5 Lftm in Pink anschl und mit 1 Kettm zum Ring schließen. In den Ring 15 Stb, in der folg Rd in jedes Stb 2 Stb häkeln = 30 Stb. Nun noch 1 Rd Stb, 1 Rd fe M und 1 Rd Kettm häkeln.

Torero-Ken

Material: Je 25 g Gold (Fb 12), Schwarz (Fb 15) und Gold-Schwarz (Fb 14) Qual „Gold + Silber" (80 % Viskose, 20 % sonstige Fasern, LL = 95 m/25 g) von SCHOELLER/ESSLINGER. 3 cm Goldborte mit Tropfen, 2 Spitzenriegel in Gold mit Straßsteinen, 4 Quasten in Gold, ca 30 cm Litze in Rot-Gold und ca 40 cm Goldlitze, jeweils 0,5 cm breit, 2 Druckknöpfe, etwas Fadengummi in Schwarz, 3 kleine ovale Goldperlen, Rest roter Seidenstoff (ca 21 x 21 cm). 1 Paar PRYM-Schnellstrick-N Nr 2 und Häkel-N Nr 2.

Rippenmuster: 1 M re, 1 M li im Wechsel.

Gl re: Hinr re M, Rückr li M.

Gl li: Hinr li M, Rückr re M.

Großes Perlmuster: 1. R: 1 M re, 1 M li im Wechsel. 2. und 4. R: M str, wie sie erscheinen. 3. R: 1 M li, 1 M re im Wechsel. Die 1. - 4. R stets wdh.

Pulli

Vorderteil: 30 M in Gold anschl und nach dem Zählmuster arb. Die Abnahmen für den Halsausschnitt wie eingezeichnet ausführen. Die restl je 10 Schulter-M nach 7 cm ab Anschlag abk.

Rückenteil: 30 M in Gold anschl und im großen Perlmuster str. Nach 7 cm ab Anschlag alle M abk.

Fertigstellung: Beide Teile ringsum mit 1 R fe M in Schwarz umhäkeln. Die Gl-Re-Raute des Vorderteils mit Kettenstichen in Schwarz umstricken. An den vorderen Halsausschnittrand die Goldborte mit Tropfen nähen und fixieren. Seitennähte schließen, dabei für den Seitenschlitz unten je 1 cm und für die Armausschnitte die oberen 3 cm offen lassen. Schulternähte mit Druckknöpfen schließen.

Hose

Re Hosenbein: 28 M in Schwarz anschl und für den Bund 4 R kr re str. Anschließend gl re weiterarb, dabei zur Formgebung 6 x in jeder 10. R je 1 M zun = 40 M. Nach 13 cm ab Anschlag beids 3 x in jeder 2. R je 2 M abk = 28 M. Nach 17 cm ab Anschlag noch 1 cm im Rippenmuster str, danach die M abk, wie sie erscheinen.

Li Hosenbein: Genauso arb.

Fertigstellung: Entlang der Mitte beider Teile die Goldlitze aufnähen. Innere Beinnähte und Schrittnaht schließen. Durch den Rippenbund einige Fadengummis einziehen.

Jacke

Achtung! Vorder- und Rückenteil werden in einem Stück gestr.

Ausführung: 51 M in Gold-Schwarz anschl und in folg Einteilung arb: Rdm, 2 M gl li, 45 M gl re, 2 M gl li, Rdm. Nach 4 cm ab Anschlag für die Armausschnitte die 12. - 17. und die 35. - 40. M abk und alle drei Teile getrennt hocharb. Das mittlere Teil bildet das Rückenteil, die beiden Seitenteile die Vorderteile. Nach 7 cm ab Anschlag alle M abk.

Ärmel: 24 M in Gold-Schwarz anschl und 4 R kr re str, danach gl re weiterarb. Für die Armkugel nach 7 cm ab Anschlag beids in jeder 2. R 1 x 6 M und 3 x je 1 M abk. Die restl 6 M nach 9 cm ab Anschlag abk.

Fertigstellung: Schulter- und Ärmelnähte schließen. Ärmel einsetzen. Einen Spitzenriegel in der Mitte teilen und die Hälften jeweils auf die Schultern nähen. An die vorderen Kanten und die Ärmelkanten Litze in Rot-Gold nähen. Von dem 2. Spitzenriegel eine Spitze abtrennen (für den Hut) und den Rest auf das Rückenteil nähen. An die vorderen Ecken und die rückwärtige Mitte der unteren Jackenkante je 1 Quaste nähen.

Hut

Ausführung: Für das Kopfteil 42 M in Schwarz anschl und im Rippenmuster str. Nach 1,5 cm ab Anschlag die M abk. Für das Dreieck (2 x arb!) 20 M in Schwarz anschl und gl li str. Zur Formgebung beids in jeder 2. R 10 x je 1 M abk. Nach ca 5 cm sind alle M aufgebraucht. Dreiecke re auf re liegend mit Gold und 1 Rd fe M zushäkeln. Rippenmusterstreifen zur Rd zusnähen und auf das Dreieck nähen. Die Goldperlen als Dreieck angeordnet vorne auf das Kopfteil nähen. An die untere Kante des Kopfteils in der hinteren Mitte eine Quaste annähen. Das abgetrennte Teil des Spitzenriegels auf den Hut nähen.

Zählmuster

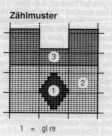

1	= gl re
2	= gl li
3	= großes Perlmuster
□	= 1 M und 1 R

Fig. 8.4   Who Makes Barbie's Clothes: *The Mattel company sold the Barbie doll inexpensively, profiting mainly from her expensive clothes and accessories. Enter the many girls and women worldwide, who used their considerable skill to design and make Barbie's trousseau. This photo [Page 287] is the image on one side of a paper pattern for a homemade Barbie dress. On the pattern's flip side [Page 286] are knitting instructions.*

Marina und Ken in Spanien

Anleitungen Seite 28 + 29

Puppen und Schuhe: Barbie-Moden, Entwurf: G. Papperitz-Ossenkopp, Text: Janne Wedell

company, evoking folkloristic images, designed a re-enactment of the American middle-class tourist experience of traveling around the world. Barbie, like American tourists with their high-valued dollars, visited Japan, Mexico, and Hawaii, and toured Europe being sure to stop in Switzerland, and the Netherlands, where the company claimed its third largest fan club (24,000 members). Mattel designers offered a West European experience of travel and international exchange across borders. Newsletters encouraged Barbie fans to get a pen pal, learn languages, and become ideal hostesses to their future husbands.[28]

Not all girls accepted their role as passive consumers. The history of the Barbie doll is also a story about the girls, mothers, and grandmothers who invested resources, talent, and technique to sew clothes to fit the sexy plastic figure and her handsome boyfriend. In Barbie's make-believe world, the doll visited fashion houses and bought ready-made clothes rather than making her own. In Europe's real world of girls, and their mothers raised in the thrifty ways of the 1950s, however, Barbie's outfits were too expensive. Home-made clothes offered a creative and money-saving way out. The pricey brand-name clothes and accessories prompted a rich subculture of designing and making one's own clothes for Barbie. In the nineteenth century, magazines had already begun to include information on how to make dolls' clothing. By the 1940s, there were paper patterns on the market for plaster dolls, like the 1949 Butterick sewing kits and McCall's Peggy doll in 1949. Although Mattel expected girls and their parents to buy ready-made clothes for the Barbie doll, in response to demand, the company issued its own paper patterns in 1962. The paper pattern provided sewing instructions for four ensembles with six outfits each. Department stores followed: J.C. Penny's sold patterns for Ken that included two ensembles with a total of twelve outfits. Sears marketed a line of kids' clothes with the tagline, "Dress up like Barbie." Well into the 1990s, paper-pattern companies like Advance, Butterick, Simplicity, Vogue, and German Burda offered detailed sewing instructions for making Barbie doll clothing—each with its national affiliations. In the Netherlands, women's weeklies like *Margriet* and pattern magazines like *Ariadne* and *Burda* carried paper patterns for Barbie.[29]

The wool firm *Neveda* produced several knitting patterns for teenage doll fashion for an international market, including Western Europe and Canada. Women and girls were instructed to use leftover yarn and special knitting needles (no. 2) that were suited to the delicate work. The producer imagined a European world where Barbie went on ski vacation to Austria in the winter and visited the Acropolis in Athens in the summer—in woolen clothing! Grandmothers and mothers created entire Barbie wardrobes based on paper patterns, fashion-magazines cutouts, and samples taken from their own wardrobes. Barbie's vibrant culture of homemade clothing bridged a culture of thrift in which generations had been raised (supported by government-sponsored programs during the war), and a new American culture of abundance and consumption that beckoned.[30]

Blue Jeans & Cold War Countercultures

The 1950s' commercial culture of abundance and urban elegance that Barbie represented provoked a passionate countercultural response. In real life, players in the U.S. fashion market presented the most fundamental challenge to the Paris-based fashion game. Within the United States, California anchored opposition to the high-end fashion system. Taking the working class; the world of farmers; and the sports field for inspiration, California manufacturers managed to capture consumers' changing moods with batch production and quick design responses. The most successful—and iconic—of its challenges to European fashion were blue jeans—a product with European roots that was transformed into an American symbol of a rancher culture. Originally, the blue trousers had been associated with Genoese sailors (thus the phrase *bleu de Gênes*), their indigo-colored fabric had come from the French town Nîmes (hence the phonetic "denim"), and a German-Jewish immigrant, Levi Strauss, had brought them to the United States in the mid-nineteenth century. Strauss' innovation was to adapt his trousers to the needs of California's gold miners and agricultural workers by adding copper rivets for sturdiness. Blue jeans' durability also attracted the U.S. Army.[31]

In the interwar period, other groups discovered blue jeans, giving them a new connotation. In this period, blue jeans came to symbolize struggling workers, sharecroppers, and dustbowl farmers. New Deal photographers like Margaret Bourke-White and Walker Evans immortalized the pants, and the 1940 movie based on John Steinbeck's novel, *The Grapes of Wrath*, did the same. Members of artists' colonies started wearing them as an anti-fashion statement. Some left-wing intellectuals, such as the German playwright Bertolt Brecht (during his U.S. exile), also sported jeans. Blue jeans had finally entered popular culture, where they symbolized youthful resistance. Rockers like Elvis Presley and Eddie Cochran wore them, as did young Hollywood movie stars like Marlon Brando in *The Wild One* (1953) and James Dean in *Rebel without a Cause* (1955). Through music and film, then, jeans began a global tour, defying class, gender, and geographical divisions. They returned to Europe with U.S. troops after the Second World War and were sold on the black market in the Allied zones of Berlin.[32]

The new art of being cool and wearing jeans defined a counter-culture with its own hierarchies and intricate rules of connoisseurship. Using the practices of crafting and alteration, its members appropriated blue jeans to meet their own needs and desires. In doing so, they redefined themselves: instead of being simple buyers or consumers, they turned into tinkering users. An authentic 501 model Levi's had to be washed five times to achieve the requisite lived-in feel; it had to be put on wet to fit correctly when dry. In the 1950s, the turn-up at the foot was essential. By the early 1960s, British "mods" sewed them inward to fit exactly on the instep of the foot; tapered the pants for a sharper look; removed the stitched patterns on the back pockets; deleted the company logo; and got rid of the little red Levi's tag. Colin Wild, a London tailor, designed bell-bottom flared trousers that became all the rage with hippies—who in turn copied the design. In a burst of do-it-yourself creativity, members of the hippie culture added embroidery and wedges of fabric in vibrant colors, as well as seams and patches with peace symbols, stars, hearts, and drawings. A decade later, by the time that punks had set a new trend with a tight-legged, tapered look, no one wanted to be "caught dead" in wide-bottomed jeans. The sense of proper fashion required a great deal of connoisseurship on the part of young adults in the United States and Western Europe.[33]

Blue jeans also challenged the Soviet bloc's establishment. In an ironic twist of history, the iconic article of American proletarian and farmers' clothing became a symbol of resistance against the straightjacket of communism in the Soviet Union and Eastern Europe, the nations of "workers and farmers."[34] The irony was not

Fig. 8.5 Flower Power to the People: *In the 1960s and 1970s, young men and women protested politically—and in their fashion choices -- here in Paris in 1967 at New Jimmy's. Paris no longer dictated what people wore. From liberal Britain to socialist Hungary, young people bought mass-produced, often U.S.-inspired clothing. Enter the tinkering process: wearers constantly re-crafted their fashions, adding everything from patches and slogans to beads and embroidery. This tweaking was one part protest against the older "establishment"; one part declaration of your subculture identity; and one part a statement of individual identity.*

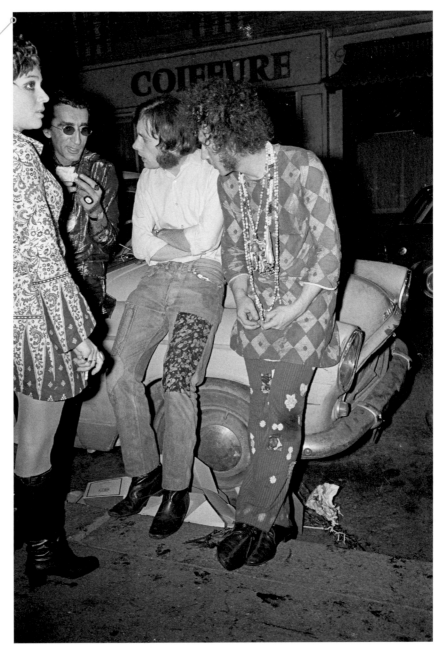

lost on party leaders, who suppressed blue jeans at times; regulated them at other times; co-opted them in other circumstances; and finally embraced blue jeans reluctantly. The appropriation of blue jeans in socialist countries required knowledge of fashion taxonomy (branded versus brand-less, fake or original, denim or jersey) and a great deal of ingenuity and resourcefulness on the part of teenagers and their families. Two years after Dean had worn his jeans in *Rebel without a Cause*, reports of blue jeans in the Soviet Union had already appeared at the Sixth Youth Festival in Moscow. This was not a local event, but an international film festival that included 180 movies from thirty countries, visited by more than a million people. Women students are reported to have made drawings of the outfits they saw, while youth delegations from other Soviet bloc countries and Western Europe illegally bought and sold suitcases full of clothes in hotel rooms and lobbies to satisfy the thirst of Russian youth.[35]

In the era of de-Stalinization, the appropriation of U.S. jeans in the Soviet Union and other East European countries happened in different ways. East Germany and the Soviet Union remained ideological strongholds whereas Poland, Hungary, and the Baltic states tried to mediate between East and West. Official East German policy considered blue jeans the trademark of a teenage dropout culture and denounced them as politically taboo and anti-Socialist. Teenagers in East Germany nevertheless found ways of getting their hands on American jeans, through personal and familial contacts across the occupied zones; and through mail packages from relatives and Western-currency stores. Before the wall went up in 1961, East German teenagers could go to the movies in the Western part of the city. In Gerhard Klein's East German movie, *Berlin—Ecke Schönhauser* (1957), the protagonist, Dieter, asks his beloved Angela how he should look in order to appeal to her: "like Marlon Brando," she answers. By contrast, Polish teenagers with fewer immediate personal contacts and no direct access to Western cross-border television had a more difficult time getting their first pair of jeans. They acquired them via packages sent from Polish-American communities in the U.S.; through aid and church organizations; and by direct purchase in private boutiques.[36]

The 1956 political upheavals in Poland and Hungary affected the world of fashion. Under the new party leader Władysław Gomułka, Polish consumer culture experienced a wave of openness and acceptance. The Polish youth fashion journalist, Barbara Hoff, could now report freely on American, Italian, and French fashion in Krakow's well-respected weekly *Przekrój*. By 1958, Hoff had gathered enough material to publish instructions and detailed patterns on how to design one's own jeans and other pieces of clothing with fabric and materials available in Polish shops. Hoff, who eventually designed and marketed her own trendy and hugely popular fashion label *Hoffland* for Polish mass production, managed to work both within Socialist institutions and outside of centrally controlled commerce.[37]

In Hungary, party leader János Kádár introduced a less strict form of socialism, the so-called "Goulash Communism." This was in response to the bloody suppression of the 1956 rebellion. In exchange for their political acquiescence, Hungarian citizens gained greater access to consumer goods. In turn, Hungary became a welcome tourist destination for citizens of other Soviet bloc countries in search of Western goods. The first jeans entered Hungary by chance or by mistake via the charity-clothing bales sent by American relatives and church groups as aid in response to the 1956 uprising. A sign of their ambiguous status, the blue jeans were alternatively called "cowboy pants" (*kobvojandrag*) "American gear" (*amsci*), or "many-pocketed pants" (*sokzsebes*). One Hungarian recalls: "The cowboy pants, for example, came through the Lutheran Church. I attended bible classes at the local church, and when they received parcels from Western connections, they distributed them among the people who attended the church."[38] Enterprising tailors like Kecsks in Budapest sold proprietary, custom-made jeans. Kecsks himself was repressed temporarily in 1959—only to achieve a breakthrough several years later when the children of the communist leadership cadres started to wear them.

For teenagers in the East, just like in the West, "lying in bath in the act of preshrinking" was an important ritual. It had its own word: *beavatás*. In Hungary, and probably elsewhere in Eastern Europe, the outcome of this watery preshrinking process was often more precarious. The moment of truth came when the jeans sold as authentic on the black market turned out to be fake or of dismal quality. Traders at the *Ecseri* second-hand flea market on the outskirts of Budapest sold blue jeans smuggled in by truckers, train conductors, and others with access to foreign countries. The ability to distinguish between real and counterfeit was a key skill. One person later recalled the kind of rigorous authenticating procedures for jeans that teenagers enacted at school: "First [the jeans were] rubbed with paper tissues to see if the paper becomes blue. Then followed the cigarette test on the bottom of the pants leg to see if it gets burnt or not. The last one was the mechanical test that was lifting the jeans owner by the belt holders of the pant."[39] Those who could not afford the "real thing" relied on the black market for fake Western labels; others found labels, Western emblems and symbols in fabric stores to embellish their blue jeans in their search to bolster the appearance of authenticity of their counterfeits.[40]

In an attempt to co-opt the blue jean from the subculture, the Hungarian state media endorsed the term *famernadrág*, farmer-trouser, to suggest a direct pedigree to the 1930s socialist writer John Steinbeck and American progressives. In West Germany, the garment industry was instructed to employ the term "double-stitched fell-seam pants" instead of blue jeans. These sartorial semantics failed to socialize the blue jean in an official way. In fact, the counterculture movement only intensified. Despite jammed airwaves, Hungarian teenagers started listening to rock-and-roll music on Radio Free Europe and Radio Luxemburg; they fabricated their own electric guitars; and continued to adapt foreign fashion to their own styles. By 1967, Kádár gave in to the mounting pressure from below. He could not help but notice the "wild-west pants,

Fig. 8.6   Black-Market Fashion: *Inspired by the clothing of workers and cowboys, blue jeans and flannel shirts became fashions icons—and American icons, at that. Authorities in Eastern European countries detested the trend toward blue jeans and colorful shirts, because they were associated with the U.S. and youthful rebellion. To get their hands on such fashions, consumers turned to the black market. In this 1992 photo, Polish consumers at an improvised street market negotiate the price of shirts, probably smuggled.*

beards, [and] hairstyles." Even though he observed that Western fashion and norms of cynicism had spread, Kádár declared in a speech to the Organization of Young Communists: "What's important here is that the Party, the Youth League, is not a fashion designer or a hairstyling salon, and does not need to deal with such things."[41] A year earlier, the Minister of Light Industry had already relented, announcing at the Soviet Central House of Fashion that "mini-skirts are okay for communism too—so long as they aren't too mini."[42] Such officially sanctioned signals of acceptance from Communist Party leaders in the Soviet bloc were misleading, however. One Hungarian, for example, remembered when, as a 16-year-old in 1968, his blue jeans provoked violent aggression: "My uncle—with a military record and military sensibility— ordered his son and me to take off our jeans (embroidered with flowers, hippie-style), cut them to pieces with scissors, throw them into the garden toilet... and to shit on them, one after the other."[43]

In the 1970s, blue jeans as a symbol of personal independence and freedom from the state could no longer be suppressed, controlled, or considered a mere fad. Cooptation was the next step. In this concerted effort, socialist authorities first responded by establishing their own blue-jeans labels. Next, the authorities forged license agreements with U.S. and Italian firms. In fact, the Polish government had already established domestic jeans production in 1961, through the Krakow textile manufacturer *Rekord*. In 1974, the East German regime followed the example of its socialist sister countries, embarking on manufacturing its own jeans.[44] A year later, the Soviets followed suit. In May of 1975, the Minister of Light Industry, Nicholai Tarasov, announced the plan to produce 17 million pairs of Soviet jeans. This was after two articles had appeared in the Communist Youth Journal (*Komsomol'skaya Pravda*), criticizing the lack of fashionable clothes. The Journal followed up with the Minister and asked a year later: "Where are the jeans you promised us?" The resulting Soviet versions of the blue jean were treated with disdain by the youthful rebels, who disliked the design, the quality, and the fabric. A write-up in the weekend supplement of the newspaper *Izvestia* that year went further still in its critique, calling the Soviet pants "a parody of jeans."[45]

This time, in response, the Ministry promised to pay more attention to the blue jeans' stitching, buttons, labels, fasteners, and zippers. Given that it had industrial sewing machines that could handle the tough denim fabric, Moscow's Workers' Clothing Factory (*Rabochnaja Odezhda*) was chosen to manufacture an all-Russian denim called *Orbita*. When, in 1979, the Soviet Union's young consumers still could not be appeased, the Ministry took its cue from Hungary, which had already signed a licensing agreement with the Levi Strauss Company. Just after the superpowers signed the Helsinki peace accords, the Soviet government sent a delegation on a fact-finding mission to the United States. Their aim was to find a suitable licensee for the Soviet jeans. Predictably, Levi Strauss won the bid.[46]

Part of the deal was the idea to dress all 23,000 Soviet employees at the 1980 Moscow Olympic Games in blue jeans and white T-shirts. But the announcement of the U.S. boycott of the Olympic Games (in response to the Soviet invasion in Afghanistan) put an abrupt end to these ambitious plans. Instead, the Soviet government decided on the so-called Jesus Jeans as the official outfit, commissioning a Turinese company, *Maglificio Calzificio Torinese*, to produce it. A former manufacturer of army uniforms, the Italian firm had started to produce blue jeans in 1973. The "Jesus Jeans" advertising campaigns followed. With the tagline, "Who loves me follows me," the message was aggressive and blasphemous, cleverly transgressing social norms to boost sales among teenagers. The cachet of the Italian brand was lost in the Soviet act of co-optation, however. The trendy Italian brand name was jettisoned in favor of the name *Tver*, which was printed in Cyrillic on the blue jeans' label. This seriously devalued the blue jeans in the eyes of the teenage Soviet citizens. Instead of snapping up the pants, adolescents continued to rebel against the state, relying on the black market for their jeans. Indeed, in port cities from Leningrad to Tallinn and Riga—and even on the edges of the Soviet empire, from Georgia to Murmansk—the black market boomed. Counterculture and all, users had won: jeans, T-shirts, and records—either authentic versions smuggled directly from the West or counterfeits crowned with "Made in the USA" labels—were everywhere.[47]

Copying, Pasting, & Sharing Technology

America appropriated Barbie and blue jeans, both having European origins, and turned them into successful global commodities. Their history exemplifies a new form of a rebellious youth culture. A similar scenario unfolded in computing. Computers were wrested from official channels by the hands of consumers and users, tinkerers and rebels. The young men who, in previous decades, had been loyal sons of corporations, governments, and churches broke out of their institutional confines and radicalized the computer industry. Creative and rebellious boys and young men transformed computers from machines for technocrats into personal toys and popular collective tools.

American government and industry failed to see the big wave of computing coming. In 1983, a U.S. market-research report scanned the European horizon for commercial opportunities vis-à-vis personal computers; they saw virtually nothing. First of all, the report focused only on the professional and science markets, ignoring computer use in the home setting. Second, the study excluded low-priced—and therefore more accessible—computers like the British Sinclair. Third, the report's authors dismissed the "smaller country markets" as immature. Fourth, schools were ignored as potential users. The report could not have been more wrong; the computer wave reached Europe just as soon as the ink on the American report had dried.[48]

Corporate America's bungled assessment of Europe's computer markets had another effect: it left users on their own to make the machines work properly. For example, when introducing its first personal computer in 1981, IBM divided its target countries into three categories: the first category comprised the United States, Germany, France, and Britain; the second the Netherlands and Belgium; the third included Greece, along with countries like Uganda, Zimbabwe, and Namibia. As Theodore Spinoulas, then-editor of the Greek computer magazine *RAM* remembered: "The first category had everything: local language support, books, people and everything needed to enhance IBM's sales. The second category only received part of these services."[49] To Spinoulas' dismay, in countries like Greece, "IBM did not provide anything. Computers were sold, but the user could be long dead and gone before Greek characters would ever appear on his screen or printed on paper—not to mention that a manual in Greek was a distant dream." Where IBM and other multinationals failed to offer service, small computer stores and journals functioned as de facto community help desks and closed the gap. One Greek writer humorously nicknamed the computer stores Pirates, S.A. Six years after IBM launched its personal computer, Greece's most popular and

Fig. 8.7 Sharing PC Knowledge in Europe: *Using a personal computer— let alone owning one— was still a distant dream for most Europeans in the mid-1980s. At the festive event pictured here in 1987, 15,000 people gathered in Athens' most popular soccer stadium in the hope of winning a PC in a lottery. The event was organized by a Greek home-computing magazine,* Pixel, *whose editors in effect mediated between U.S. manufacturers and Greek users. In attendance were mostly young people and families.*

authoritative computer magazine, *Pixel*, hosted a veritable user-fest. The magazine hosted 15,000 Greek computer enthusiasts gathering in the famous Panathinaikos soccer stadium for an annual lottery to win a computer. The event was foremost a testimony to the success of stores and magazines in attracting large numbers of computer users—and in providing support where IBM failed.[50]

Originally, computers were built as utility machines to solve complex mathematical problems in the military and academia;

computers also found their way relatively easily and quickly into growing government bureaucracies like census, tax, and intelligence agencies. Integrated circuits and time-sharing enabled users of large computers to handle increasingly more data. In the 1970s, the microcomputer opened up fundamentally new and unforeseen areas of application. Quickly realizing the liberating potential of this technology, electronics geeks formed local groups and national networks. The basics for the emerging do-it-yourself computer were a microprocessor, a video monitor, a keyboard to enter data, and a unit for storing information. Saving data took different forms: first the tape recorder, then the floppy disk, and later the hard drive did the job. The first personal-computer users bought ensembles of these components as a hobby, just for fun.[51]

It was not only the American computer industry that misjudged the European market. Virtually all large multinational computer firms focused on business users and failed to envision other applications and opportunities. To some extent, governments picked up where corporations left off in imagining new kinds of users. Even so, European states on both sides of the Iron Curtain were ambivalent about what these new users should do and how they ought to behave: government officials tried to formulate research and development programs, technology procurement plans, and policies to improve public access to computers. European officials took various initiatives to limit what they conceived as U.S. corporations' undue domination. Governments rejected the idea of computer use for sheer pleasure; personal computing should just prepare the young for the incipient digital revolution.

In the late 1970s, several European nations started computer literacy programs, spurred by economic fears and rationalized by the argument that the coming "information society" would force citizens to acquire computer skills or risk being left behind. In 1978, the BBC documentary, *Now the Chips are Down*, predicted that the microcomputer revolution would cause massive unemployment. A government report advanced the idea that a future economic crisis could be averted only if the British mastered the new technology and became computer literate. In 1978, the Dutch government commissioned a similar report that became the basis for bringing computers into the classroom. By the early 1980s, with the economy deep in crisis, government programs in social-democratic countries like Britain and the Netherlands—as well as socialist countries like Poland and Yugoslavia—assigned teachers the task of increasing information equality with their students. In Sweden, a social-democratic culture that emphasized education and equal opportunity joined forces with conservative governments. Together, they embraced entrepreneurship and technological enthusiasm to combat Europe's economic malaise. By supporting computer use in libraries, schools, and community centers, governments tried to harness "personal" computing for the public good.[52]

In various European countries, governments and business developed different models for educational purposes. In Britain, public broadcasting responded to the government's call to catch up with digital technology. Almost eight million viewers watched the BBC's TV series "The Computer Programme," while many bought a computer that the public broadcasting service developed. In the early 1980s, even before becoming the British school standard, this so-called Acorn machine also found its way to the United States and West Germany.[53]

The Dutch government sponsored specialized computer-literacy programs for women, children, and the elderly, and in 1985 Dutch companies received a tax break when they subsidized employees' home computers as an inexpensive alternative to on-the-job computer training. On the commercial side, the Dutch department store V&D competed directly with mail-order houses and hobby clubs, selling the Commodore 64 in 1984 cheaply and establishing a computer club for support. Two years later, the store's club claimed as many as 60,000 members.[54]

In 1981, the Swedish administration also decided to support a national personal-computer project for school children. The Swedish Compis computer represented the efforts of a political collaboration between industrial and public interests, just like in Britain and the Netherlands. The collaboration boosted Swedish research and development in general and the Swedish electronics industry in particular, while at the same time promising to fulfill social-democratic promises of equal access to computers. Mathematics and computer-science teachers participated in the project's group meetings to formulate a set of user requirements for a (yet to be developed) product that combined software flexibility and uniformity. This was part of their vision for a user-friendly school computer that teachers could handle and students could easily learn. The acronym Compis ("Computers in School") conveyed the notion that school children needed to develop a certain familiarity with modern computer technology. The name

Compis was also a homophone: the Swedish word *kompis* means "pal" or "buddy."[55]

Socialist governments in Europe also subsidized computer clubs to introduce their citizens to the new technology in the hope that clubs would spur socialist progress. In Yugoslavia, although the federal government did not have a coherent policy on the subject, the country's republics established their own computer-education projects for primary schools. Federal law required all computers to be domestic products, which stimulated the fledgling electronics industry. In 1985, the Ivo Lola Ribar Institute, specializing in command-and-control computers for military uses, offered the Lola-8 home computer for educational purposes. In the late 1980s, Poland started computer clubs and computer-literacy programs under the auspices of socialist youth organizations, local housing associations, state-owned companies, and schools. The broad coalition to popularize computing included members of the Polish Association of Computer Science of the Polish Association of Informatics (*Polskie Towarzystwo Informatyczne*, PTI) founded in 1982; activists from the Association of Polish Socialist Youth (*Związek Socjalistycznej Młodzieży Polskiej*); and journalists from The Young Technician (*Młody Technik*).[56]

Polish computer enthusiasts lobbied decision-makers and encouraged children and adolescents to use computers for "serious" instead of entertainment purposes. But, as one journalist discovered in 1987, Polish youth did not see computers as a sure path towards a productive future or a strong nation state. The writer described disapprovingly primary- and secondary-school children coming to Warsaw's computer bazaar, knowing exactly "what, where, and how, how much [the computer] costs and if it is modern... They know how to use computers; they are knowledgeable about technical details. Moreover, they are building closed clans of 'Atarians' [*atarowcy*], 'Spectrumians' [*spektrumowcy*]. Their free time is dominated by computers."[57] By 1993, out of 3,800 state-sponsored cultural centers, more than eight hundred had computer clubs with access to some four hundred computer rooms catering to 15,000 members. The local industry also sought to manufacture its own computers, such as the Meritum (a TRS-80 clone) and the Elwro Junior (a ZX Spectrum clone) for selected schools. But the presence of computers in schools by no means guaranteed their use for educational purposes. A 1989 editorial in the computer journal *Bajtek*, for example, criticized computer games for failing to advance the national economy. European social-democratic and socialist educators alike worried that computers were being used inappropriately, concerned that gaming subverted the more serious and important goals of education.[58]

On both sides of the Iron Curtain, then, governments linked educational initiatives to solving the 1980s economic crisis by stimulating their national electronics industries. This national computer initiative aimed to pre-empt multinational corporations and Cold War power plays. The obstacle was that after purchasing computer equipment, governments generally failed to allocate enough funds for maintenance and support. Computing standards changed rapidly, and educational software suitable for classroom use was frequently lacking. With the exception, perhaps, of the successful BBC Micros (Acorn's successor), which had a relatively long life in British schools, the various national projects—the Dutch Tulip, the Swedish Compis, the Yugoslav Lola, and the Polish Meritum—competed futilely with the global IBM. Everywhere, official policy was caught between the delayed corporate push towards monopoly and standardization and a vibrant culture of young users who pushed the technology in different directions. This was a remarkable turn of events.[59]

The story of IBM's neglect of Greek users is telling. When American personal computers and their clones hit the European market, they were unfinished products without a clear script—or, in some cases, even a manual—for how they could be used elsewhere in the world. Young users and their allies came to close the divide between large infrastructures and individual use, between product and consumption. Young hackers helped revolutionize computers by turning them into toys, end-uses for which they were never intended; social activists forced computer access for all citizens by bringing them into the public domain. Users from Athens to Belgrade were confronted with computers' incompatibilities—between, for example, imported computer hardware and local software needs. They challenged manufacturers' claims that computers were universal machines that could be used anywhere. They changed technical standards that had little to do with local circumstances, developing applications using Cyrillic and Hellenic alphabets. Through piracy, copying, and tinkering, hacker communities facilitated the appropriation of personal computers. Rooted locally in clubs and stores, vibrant, often anarchistic, subcultures

Fig. 8.8   Closing the "Digital Literacy Gap": *In the 1980s, mainframe computing dominated. The 1980s was also a decade in which all kinds of users began to discover the personal computer. Government outreach programs, for example, sought to bridge the so-called digital literacy gap. Personal computers were introduced for educational purposes—with varying degrees of success. In this 1985 photo, members of the Women's Housewives Association of Amsterdam attend a computer session, which took place on a refitted bus.*

connected with each other transnationally through mail orders and magazines that nurtured collective use of technologies.[60]

In the early days of the personal computer, computer users who modified programs rarely got into trouble with companies or the police. Software was seldom legally protected. There were few restraints, only welcome challenges to sharing insights and resources. Given that computer manufacturers focused on business users, the emerging subcultures were largely ignored, computer manufacturers having written them off as commercially uninteresting. William (Bill) Gates was a notable exception. In a famous 1976 letter, Gates accused American hobbyists of theft and demanded that they pay for the software they had acquired. Despite Gates' seminal advocacy of companies' intellectual-property rights, it would take two decades of arm wrestling between corporations and users on both sides of the Atlantic before the economic argument took hold. Only in the late 1990s did governments, under pressure from corporations and intelligence agencies, begin to issue anti-piracy laws.[61]

From the moment of Gates' public plea to the time of the first software protection laws and beyond, a vibrant digital culture of playing, copying, and designing thrived. Teenagers and university students—from Warsaw and Turku to Delft and Berlin to Athens and Zagreb—worked busily to make computers function in their local contexts. Hobby clubs, magazines, stores, trade tourists, and vendors in computer bazaars on the street bridged the gap between production and use, between American (most often) software and local clones. Sharing and pirating was the unquestioned norm across Europe. A youthful European computing counterculture developed that was both shared and diverse. Computing practices spread through the endless copying of computer codes and listings, which turned play into useful innovation.

The unacknowledged history of computing is a transnational story of young Europeans appropriating a U.S.-based technology. The relationship between Britain and the Netherlands serves as a case in point. In both countries, hobbyists, teachers, TV directors, and students of electrical engineering designed standards and protocols to help users navigate between different computers to exchange information more easily. Britain led the way with the Amateur Computer Club (ACC) in 1973. Dutch hobbyists soon followed, founding an association that would soon grow into Europe's largest club, with 67,000 members. In the late 1970s, through the British Association of Microcomputer Users in Secondary Education (MUSE), teachers developed a common, user-friendly protocol for BASIC; this was another step in standardizing and appropriating a computer language. Dutch teachers elaborated on their British colleagues' protocol in 1982. For its 1984 *Chip Shop* program, in turn, the BBC adapted the version the Dutch radio program had developed together with Philips and hobbyists. In the next round of tinkering, the West-German channel WDR3 followed in adopting Basicode—as did their Belgian, Danish, Australian, and East German public-sector counterparts. Thus, Europe's public radio and television enjoyed a rich exchange of borrowed and redesigned offerings. More generally, these exchanges illustrate the variety of circulation channels that the activists and professionals devised.[62]

In computing's private sector, the British served—as they had for European youth's music and fashion scene—as a transatlantic bridge-head for U.S.-based computing. British entrepreneur Clive Sinclair created an inexpensive and user-friendly alternative to American models designed for educational purposes. In 1980, Sinclair offered the mean and lean Z80 ready-built, based on the American Zilog processor, at the bargain price of £99.95. Targeting a non-technical, non-hobbyist clientele, Sinclair quickly sold 50,000 units. Users simply had to plug the Z80 into their TV set and they were able to store data on standard tape recorders. Sinclair's machine sold outside the specialized hobby channels, notably through the book and stationery chain W.H. Smith. Before gaming turned into the main activity among users, the Z80 was a cheap educational computer designed for a computer-illiterate clientele. The Sinclair became particularly popular among computer clubs operating in public spaces like pubs and libraries; by 1983, more than 200 clubs populated the country. A year earlier, during her state visit to Japan, Prime Minister Thatcher had touted the Sinclair as a genuine British product. Thatcher praised the affordable computer as a forward-looking product that expressed the country's new entrepreneurial spirit—in contrast to the old, if not ossified British society she so loathed.[63]

The U.S. market-research institute that had originally misjudged the European market claimed the cheap Sinclair was a spoiler; they were wrong again. Technophiles found the educational computer unsophisticated. The user-friendly plug-in computer nevertheless attracted devotees who modified the machine in innovative ways. Indeed, the Sinclair became the European answer to the American Commodore. The British computer became popular in the rest of Europe, gaining the greatest traction in Greece and socialist countries, including Poland, Yugoslavia, and the Soviet Union. A successful translation of U.S. technology into social-democratic and socialist Europe, the Sinclair's success in turn prompted the American Commodore company to produce a low-cost computer, the VIC-20.[64]

During the Cold War, Finland, Poland, and Yugoslavia functioned as computer-knowledge distribution channels within Europe, which was divided politically and economically. Delicately negotiating the tensions between East and West during the Cold War—and with no piracy laws in place—Finland gained a reputation as the "wild west" of software piracy where "bandits and hooligans could easily rampage without fear of law."[65] Polish tourist-traders had a reputation as the modern-day "Phoenicians of the Socialist Bloc."[66] Polish open-air markets served as the go-betweens for those in the region who faced tougher travel restrictions—most notably East Germans and Soviets. Computers from the West entered Poland through illegal and legal routes. Polish guest workers and tourists with travel permits returned home with British Sinclair computers or American Commodores, often selling them at local bazaars like Grzybowska Street and the Persian Bazaar in Warsaw. The official *Pewex* and *Baltona* stores offered a legal distribution channel for Western computers. According to a survey of the time, four out of ten Polish computer owners during the Cold War had bought their machines abroad, using foreign currencies. To the dismay of the regime, Polish citizens would travel, for example, to the West German city of Hamburg to buy British Sinclair computers for 120 Marks, only to sell them in Poland at four times the price. Also popular were Commodores (64 and Amiga), equipped with the British QWERTY keyboard or the cheaper, but more cumbersome German QWERTZ keyboard with diacritics. One third of Polish consumers had bought their imported computers domestically, in zloty, on the black market. Only a minority (a quarter) chose the officially sanctioned route of buying computers with U.S. dollars at the *Pewex* or *Baltona* state-run retail stores. Imported computers came with manuals in English or German, languages most young Poles had not mastered (the required foreign-language education was in Russian). To capitalize on the language gap—and the gap in local users' knowledge of the foreign computers—entrepreneurial Poles sold bootlegged translations of manuals; wrote computer books; and copied and cracked software.[67]

Surfing the wave of the music scene, personal computers reached the shores of Tito's Yugoslavia even earlier than other countries. The first imported PCs came to Yugoslavia through non-standard and creative routes. Disassembled Sinclair Z81 or Commodore 64 home computers bought in Germany usually found their way into Yugoslavia via the Italian border town of Trieste, and the black markets in Sarajevo. Zoran Modli, a Yugoslavian pilot, radio personality, amateur musician, and leading figure in the New Wave music scene, hosted an iconic radio show called *Ventilator 202* ("Electric Fan 202") on Radio Beograd. Mixing music hits and demo tapes of local new-wave bands, Modli started to broadcast tape recordings of software programs for different types of personal computers in the wee hours between 2 and 4 a.m. He saw home computing as an opportunity to give the power of technology to the "ordinary guy on the street." In the first computer-related broadcast, in autumn 1983, he played the application "Paginator" by transposing the binary codes "0" and "1" into sound frequencies (1,200 and 2,400 hertz) for the British import Sinclair Spectrum Z81 on air—a technique invented by hackers in the U.S. state of Kansas. Yugoslavian radio listeners responded with passion: enthusiasts began submitting their own software for other types of personal computers and wrote game software especially for his program.[68]

Evidence of a homegrown Yugoslavian computer culture dates to 1983, if not earlier. That summer, Voja Antonić, a young electrical engineer teamed up with the computer enthusiast-cum-journalist Dejan Ristanović to design a do-it-yourself computer that could be built from off-the-shelf parts available at local electronics stores and foreign mail-order houses. The two also redesigned cheaper and simpler graphics cards. When Ristanovic's popular-science magazine *Galaksija* published diagrams and detailed instructions on how to build one's own computer, he had to print 120,000 extra copies. Within months, 8,000 people had ordered the do-it-yourself kit from Antonić; another 12,000 had built their own *Galaksija* based on the drawings. Firmly rooted in a youth counterculture, and oriented towards the West, Yugoslavian personal computing was carried along by a user movement that held in high esteem the values of sharing and tinkering. Radio and music paved the way.[69]

In short, young people all over Europe appropriated American-based computer technologies. These pioneers found ways to adapt computing to their local circumstances, their own needs and desires. Participating in a transnational culture and cutting through Cold War barriers, they redesigned computers; introduced computer "Esperanto" (Basicode); converted alphabets (Latin to Greek and Cyrillic) and keyboards; and translated information like manuals into users' native languages. Together, these European users changed the landscape of personal computing.

Demanding Internet Access for All

In the 1990s, the European subculture of hacking went public. Politically active hackers throughout the Continent sought to make computers and the Internet universally accessible. In Hamburg, Amsterdam, and Prague, activists sought to put these digital technologies squarely in the public domain, beyond the exclusive control of corporations, the military, and academia. The hackers' expressed goal was nothing short of changing the world. Many first-time users operated in collective settings, at times in an explicitly political fashion. The "counterculture" in West Germany, the port city of Hamburg in particular, played a unique role. In the birthplace of the *Lilli* doll, and where the squatting movement thrived, a local computer subculture helped to create unique niches in the digital network that were neither private nor public and neither state nor market controlled. In this environment, as well as in the divided city of Berlin, the German Computer Chaos Club (CCC) emerged. Inspired by countercultural currents in California and deeply rooted in anarchist thinking, the CCC was established in 1983. Five years later, in a brazen stunt, some of its members hacked the German Postal Service, which functioned as a bank as well, wired 100,000 German Marks into their own account, and presented their bold act to the press the next morning. The CCC argued that the action proved the institution could not be trusted to protect citizens' privacy. In a country highly sensitized to issues of individual rights, the argument resonated and garnered broad public support. The Computer Chaos Club was challenging the established social-democratic forces and corporate interests—and promoting grassroots democracy in their place.[70]

Amsterdam followed in the quest to proclaim the Internet a public space. Committed to community ideals, Dutch activists sought to make the new technology accessible for all citizens. Their initiatives grew into a city-wide and national user movement. They claimed the Internet was for the public rather than for commercial purposes. Impressed by these activities, Spanish sociologist Manuel Castells in 2001 singled out Amsterdam's digital city as a broad-based user movement that had created "a new form of public sphere combining local institutions, grassroots organizations, and computer networks in the development of cultural expression and civic participation."[71] What would come to be known as the Digital City had four practical foundations in Amsterdam. First, Amsterdam became a transatlantic gateway for Internet hardware for Europe. Second, the local hobbyist and hacker community had been large and internationally well-connected from the start. Third, the city nurtured a host of independent media and cultural centers connected to the squatting movement. Fourth, unlike their German counterparts, the city's sub-cultural community and its social-democratic government were mutually supportive rather than antagonistic.[72]

The Amsterdam story is intriguing. At the conference entitled "Hacking at the End of the Universe" in summer 1993, a group of international hackers gathered at a campsite in the Dutch polders to exchange ideas about the latest innovations, social engineering, and hacking tricks. The meeting inspired Marleen Stikker, a woman activist with deep ties to the squatter movement and the hacker community, to find ways to make the personal computer and the emerging Internet useful as political and community tools. While women played several key leadership roles, the project did not articulate an explicit feminist agenda. Rather, its examples came from other user-based digital projects in the U.S., the American Well and the North American Freenets, yet were translated to the local context on the eve of local elections. With subsidies from the local social-democratic government and in alliance with community organizers and the hacker movement, the Amsterdam group launched the Digital City (*de Digitale Stad*) in 1994. During its first week, 3,500 people registered, and before long 10,000 visitors had signed up. The ten-week project matured into a permanent presence in the city and on the Internet, maintaining twenty phone lines 24 hours a day to support the booming traffic and help users navigate their way through the new digital protocols. Most important, the experiment became the entry point to the Internet for many first-time Dutch users outside the academy. Through the

Fig. 8.9 Envisioning the Internet: *It was 1994: the emerging Internet was still clunky, still steeped in the culture of the military and engineering, from which it came. Enter the Dutch "Digital City" project, an alliance of activists, squatters, and members of local government. These pioneering users broke with engineering-based interfaces of letters and code. This is Digital City's first graphic design for a user-friendly interface. This image— and Digital City's approach—imagines the Internet as a city unto itself. Each octagon on the screen represents a different activist group in the city, from Gay Square to news-junkie politicos.*

Digital City's efforts, these new users now had e-mail addresses and a web presence in their native language—a forum for dealing with local issues.[73]

The project's cultural and media centers involved users as a driving and urgent force behind shaping the Internet. During a 1990 festival, for example, a coalition of users established a temporary network connection with the AIDS wards in hospitals in Amsterdam; Ithaca, NY; and San Francisco. The agenda was explicitly political. Amsterdam set up the network to enable the participation of people with AIDS and HIV whom the Reagan administration had prevented from traveling abroad. Soon after, Amsterdam Internet provider XS4ALL—the domain abbreviation for "access for all"—played a crucial role in organizing an Internet connection between the besieged Sarajevo and radio station B92 in Belgrade. Additionally, the Amsterdam City Council, as the owner of a local cable infrastructure, supported a local television and radio effort enabling activists to acquire skills and connections.[74]

The Amsterdam group's most important innovation was, perhaps, their mobilization of the community metaphor of the city as a representation of political citizenship. While connected transnationally and committed to the squatting movement's progressive grassroots politics, the image of the city brought the innovative Internet closer to citizens' daily practices. The software was originally text-based and modeled after the American Freeport, leaving a great deal to the imagination of the first users who visited the site. Users were expected to visualize "the city" based on the simple labeling like "cafés," "kiosks," "library," and "shops." Given that experienced users found the interface rather disappointing, the designers made the crucial decision to adopt the brand new CERN www protocol. That graphic interface better matched the initiators' vision of an electronically connected community built around the metaphor of the city. Most important, however, were the ways that non-technical user communities filled the space with content. The alliance with communities of hackers, artists, and local activists accounted for the immediate and lasting success of the project. Also instrumental in the success was the Digital City's collaboration both with the famous music temple of pop, *Paradiso*, which attracted members of the CCC to Amsterdam and with the Philips research project "New Topia" (*Nieuw Topia*).[75]

From the start, the Digital City project was simultaneously embedded locally, connected internationally, and operating transnationally. By the second half of the 1990s, the greatest threat to the original vision became the interest of corporations and the government's retreat from supporting such initiatives. And so, the counter-cultural spirit of California, Hamburg, and Amsterdam moved to Prague, to build access for all citizens.

At the very start of the new millennium, activists in Prague built visions of interconnectedness and shared use through a local product called RONJA, which they saw as a cheaper and technically superior alternative to Wi-Fi. The first wireless networks in Prague had been established within the squatting movement. Like in Amsterdam, the goal of creating a community computer network was, in part, to build alternative media outlets. Publishing fanzines and broadcasting with pirate radio and street TV were means of establishing independent communication infrastructures that could circumvent and replace corporate and state cables and satellites. The RONJA project—among many other similar initiatives—fell on particularly fertile ground. The state's telephone company was in no hurry to offer users high-speed ADSL connections. Moreover, the Czech technical culture of self-reliance readily stepped in where state and corporate institutions neglected to provide services. "RONJA" stood for Reasonable Optical Near Joint Access—cheap, do-it-yourself hardware equipment for sending data through a visible red light. RONJA became the fastest, most reliable method of connecting computers for Czech citizens. RONJA also helped users without knowledge of electronics to build their devices from widely available, off-the-shelf components. Improvements in the technology were discussed among a community of developers and users. As a "user-controlled technology," its designs and schematics were published as free software.[76]

The Czech RONJA system was far more stable than the Wi-Fi systems. It was also deliberately low-tech and inexpensive. In experimenting with optics to connect computers into a LAN net, its designer, Karel Kulhavý, deliberately selected cheap parts like a Chinese-made magnifying glass sold at Prague's flea market. This was to allow other users to easily reproduce the system (some enterprising users even tinkered with pineapple cans to turn them into antennas). The technology allowed users to maintain full control over the system. By 2008, Prague had an estimated 250 independent wireless networks, some involving a handful of

Fig. 8.10
Do-it-yourself
WiFi: *Computer activists shun commercial closed technologies (like Apple products) that are invulnerable to tinkering. Instead, active computer users prefer open-source solutions that allow for change and even redesign. Here, perched on a Prague rooftop, is the Czech RONJA system, a counterculture alternative to off-the-shelf WiFi box. RONJA, was built in 1997. The system helps consumers skirt computer copyright and licensing issues created by the struggle between users and corporations.*

friends, others hundreds—or even thousands—of members. With more than 8,000 users, the city of Plzeň claimed the largest single, non-profit wireless community network in the Czech Republic. Regional activists succeeded in coordinating their efforts with neighboring towns and villages by connecting separate wireless networks through high-speed links, establishing a local intranet of an estimated 20,000 users.[77]

RONJA's wireless system could thrive in part because corporations and the state had chosen to neglect individual users. Currently, in the early 2010s, however, corporations and the state are engaging in an aggressive push-back where hackers are concerned. Businesses have sought to stop user activism through political and legal means. Initially, governments were more likely to support corporations rather than users. This favoritism changed radically in 2012, when Internet activists started to successfully lobby politicians to rescind their support of corporations. In January 2012, a group of Polish Parliamentarians protested the ratification of the Anti-Counterfeiting Trade Agreement (ACTA) on behalf of European Internet users at large.

The controversy symbolized the core conflict between corporations, mostly U.S.-based, and European Internet users. Corporations claimed that the act of copying and pasting violated the rights of intellectual property owners. Their lawyers sought to criminalize the act of copying and pasting; the firms lobbied the European Union to approve the powerful "Anti-Counterfeiting Trade Agreement," also known as ACTA. In the first round of legislation for which corporations lobbied, the European Parliament ratified the Anti-Counterfeiting Trade Agreement. But users revolted, defending their right to copy, paste, and create. In more than two hundred European cities, hundreds of thousands of citizens took to the streets to oppose ACTA. The rebels declared their freedom from corporate and state control of the Internet. The issue had special resonance in East European countries like Poland, where protesters compared ACTA to the censorship and surveillance of former socialist regimes. Similar protests erupted in the United States concerning two measures that also sought to provide a legal framework against purported piracy: the Stop Online Piracy Act (SOPA) and the Preventing Real Online Threats to Economic Creativity and Theft of Intellectual Property Act (PIPA).

This first round of grassroots user protest was, in fact, well-organized and effective in Europe. In the confrontation with globally-acting software corporations, users saw themselves as defending free access to the new technology. Internet users proved their ability to unify; to leverage their power; and to mobilize in the mainstream political arenas of national and European parliaments. For example, in Slovenia, diplomat Helena Drnovsek-Zorko, who had signed the treaty on behalf of her country, apologized to Internet users at large for having failed to serve the public's interest when she voted pro-ACTA. In Poland, parliamentarians protested the approval of the trade agreement by sporting Guy Fawkes masks. Brussels saw protests in the European Parliament—on the very day that ACTA was adopted: Kader Arif, a French member of the European Parliament, resigned as *rapporteur* on ACTA to signal his protest in the strongest possible terms. He called the new legislation "wrong in both form and substance."[78] To his mind, ACTA negotiations had misled legislators and ignored the needs of ordinary software users: "Voilà, that's the masquerade that I denounce," he exclaimed.[79]

Activists believed that national and European-Union anti-piracy laws were "harmonized" to serve U.S. media corporations' schemes to protect their market. In 2006, when running for national parliament—and before his election for the European Parliament—the Swedish Pirate Party leader Rickard Falkvinge had resolutely moved the defense of computer user rights out of the subcultural backrooms of hackers and into the official political arena. This tactic was successfully reproduced in Germany, the Czech Republic, Spain, and Switzerland, where pirate-party candidates had begun to make traditional political parties uneasy. In Poland and Hungary, liberal parties subsequently took on the cause. In ways that shocked Europe's political establishment and U.S. corporate interests, the pirate parties mobilized a young electorate. These youthful Internet users felt that the legal measures initiated by the U.S. record and film industries threatened their innovative practice of crafting, copying, and sharing, tinkering, and tweaking. This most recent incarnation of a technology user movement has rallied against corporate power and state control in the field of computing.

The defense of users' rights coalescing around technology was not new. European history includes numerous examples in which the practices of piracy (to its opponents) and sharing (to its proponents) have been contested. In the late nineteenth century, the French *Syndicat de la Haute Couture* sought to protect its sartorial monopoly by opposing the copying of fashion. Their tactics— inveighing against the home industry's use of sewing machines to recreate dress patterns—remind us of the freedoms at issue in the ACTA case. The French fashion industry has continuously targeted key copyright offenders through court cases and publicity campaigns. Nineteenth-century French *haute couture* ateliers and twenty-first-century U.S.-based software companies alike have despised piracy. On the other hand, some companies have realized that the practices of technology user movements can *also* generate value by functioning as a marketing wedge. That means users can participate in setting standards; users' deep engagement with the technology can lead to dependency on that technology; and, ultimately, users' involvement can generate greater corporate profits. Indeed, protesters have succeeded in winning the support of companies such as Google and its subsidiary, YouTube, both of which expect higher turnover from an open-access Internet.[80]

Similarly, consumers, tinkerers, and rebels embraced—and were also embraced by—the French fashion establishment. While enterprises used legal measures to thwart competitors, the industry did not protest the vibrant international women's culture of personal tinkering with *haute couture* on a small scale and for private use. Some couturiers realized that this copying of French styles in cities from

Madrid and Budapest to Istanbul and New York only enhanced the reputation of Paris as Europe's—and indeed the world's—fashion capital. Like the young computer users of today, ordinary women exhibited enormous ingenuity and skill in copying French fashion; users adopted the novel techniques of the fashion plate, the sewing machine, and the paper pattern. Indeed, the nineteenth-century Syndicat established Paris as the capital of fashion not because of, but despite the many rules aiming to protect the monopoly against copying by using paper patterns. The same was true for U.S.-based personal computing in the late twentieth century.[81]

Shortly after blue jeans had become an icon for a rebellious generation, computers came to symbolize a subculture of young enthusiasts and hackers, who toyed with—and ultimately subverted—the roles that corporations and states had defined for them. Their creativity thrived outside of—and in opposition to—both state-sanctioned politics and high-culture prescriptions. On both sides of the Iron Curtain, young men and women readily adopted clothing styles and computers from the Anglo-Saxon commercial world. They did so in innovative and ironic ways. They altered American jeans to fit the latest trends; they tinkered with Barbie clothing; and they modified U.S.-based computer software to suit their own needs. Many of their "playthings" were U.S.-based, imported from powerful U.S. corporations like Mattel, IBM, and Levi-Strauss. Young Europeans were not equal partners in the transatlantic exchange, yet they were important in tweaking the messages to suit local uses. In fact, a distinct imbalance of power existed between different countries, an imbalance exemplified by the German *Lilli* doll losing her trademark infringement case; by the commercially booming Barbie clothes being sewn in a satellite country, Japan, for low wages; and by IBM's dominance in setting a global standard. Nevertheless, users were vital. Only when employed creatively, over and over again, do technologies become useful and successful. Computers are, perhaps, the best case in point. In a politically divided Europe, young Europeans ignored national boundaries, connected with one another—and empowered armies of users. They felt part of an imagined, virtual, transnational community of youthful rebellion. In the 1990s, these subcultures—ranging from the crafting counterculture of blue jeans to the hacking culture of geek programmers—turned into markets in their own right, markets that encompassed both work and play.

Conclusion

Henri de Saint-Simon expected science and technology to be the main forces in shaping Europe into a political union. Engineers would make plans, and scientists would assess them. Enlightenment-era France and liberal Britain figured prominently in Saint-Simon's thinking. But what has become of his vision? Have scientists and engineers—inspired by politicians and philosophers like Saint-Simon himself—been the lead actors in shaping and unifying Europe?

As the visionary had hoped, new technologies *did* help to unite the people of Europe. For example, technologies like fashion plates and paper patterns gave rise to a cosmopolitan Europe, a region that was part of a global and colonial world. When it came to fashion, Paris was the mecca of the industrializing world—thanks to the legions of creative women across national and regional borders who copied and tinkered with Parisian styles. These users' activities transcended the rules and individual cultures of nations. In each individual act of appropriation, the prestige of Paris as the center of fashion was further enhanced. The technology of the bicycle also contributed to a shared European culture. Youthful cyclists traveled in Europe and beyond on their "iron horses." Forming an early user movement, bicyclists rebelled against the Europe of the aristocracy—the rarified realm of horseback riding, stuffy parlors, and elaborate codes of fashion. At once local and transnational, the bicyclists' urban world was modeled after liberal London—instead of aristocratic Paris or imperial Vienna. The bicycle movement's members guided each other in forming a pan-European fellowship exemplifying middle-class liberalism. It was also the bicycle movement that shaped one dimension of Europe's tourist infrastructure. This enabled the middle class to navigate the countryside—and escape the unpleasant sights, smells, and sounds of the industrializing cities. Tourism for the bourgeois, thus, balanced precariously between the cosmopolitan culture of the aristocracy and the urban culture of the working classes.

In some ways, The Long Twentieth Century's technologies served to erase social differences and bring people together. In other ways, those very same technologies divided people by reproducing class, gender, and ethnic differences. For example, the railroad tracks that crisscrossed Europe exemplified the kind of pro-integration European force that Saint-Simon had in mind. Yet Europe's social classes felt insecure about this new, increasingly democratic world of trains. Entrepreneurs and engineers eventually designed first-, second-, third-, and even fourth-class tickets; amenities; and hotels. These practices accepted society's demand to maintain social differences. The migrating rural poor escaping Eastern, Central and southeastern Europe for the cities and for the United States were subjected to a grueling transport regime—including strip-searches and physical disinfection. Such treatment was a stark contrast to the luxury dining cars and first-class service enjoyed by the vacationing *haute bourgeoisie*. Indeed, fundamentally and profoundly, before the First World War, Europe was built on separate tracks to create transnational, national, and regional corridors—and distinct class corridors, as well.

The train example, however, raises vital questions about how much leeway users had in directing the technologies of their time. The governments of Prussia, Russia, Austria-Hungary, and the city-state of Hamburg privatized transmigration through Germany because they lacked the capacity to manage the migration. Specifically, the movement of so many immigrants looking for a better life in Western Europe and the United States had brought with it the need to patrol the Continent's eastern borders. It became the responsibility of businesses and user associations—namely, steamship companies and transnational humanitarian agencies—to deal with the stringent U.S. and Prussian migration laws. In this separate corridor of trans-European and transatlantic migration, a network of shelters emerged. Missionaries and self-help groups guided migrants through the obstacles and on to their new destinations and new lives. Yet, the one hand, migrants had little opportunity to affect railway technology. But, the other hand, migrants' advocates were indeed able to modify railroad structures, because migrants' huge numbers represented a lucrative market: they were twenty-five million strong, and steamship companies depended on their business.

Consumers, Tinkerers, Rebels presents a provocative view of history in which people co-produced a shared European experience. Most experts and most individuals did not foresee such a process. The shaping of Europe transpired not primarily in parliamentary chambers and ministries, science labs and drafting departments, but far, far beyond. Europe was created on the pages of popular magazines and newsletters; in working-class kitchens and nurseries; in computer-club rooms and recycling yards; in train compartments and bicycle lanes. These were some of the places where technology's champions—and its critics—repeatedly met. And these were the places where regulators encountered consumers and users determined to, by turns, appropriate, tinker with, and, at moments, reject available artifacts and systems. Sometimes user groups created their own technologies or collaborated with companies. At other times, they cooperated with the state. In still other instances, they found experts acting in their place. European history is rife with examples of users questioning, protesting, and even undermining the intentions of companies, states, and experts.

What room for negotiation did technology users really have in shaping Europe throughout The Long Twentieth Century? The period has seen at least two major turning points: the First World War and the 1960s. As we showed in Part I, *The Nineteenth Century: Shaping New Technologies*, in the 1800s, users either created their own worlds or appealed to the government for protection against ruthless business practices. This was the time when companies expanded their markets globally and experts created new knowledge domains. The First World War, when the state stepped in, was a watershed. Together, European governments began to more actively direct new technological systems. Often, coalitions involving government and businesses sought to incorporate and co-opt users for their own purposes. The second major turning point was the rebellion of the 1960s. Colonies revolted against European imperialist governments. National governments folded before users' activism. Corporations under pressure changed their tactics. And the European Union began to demand its place on the world stage.

To recap: the year 1919 signaled the first turning point, when the working classes gained national access to political power in many European countries. The political shift had a profound impact on users' power to direct new technologies. In the nineteenth century, users, experts, and companies interacted in directing these new technologies; for the most part, states did not take an active role, let alone the initiative. After the Great War, power struggles ensued over how innovations were to be developed and who should have access to them. In Europe's political chambers, all governments sought to shape technologies for a classless Europe, but each in its own image—whether liberal, socialist, welfare, or fascist. To this end, states drew on new expert groups when providing citizens with housing, leisure activities, and food. Engineers and architects, home economists and nutritionists, worked to shape the new century's large projects. The role of these expert advisers, with their influential visions and hopes, is the subject of two companion books in this series—*Building Europe on Expertise* and *Writing the Rules for Europe*—which readers interested in these issues might consult.

After the First World War, the state and its experts stepped in. As shown in Part II, *After the Great War: Who Directs Technology?*, debates raged over traffic control, housing, and industrially processed food. These events spotlight the power struggle between the state, experts, and users over how the many ought to live their lives given the priorities established by the few. Professionals, often in collaboration with the government, came to intervene and act in the name of user groups. The technologies of the bicycle, food production, and kitchens illustrate this point.

In the case of the bicycle, we showed how, for a brief but crucial period before 1900, middle-class bicycle clubs—often supported by companies—were hugely successful in helping to build road and tourist infrastructures. After the First World War, the bicycle clubs lost their power to traffic experts and the state. At this point, bicycling shifted from a bourgeois to a working-class movement. When the technology of the bicycle became affordable, workers and small entrepreneurs adopted cycling to transport goods and commute to work. At the same time, the upper middle class domesticated the automobile with their user activism. Former cyclists were now motorists. A conflict ensued between working-class cyclists and bourgeois motorists over how the streets should be used. The struggle took place within city councils and national parliaments. Designing fast lanes for cars and trams, politicians and experts made rules that either pushed working-class pedestrians and cyclists out of the way or directed them to take the more expensive route of public transportation, if available.

The case of industrially processed food also highlights the new roles that the state and its experts assumed in building large technical systems. Prior to the First World War, consumer groups on both sides of the Atlantic rallied against high prices, food scandals, and inhuman working conditions. Trying to regain control over their own food supply, middle-class consumers developed skills and appropriated technologies like intricate home-canning kits. Consumer and producer cooperatives offered alternative ways to pool resources and circumvent middlemen. Again, the First World War proved important: to deal with pressing concerns about how to feed the many, an alliance between the state and food experts proved crucial. Dairy cooperatives, national governments, and medical experts collaborated in bringing allegedly healthy food— milk in particular—to workers and school children.[1]

A coalition emerged among the state, experts, businesspeople, and users. Together, they created large technical systems like the built environment and the food chain. This coalition was laced with tensions. As the companion *Making Europe* volume *Communicating Europe* illustrates, such tensions also pervaded mass media. In our work, some of the most visible tensions played out in large-scale housing projects. Middle-class, modernist architects with a social-engineering mindset, insisted on designs that separated spaces according to their functions: the kitchen was for preparing food, the dining room for eating, the bedroom for sleeping, and so on. Modernists also abhorred clutter of any kind. Residents often responded in subtle (and sometimes less-than-subtle) ways. Many occupying state-subsidized housing celebrated their working-class and rural traditions, transforming the single-purpose kitchens into multipurpose living rooms. They took their fates into their own hands by, for example, tearing down kitchen walls and moving tables—and even beds—into kitchens, where they were not "supposed" to be. In a similar spirit of self-determination, Hungarians, neglected by the state, rebelled by becoming do-it-yourself builders of their own houses. These were small, individual acts of rebellion, rather than the outcome of organized user movements. Still, their acts of resistance were historically significant.[2]

More often, tensions were distributed among the lead players in housing—the state, experts, and builders. Users were included only marginally: when it came to creating housing designs, user needs and preferences were incorporated by imagining ideal residents—what social scientists call "projected users." Other professionals sought to mediate between the ideal user and the users they encountered in real life. Home economists, in particular, worked to link state officials, architects, and users. By listening to residents' opinions, home economists tried to offer practical solutions to diffuse tensions between designers and users. Home economists succeeded—at least in part. For example, they championed dish racks and easy-to-clean surfaces; they dismissed high-tech and expensive household solutions; and they rejected elitist, poorly-grounded interventions on the part of architects. As professionals who positioned themselves as the sole spokespeople on behalf of user groups, home economists defended their knowledge vis-à-vis architects and politicians.

This book also has depicted a critical, but less well-known pivot point: the youthful revolt of the 1960s. This era saw the empowerment of citizens, consumers, and users. In Part III, *Beyond the 1960s: Users Empowered?*, we showed how rebels protested the powerful alliance between the state and big business. To break this alliance, user movements mushroomed. They questioned, confronted, and sometimes even undermined the strong state and powerful corporations. In Eastern Europe, the act of rebellion took the form of wearing—and tinkering with—blue jeans. In Western Europe and Scandinavia, many user movements sought to mobilize older technologies like glass bottles, bicycles, and windmills. The mission was to help save planet Earth rather than the nation state and the (Cold) war economy. Hackers throughout Europe effected an almost magical reversal: business machines became personal computers, and serious work turned into play. The rebellion of user movements turned the world upside down. The injunction to "Think globally, act locally" epitomized the idea that activists were no longer tied to city or nation. Users began to act locally, nationally, transnationally, and transatlantically—all at the same time.

The new user movements also challenged directly the knowledge of experts. In place of professional expertise, users offered their own brand of counter-expertise: knowledge that was experience-based, often informal, and deliberately collaborative. Sharing knowledge is a key aspect of user movements. To this end, users from Germany to Greece arranged meetings and issued newsletters or journals. Hackers and self-professed geeks regularly organized events in which teenagers not only traded hardware and software, but taught each other the latest cracking techniques. Similarly, for the Danish windmill builders, sharing technical information, crafting legal frameworks, and lobbying utility companies was part of building a more equitable, sustainable community. Sharing technical knowledge and skill became a way of life. These actors prefigured the knowledge society to come: boundaries were increasingly blurred between professional designers and lay users.

Having examined the interplay between the state and experts, business and users, this book concludes by offering a new scenario of how Europe and its technologies came into being. Europe is a complex social and cultural entity that cannot be reduced to a system of GPS coordinates or elevation lines; nor is Europe equivalent to the European Union. Only as a first approximation is it appropriate to treat Europe as a three-dimensional, physical space that both divides and unifies people. It is true that Europe and the rest of the world have developed unifying physical infrastructures, such as harbors and roads. Other unifying infrastructures of The Long Twentieth Century included the telegraph, telephone, and wireless networks as well as the expansion of airports. Our case studies show, however, that structures were also erected with the intention of being divisive. This includes trenches, checkpoints, and walls. These structures separated nations. They marginalized migrants. And they cut off family members from one another.[3]

The historical topography of Europe cannot be reduced to a two-dimensional map, true to scale, with straight, connecting lines and jagged borders. Europe's working spaces then, as now, depended on technological systems that connected distant (and nearby) places. For example, while Paris and Vienna were easy to reach by rail, San Marino and Moldova were not. Small villages were simply passed over by the new railroad lines. What emerges from these examples is a mosaic of centers versus peripheries, main routes versus back roads. Users' access to European places was defined by the transportation infrastructure—not by Europe's physical geography. Transportation systems brought citizens of urban centers closer to each other, but these same systems also further marginalized people at the peripheries. Thus, the technical systems redrew the geography of Europe.

Defining Europe in terms of the social relationships and technological connections highlights its contribution of transnationally networked countercultures. Instead of treating "Europe" as a physical, three-dimensional space, we have worked to show that distance and proximity may not only be measured in miles or kilometers, but also in degrees of inclusion and exclusion, linking and delinking. Physical proximity does not guarantee inclusiveness, just as physical distance does not necessarily mean that individuals are cut off from each other. In many turn-of-the-twentieth-century apartment houses, for example, well-off families lived in the same building as—and in close proximity to—lower-middle-class residents. The cheaper flats could be found in attics, basements, and courtyards. The most expensive and largest apartments were situated just above—on the first floor facing the street. The physical proximity of the social classes did not foster empathy between them. The different classes did indeed live different lives—and each experienced Europe in a different way. Seen in a larger frame, then, this book contributes to transnational history by parsing these divergent experiences of integration and divisiveness, harmonization and tension.[4]

In his original vision, Saint-Simon expected that industrial and liberal Britain would lead the way to a unified Europe. In the end, however, the United States rather than Britain came to shape a shared European culture. Throughout The Long Twentieth Century, "America"—as a vision—loomed on Europe's horizon before imparting to Europe a corporate culture. In developing technologies like trains and sewing machines, living rooms and bicycles, industrializing Europe and the United States were part of the same transatlantic world. Often, European inventions were produced and marketed on a grander scale in the United States before finding their way back to Europe. Still, the two industrializing continents faced the same issues. And they found similar solutions, often inflected, though, by local circumstances.

In and about the 1940s, the paths of Europe and the United States began to diverge. For one, the close collaboration between experts and the nation state was now recognizable as a distinctly European feature. Europe's welfare and socialist states, for example—supported by many experts—offered sober versions of how to live. Home economists, acting on behalf of users, demanded low-tech, practical home solutions. The United States, in contrast, proffered a corporate culture—rather than a welfare state in which consumers, as well as experts, had key roles in shaping public policy. "America" promised abundance, typified by refrigerators filled with food, and an attempt to force-feed modernist and gadget-packed kitchens to the world at large. Despite this divergence, the profound importance of the United States in the history of postwar Europe cannot be questioned. After the Second World War, the U.S. had become a superpower. And by the 1990s, it had once again become clear that the United States had the power to shape Europe's future. The narratives of Europeans appropriating Barbie dolls, blue jeans, and personal computers support this viewpoint: we see an integrated transnational market, ruled by U.S. foreign policy and American-style commercialism and multinational corporations, in which users tweaked, tinkered with, and created their own cultures.

Saint-Simon was certainly canny in his vision that technology could cultivate European integration. Technology has indeed helped to rally user groups in Europe. But what The Long Twentieth Century has taught us is this: that any sense of unity is based on not only *national* and *regional* identities, but on *technology-user* identities as well. This means that a hacker in Amsterdam, for example, has more in common with a fellow hacker in Prague or Silicon Valley or Bombay than with fellow (non-hacker) Europeans. The core identity of modern consumers, tinkerers, and rebels is based on their common—and often transnational—user identities. The authors hope that this book will inspire future researchers to investigate, in even greater detail, how users perform while working and playing; what users do with particular innovations; what drives their behaviors; and in which political arenas they prove most effective.

What can be concluded about European users' role in shaping future technologies? As this book goes to press, the question is still under debate. Specifically, anti-piracy laws continue to be contested in Europe. The debate concerning ACTA (the Anti-Counterfeiting Trade Agreement) reminds us that competing visions of Europe

Fig. 9.1   Rebels in Parliament: *In the 2010s, users rebelled against corporations and state control by taking technology into their own hands. Claiming the right to copy material from the Internet, Europeans challenged the Anti-Counterfeiting Trade Agreement (ACTA). Under pressure from rebel "pirate" parties, Europe's politicians defended the rights of consumers in general and Internet users in particular. In this memorable scene, from a Polish parliament debate in January of 2012, parliamentarians hold up masks to protest against the European Union's ratification of ACTA. The masks reference Guy Fawkes and his legendary protest of 1605.*

sometimes intersect—precisely when individual users decide to act collectively. In one reading of the ACTA story, the anti-piracy laws are a plot against consumers by U.S.-dominated multinationals. As this reading goes, the multinationals vie for competitive edge by setting a legal standard. They circumvent political transparency and threaten users with fines and jail terms. But the corporations are stopped dead in their tracks by the European Parliament and national governments.

In another reading of the narrative, American and European actors—corporations and countercultural movements—operate transatlantically and even globally. Here, the ACTA story calls into question the extent to which the shaping of technologies has been an exclusively European affair. What we now know definitively is what *kind* of user is most effective in shaping technology: it is the *user-citizen*, the consumer who is able to leverage the political power of his or her particular nation state—if not the political power of the European Union. Specifically, when individual users of consumer goods transform themselves into user-citizens, they become a force to be reckoned with.[5]

Indeed, the process of shaping technologies—and cultures—is never ending; so is the struggle over who has the power to govern such shaping. That conflict continues: the outcome of the ACTA protests, still ongoing, will determine how a formidable force, the "pirate" user movement, fares in challenging the computer industry's corporate and state power—and in shaping Europe in the process. There are two competing scenarios for Europe: users acting as mere consumers, versus users acting as citizen-consumers. Which scenario will be played out remains an intriguing question.

Endnotes

Introduction

1   Saint-Simon, *De la réorganisation de la société européenne*. For an extensive discussion on Saint-Simon and the belief in technocracy, see: Kaiser and Schot, *Writing the Rules for Europe*. For the role of experts: Trischler and Kohlrausch, *Building Europe on Expertise*.
2   Arrighi, *The Long Twentieth Century*.
3   Oudshoorn and Pinch, eds., *How Users Matter*.
4   The term "technology-in-use" comes from: Edgerton, *The Shock of the Old*. For a consumer approach to the American history of technology, see: Corn, *User Unfriendly*. For a history emphasizing the pleasures in using American technologies: Maines, *Hedonzing Technologies*.
5   Giedion, *Mechanization Takes Command*.

1   Poaching from Paris

1   The authors would like to thank Eileen Boris and Emanuela Scarpellini for their comments and help.
2   The well-researched article by Ruether, "Heated Debates over Crinolines," 359, quoting *Berliner Missionsberichte* (1876), 134–36.
3   Ibid., 363, quoting Evangelisch-Lutherisches Missionswerk in Hermannsburg, Leitung und Verwaltung in Bethanie, Behrens July 3, 1865.
4   *The Englishwoman's Domestic Magazine,* February 1876, 133, as quoted in Breward, "Patterns of Respectability," 24.
5   Ross, *Clothing*. Hoganson, *Consumers' Imperium*, 58, 61–65, 73–75.
6   Ross, *Clothing*, 54–55.
7   Green, *Ready-to-Wear*.
8   Le Pay's study is discussed at length by Crane, *Fashion and its Social Agendas*, chapter 2.
9   Ibid., 28–29.
10 Green, *Ready-to-Wear*.
11 Hoganson, *Consumers' Imperium*, chapter 3.
12 Ross, *Clothing*, 71–76; Hoganson, *Consumers' Imperium*, 58, 61–65, 73–75.
13 For the establishment of the Paris fashion system see: Steele, *Paris Fashion*.
14 Ruane, "Clothes Shopping in Imperial Russia," 766; Ruane, *The Empire's New Clothes*. Gemeentemuseum, *Voici Paris!*, 23–24.
15 Troy, *Couture Culture*, 262–63.
16 Ruane, *The Empire's New Clothes*; Pouillard, "In the Shadow of Paris?," 67–72. Gemeentemuseum, *Voici Paris!*, 72–74; Green, *Ready-to-Wear*, 120; Troy, *Couture Culture*, 260–65.
17 Gemeentemuseum, *Voici Paris!*, 24; Ruane, "Clothes Shopping in Imperial Russia," 766–70.
18 Miller, *The Bon Marché*, 61–64; Hoganson, *Consumers' Imperium*, chapter 2; Ruane, *The Empire's New Clothes*, 134–38; Merlo and Polese, "Accessorizing, Italian Style."

19 For an exhaustive and detailed description of the transnational circulation of fashion magazines, see: Ghering-van Ierlant, *Mode in Prent, 1550–1914*, 83–91. For Russia, see: Ruane, *The Empire's New Clothes*, chapter 4 and 99.

20 Meyer, *Das Theater mit der Hausarbeit*, 35; Gordon, *Make It Yourself*, chapter 2; Peiss, *Cheap Amusements*, 62–67.

21 Hoganson, *Consumers' Imperium*, chapter 2; Ruane, "Clothes Shopping in Imperial Russia," 769–73; Ruane, *The Empire's New Clothes*. On sharing skills, designs, and informal copying techniques, see: Buckley, "On the Margins"; Gordon, "Home Sewing."

22 Green, *Ready-to-Wear*, 163; Godley, "Homeworking and the Sewing Machine"; Ruane, *The Empire's New Clothes*, chapter 3; Cf. Green, *Ready-to-Wear*, chapter 6 who correctly criticizes the notion that immigrant men and women brought skills with them and demonstrated rather than acquired the skills with the sewing machine on the job. Future scholars will have to square this with the international movement towards the institutionalization of sewing education. Gamber, *The Female Economy*, 98.

23 Ruane, *The Empire's New Clothes*, chapter 2; Helvenston and Bubolz, "Home Economics and Home Sewing"; Georgitsogianni, *Panagis Charokopos*, 125, 35, and 42. We thank Faidra Papanelopoulou for this reference. Aliberti, *L' economia domestica italiana*; Cosseta, *Ragione e sentimento dell'abitare*.

24 Green, *Ready-to-Wear*, chapter 6; Hoganson, *Consumers' Imperium*, chapter 2; Troy, *Couture Culture*, 262.

25 Karl Marx, *Das Kapital* Vol. I (London 1976), 601 as quoted in Ross, *Clothing*, 58. On the gendering of the sewing machine, see: Green, *Ready-to-Wear*, 164–65; Baron and Klepp, "'If I didn't Have My Sewing Machine'"; Coffin, "Credit, Consumption, and Images"; Offen, "'Powered by a Woman's Foot'"; Hausen, "Technical Progress and Women's Labour"; Perrot, "Femmes et machines au XIXe siècle." See also: Wilson, *Adorned in Dreams*, chapter 8; Levitt, "Cheap Mass-Produced Men's Clothing," 179; Zakim, "The Birth of the Clothing Industry in America." For the U.S., see: Kidwell and Christman, *Suiting Everyone*. On Italy, see: Paris, *Oggetti cuciti*.

26 Green, *Ready-to-Wear*, 34–35; Putnam, "The Sewing Machine Comes Home."

27 Godley, "Selling the Sewing Machine," 266–67, 286; Davies, "Singer Manufacturing Company in Foreign Markets," 316–19; Hounshell, *From the American System to Mass Production*, chapter 2; Coffin, "Credit, Consumption, and Images." Owning a sewing machine—as with mirrors, clocks, and pianos—increasingly became a status symbol for working-class families aspiring to a bourgeois lifestyle: the home represented not just a shelter, but a center of leisure and self-expression. Oddy, "A Beautiful Ornament in the Parlour or Boudoir."

28 Robert Davies, *Peacefully Working to Conquer the World* (New York 1976), as quoted in Godley, "Selling the Sewing Machine," 297, 285. Although these details come from the Singer's Japanese branch, they most likely apply to Europe since the company pushed a one-size-fits-all global model throughout: Gordon, "Selling the American Way," 674; 8–9. Davies, "Singer Manufacturing Company in Foreign Markets," 306. The Singer Company mobilized the home economics teachers from the beginning: Helvenston and Bubolz, "Home Economics and Home Sewing."

29 Gordon, "Selling the American Way," 691–92; Ruane, *The Empire's New Clothes*, 138–43.

30 Coffin, "Credit, Consumption, and Images"; Douglas, "The Machine in the Parlor." This point has also been made by Forty, *Objects of Desire*. Cf.: Putnam, "The Sewing Machine Comes Home."

31 Letter Robertson to McKenzie, November 13, 1880, as quoted in Godley, "Homeworking and the Sewing Machine," 161.

32 This important observation, but under-researched question, comes from Coffin, "Credit, Consumption, and Images"; Putnam, "The Sewing Machine Comes Home," 278; Green, *Ready-to-Wear*; Ruane, *The Empire's New Clothes*.

33 Maher, *Tenere le fila*; Waldén, *Genom symaskinens nålsöga*, 85–86. Waldén refers to various reports from 1894, 1907, and 1917. For France and the U.S., Green, *Ready-to-Wear*.

34 Waldén, *Genom symaskinens nålsöga*, 78–80, 112–18.

35 Gurova, "The Life Span of Things in Soviet Society," 50.

36 Godley, "Selling the Sewing Machine," 278; Gordon, "Selling the American Way," 696–98.

37 Seligman, "Dressmakers' Patterns," 95–96.

38 "Our Address," *The Englishwoman's Domestic Magazine*, May 1852, 1, as quoted in ibid., 96.

39 Ibid.

40 Emery, "Dreams on Paper," 236, 244; Seligman, "Dressmakers' Patterns," 97–80; Hoganson, *Consumers' Imperium*, 63; Walsh, "The Democratization of Fashion." See also: Ruane, *The Empire's New Clothes*, 103–08. Pattern Collection, depot mode, iconografie at the Gemeente museum, The Hague, the Netherlands and London College of Fashion, Paper Pattern Collection, London, U.K. Ghering-van Ierlant, *Mode in Prent, 1550–1914*, 92–111.

41 Emery, "Dreams on Paper," 236; Gamber, "'Reduced to Science'"; Walsh, "The Democratization of Fashion."

42 Emery, "Dreams on Paper," 244.

43 Gamber, "'Reduced to Science'"; Walsh, "The Democratization of Fashion," 770 and fn. 34; Ruane, "Clothes Shopping in Imperial Russia"; Ruane, *The Empire's New Clothes*, chapter 4 and 103–06. Burman, "Home Sewing and 'Fashions for All,' 1908–1937"; Hoganson, *Consumers' Imperium*, 62–64, 75–80.

44 Iulli L. Elets, *Poval'noe bezuie: K sverzheniiu iga mod* (St. Petersburg 1914), as quoted in Ruane, "Clothes Shopping in Imperial Russia," 771–73.

45 Berg and Eger, eds., *Luxury in the Eighteenth Century*; Brewer and Porter, eds., *Consumption and the World of Goods*.

46 Hoganson, *Consumers' Imperium*.

47 Sklar, "The Consumers' White Label Campaign"; Chessel, "Consumers' Leagues in France"; Oldenziel and Bruhèze, "Theorizing the Mediation Junction."

48 Crane, *Fashion and its Social Agendas*, 101–05, 113.

49 Ibid., 114–15.

50 Gordon, "'Any Desired Length'."

51 Ibid., 36–39; Crane, *Fashion and its Social Agendas*, 113, 118. Despite the Parisian government's regulations for women policemen to wear trousers and a feminist movement in the 1960s to repeal it, the law remains on the books. Samuel, "Women Banned from Wearing Trousers." Between 1893 and 1906, the fashion magazine *Gracieuse* featured at least twenty different costumes.

52 Schweitzer, "The Rise and Fall of Fashion Nationalism." Helene Tuchak, "Los van Paris!," *Wiener Mode* 28 (October 1, 1914): 1, as quoted in Hess, "The Wiener Werkstätte and the Reform Impulse," 126.

53 Troy, *Couture Culture*, Chapter 4 and 313–16; Gordon, "'Any Desired Length'," 45.

54 Lewis, "Cosmopolitanism and the Modern Girl"; Barlow et al., "Modern Girl Around the World"; Gordon, "Selling the American Way," 696–98.

55 The Paris fashion industry tried to strengthen its position in Europe, developing a complex net of commercial intermediaries, often in other countries, such as the Chambre Syndicale de la Haute Couture Belge. Pouillard, "In the Shadow of Paris?"; Merlo, *Moda italiana*; Gnoli, *Un secolo di moda italiana*; Steele and Fashion Institute of Technology Museum, *Fashion, Italian Style*. Hoganson, *Consumers' Imperium*; Troy, "The Theatre of Fashion." Gemeentemuseum, *Voici Paris!*, 26–27.

2   Creating European Comfort

1   Chekhov, "Peasants," 30.

2   Ibid., 52.

3   Ibid., 47.

4   Edelman, "Everybody's Got to Be Some Place."

5   Ibid., 13.

6   Ibid., 15.

7   Ibid., 13. Oldenziel et al., *Huishoudtechnologie*, 17–22.

8   Tolstoy, *Anna Karenina*, 170.

9   Ibid., 147, 142.

10 Ibid., 158.

11 Gläntzer, "Nord-Süd-Unterschiede städtischen Wohnens"; Kloek, *Vrouw des huizes*, chapter 3; De Vries, *The Industrious Revolution*, chapter 5.

12 Letter by Wilhelmine Heyne-Heeren, May 30, 1796, as quoted in Panke-Kochinke, *Göttinger Professorenfamilien*, 90; Von Saldern, "Im Hause, zu Hause," 154.

13 Marcus, *Apartment Stories*, 21.

14 Löfgren, "The Sweetness of Home"; May, "Die Villa als Wohnkultur"; Burnett, *A Social History of Housing*, 190.

15 Friedrich von Buchwald, *Oekonomische und Statistische Reise durch Meklenburg, Pommern, Brandenburg und Holstein* (Copenhagen 1786), 5, as quoted in Gläntzer, "Nord-Süd-Unterschiede städtischen Wohnens," 80; Moser, "Nord-Süd-Vergleich in Hausbau."

16 Marcus, *Apartment Stories*, 94; Ames, "Meaning in Artifacts." Anthropologist Pfaffenberger later picked up Ames's analysis: Pfaffenberger, "Technological Dramas."

17 Clarence Cook, *The House Beautiful* (New York 1877), as quoted in Ames, "Meaning in Artifacts," 42.

18 Pfaffenberger, "Technological Dramas"; Ames, "Meaning in Artifacts," 40.

19 Brewer, *From Fireplace to Cookstove*, 18; Moser, "Nord-Süd-Vergleich in Hausbau," 70.

20 Brewer, *From Fireplace to Cookstove*, 21.

21 Ibid., 24. Gloag, *Victorian Comfort*, 33; Cowan, *More Work for Mother*, 54–55; Albert S. Bolles, *Industrial History of the United States* (1879), as quoted in Cowan, *More Work for Mother*, 56.

22 Robert Kerr, *The Gentleman's House. Or, How to Plan English Residences from the Parsonage to the Palace* (1864), as quoted in Gloag, *Victorian Comfort*, 99.

23 Advertisement (1890s) by Oetzmann & Company, London, as quoted in ibid., 73.

24 Forty, *Objects of Desire*, 99–104. Heynen, "Modernity and Domesticity," 6.

25 Gyáni, *Parlor and Kitchen*, 69; Buzinkay, "Das Wohnideal des Mittelstandes," 30; Giedion, *Mechanization Takes Command*, 364. The author of the German Room was Georg Hirth and the original title *Das deutsche Zimmer der Renaissance* (1881). See: Sármány-Parsons, "Villa und Einfamilienhaus," 252.

26 Giedion, *Mechanization Takes Command*, 367–68; Gloag, *Victorian Comfort*, 42; Gyáni, *Parlor and Kitchen*, 91. For the U.S., cf. Hoganson, *Consumers' Imperium*, 16, 20–23, 37, 255.

27 Gyáni, *Parlor and Kitchen*, 90–92; Veblen, *Conspicuous Consumption*.

28 From an anonymous pamphlet, *A budapesti társaság* (Budapest 1886), 548, as quoted in Gyáni, *Parlor and Kitchen*, 103.

29 Ibid., 103.

30 Elias, *The Civilizing Process*. See, for example: Frykman and Löfgren, *Culture Builders*.

31 Anonymous contribution (italics in the original) to *Svenska husmodern* (1877), 3, as quoted in Berner, "The Meaning of Cleaning," 337.

32 Military metaphors were often used in contemporary manuals: Breuss, "Die Stadt, der Staub und die Hausfrau," 369. Tomes, *The Gospel of Germs*.

33 Mathilda Langlet, *Ett eget hem* (1891), 108, as quoted in Berner, "The Meaning of Cleaning," 324.

34 Manó Somogyi, "The Situation of the Óbuda Shipyard Workers in 1900" (ca. 1900), as quoted in Gyáni, *Parlor and Kitchen*, 176.

35 Ibid., 176.

36 Frey, *Der reinliche Bürger*.

37 Berg, *The Machinery Question*; Chadwick, *Sanitary Condition of the Labouring Population*; Engels, *The Condition of the Working Class*.

38 Engels, *The Condition of the Working Class*, 98.

39 Chadwick, *Sanitary Condition of the Labouring Population*, 422.

40 Hamlin, "Edwin Chadwick and the Engineers"; Hardy, *Ärzte, Ingenieure und städtische Gesundheit*; Evans, *Death in Hamburg*, 152.

41 Buiter, "Constructing Dutch Streets"; Hardy, *Ärzte, Ingenieure und städtische Gesundheit*, 170–82.

42 Hardy, *Ärzte, Ingenieure und städtische Gesundheit*, 420–46.

43 Rodenstein, *Mehr Licht, mehr Luft*.

44 Hyldtoft, "Arbeiterwohnungen in Kopenhagen," 208.

45 Barnes, *The Great Stink of Paris*, 49–51; Hoy, *Chasing Dirt*, 80–81.

46 Dinçkal, "Arenas of Experimentation."

47 Dinçkal, "Reluctant Modernization."

48 As quoted in Otter, *The Victorian Eye*, 103–105.

49 Frevert, "Fürsorgliche Belagerung".

50 Rodríguez-Lores, *Sozialer Wohnungsbau in Europa*, 203.

51 Power, *Hovels to High Rise*, 34; Rodríguez-Lores, *Sozialer Wohnungsbau in Europa*, 40–41, 51. Von Saldern, *Häuserleben*, 56; Berghout, ed. *Wij wonen*; Huisman et al., *Honderd jaar wonen*.

52 Rodríguez-Lores, *Sozialer Wohnungsbau in Europa*, 60–202; Vossen, *Bukarest*, 189; Piccinato, "Zum italienischen Volkswohnungsgesetz"; Guerrand, *Une Europe en construction*, 156–57.

53 The original is a text written by an anonymous local resident—a private source given by Anna M. Pock, Vienna to Alfred Georg Frei, as quoted in Frei, *Rotes Wien*, 94, 157, see also 84–88; Stieber, *Housing Design and Society in Amsterdam*, 109–13; Von Saldern, *Häuserleben*, 137.

54 *De Dageraad*, Board Minutes (1921), as quoted in Bervoets, "Mediating in the Name of Working-Class Collectivism," 109.

55 J. van de Berg, as quoted in ibid., 105.

56 Rodgers, *Atlantic Crossings*, 441–43.

57 Holm, *HSB*.

58 Nordström, *Lort-Sverige*, 372; Sörlin, "Utopin i verkligheten."

59 Nordström, *Lort-Sverige*, 21, 67–68; 337–40; Frykman, *Modärna tider*.

60 Nordström, *Lort-Sverige*, 230; Wright, *Home Fires Burning*, 195; Oldenziel et al., *Huishoudtechnologie*, 16–23.

61 Myrdal and Myrdal, *Kris i befolkningsfrågan*. Cf. Etzemüller, "Die Romantik des Reißbretts." Gunnar Myrdal is probably best known to an international audience as the author of *An American Dilemma* (1944) and as the director of the United Nations Economic Commission for Europe (1947–57). Alva Myrdal received the Nobel Prize for Peace in 1982. Hayden, *Grand Domestic Revolution*; Arons et al., *Kollektivering*; Oldenziel et al., *Huishoudtechnologie*, 37–40.

62 Myrdal, *Nation and Family*, 123; Bervoets, *Telt zij wel of telt zij niet*, chapter 1.

63 Hirdman, "Utopia in the Home"; Asplund, *acceptera*; Rudberg, *Stockholmsutställningen 1930*; Mumford, *The CIAM Discourse*; Heynen, *Architecture and Modernity*, chapter 2.

64 Mumford, *The CIAM Discourse*.

65 *Frankfurter Generalanzeiger* (August 18, 1928), as quoted in Hessler, "*Mrs. Modern Woman*," 271.

66 Statement in the local parliament, Files of "Stadtverordneten-Versammlung zu Frankfurt," Frankfurt City Archives, as quoted in ibid., 274.

3 Crossing Borders—in Style?

1 As quoted in Schivelbusch, *Railway Journey*, 88. The story and insight come from Schivelbusch's now classic book. Unless otherwise indicated, what follows comes from chapter 5 of this study.

2 *San Francisco Mail News*, January 15, 1886, 3. As late as 1893, the Poinsot murder was still part of a transnational collective memory. When the banker and former mayor of Palermo, Marquis Emanuele Notarbartolo di San Giovanni, was assassinated on a train in Sicily, contemporaries believed it was the Mafia's copycat of Poinsot's murder. Lupo, *Storia della mafia*, 121–22.

3 Emile Zola, *La bête humaine* (1890) see also for the connection: introduction to Zola, *The Beast Within*. For a sample from the international press, see: *Journal des Instituteurs* 3, 50 (December 9, 1860): 317–18; "Chronicle December"; "Laatste berigten. Paris 6 december." "Strange Murder in Paris"; "Der Mord auf der französische Ostbahn." See also: Gecser and Kitzinger, "Fairy Sales."

4 Judt, "The Glory of the Rails."

5 "Horrible Murder in Railway Carriage."

6 Schivelbusch, *Railway Journey*, 84–86. On engineering solutions, see: Dapples, *Le matériel roulant des chemins de fer*. For example, platform standardization so that passengers juggling bags could get comfortably on and off the train

or a seat designed to place hand luggage underneath. On this point, see especially: "Security for Railway Travellers." On the public responses to the Briggs murder see: Fox, "Murder in Daily Installments."

7   Lewis Morris Iddings, "The Art of Travel," *Scribner's Magazine* (March 1897), 351–67, here 353, as quoted in Richter, *Home on the Rails*, 81. Perl, "*Transports des voyageurs et de leurs bagages*," 1544. Dapples, *Le matériel roulant des chemins de fer*; *Scientific American* (October 2, 1852), as quoted in White Jr., *The American Railroad Passenger Car*, 203.

8   Welke, *Gender, Race, Law, and the Railroad Revolution*, chapter 7. Richter, *Home on the Rails*, chapter 4. Although the Anglo-Saxon common law of common carriers ruled that privately owned railroads just like coaches, telegraph operators, and inns had to provide their services without discriminating. Neither state agencies nor railroad companies put this into practice. One angry male traveler described with disdain "the woman who sails through the crowded car, and brings to beside you like a monument, looks at you as if you had no business to be born without her consent, and says in a clear, incisive voice, that cuts through you like a knife, '"I know a gentleman when I see him'." Benjamin F. Taylor, *The World on Wheels and other Sketches* (Chicago: S.C. Griggs, 1874) 54–55, as quoted in Richter, *Home on the Rails*, 106–11.

9   Schivelbusch, *Railway Journey*, chapter 5; Dapples, *Le matériel roulant des chemins de fer*.In the United States in 1880, there were 130 parlor cars; in 1890, 370; in 1900, 500; and by 1920, 1,570, but they never represented more than 2.5 percent of the total. White Jr., *The American Railroad Passenger Car*, chapters 3, 4 and 287, 658.

10 Bonzon, "Convention Internationale sur le transport des voyageurs," 4, 22.

11 Ibid., par. 6. Wedgwood and Wheeler, *International Rail Transport*, 3. Buiter and Anastasiadou, "Regulations and flows in transnational rail traffic." On this point, we disagree with the authors, who take on the railroad's perspective.

12 Leeuw, "Le transport international," 32 ; Bonzon, "Convention Internationale sur le transport des voyageurs," 49, 100; Esbester, "Design and Use of Nineteenth-Century Transport Timetables."

13 Leeuw, "Le transport international."

14 From an advertisement of the company Moritz Mädler in Leipzig (1906), as quoted in Mihm, *Eine Kulturgeschichte des Reisekoffers*, 43, 50–53. On Hare, see: Mullen and Munson, *The Smell of the Continent*, 205.

15 Leeuw, "Le transport international," 92–99.

16 Ibid., 5, 22–25. On international protection: Bonzon, "Convention Internationale sur le transport des voyageurs," par. 10; Wessely, "Travelling People, Travelling Objects." The plight of passengers is belittled by Wedgwood and Wheeler, *International Rail Transport*, 4. See also: Hammer, "A Gasoline Scented Sindbad."

17 Wedgwood and Wheeler, *International Rail Transport*, 29–31; Leeuw, "Le transport international," 104–7; Bonzon, "Convention Internationale sur le transport des voyageurs," 156–57.

18 Ackermann, *Répertoire de jurisprudence*, 13. For English and American jurisprudence, see: MacNamara, *A Digest of the Law of Carriers*.

19 Maritime and airline passengers' rights were governed respectively by the Brussels Convention and the Warsaw Convention. While the Berne International Convention on the carriage of passengers and luggage by rail (CIV) came into force in 1928, it was already nulled by 1938 (convention of Rome).

20 Armstrong and Williams, "The Steamboat and Popular Tourism."

21 Simmons, "Railways, Hotels, and Tourism," 216; Schueler, *Materialising Identity*; Mullen and Munson, *The Smell of the Continent*, 19; Ring, *How the English Made the Alps*; Heafford, "Between Grand Tour and Tourism"; Barker, "Traditional Landscape and Mass Tourism in the Alps"; Hazbun, "East as Exhibit."

22 Cepl-Kaufmann and Johanning, *Mythos Rhein*, 117; Disco, "Taming the Rhine."

23 Taylor, "From Trips to Modernity to Holidays in Nostalgia," 6.

24 Barton, *Working-Class Organisations*, chapter 3.

25 Ibid., chapter 3.

26 Quoted in Hamilton, *Thomas Cook*, 129.

27 Quoted in Pudney, *The Thomas Cook Story*, 149.

28 *Illustrated London News*, September 25, 1880, as quoted in Mullen and Munson, *The Smell of the Continent*, 35.

29 Brendon, *Thomas Cook*. Hazbun, "East as Exhibit."

30 Richter, *Home on the Rails*, 9–13. Simmons, "Railways, Hotels, and Tourism."

31 Behrend, *The History of Wagons-Lits*, 12; Sherwood, *Venice Simplon Orient-Express*, 23–25. The reinforcement of national difference by the tourist experience is the subject of vast literature, for example see: MacCannell, *The Tourist*; Urry, *The Tourist Gaze*. For the belief in the capacity of infrastructures to bring international understanding, see: Kaiser and Schot, *Writing the Rules for Europe*.

32 On the different geo-political roles of railroads, see: Divall, "Railway Imperialisms." On *Wagons-Lits*, see: Mühl and Klein, *Die Luxuszüge, Geschichte und Plakate*, 279–81.

33 Mühl and Klein, *Die Luxuszüge, Geschichte und Plakate*, 280–84. Behrend, *The History of Wagons-Lits*, 7. More generally on the railroads' role in governance, see: Kaiser and Schot, *Writing the Rules for Europe*.

34 Behrend, *The History of Wagons-Lits*, 10; Mühl and Klein, *Die Luxuszüge, Geschichte und Plakate*, 33, 41, 53, 61.

35 Hunter, "Cook Enterprise on the Nile," 41; Hazbun, "East as Exhibit." On the power of fictive travel movement, the rise of tourist mentality, and the imperial ways of seeing, see: Pratt, *Imperial Eyes*; Hoganson, *Consumers' Imperium*, chapter 4. Package tours may be analyzed by means of the concepts of "the smooth and the striated," as developed by Deleuze and Guattari, *A Thousand Plateaus*.

36 Behrend, *The History of Wagons-Lits*; Mühl and Klein, *Die Luxuszüge, Geschichte und Plakate*; Torpey, *The Invention of the Passport*.

37 See also: Forêt, "Railroad Literature." On the contrasting consumer cities of (Western) Shanghai vs. (Chinese) Suzhou, see: Finnane, "Yangzhou's 'Mondernity'." On terminus hotels: Mühl and Klein, *Die Luxuszüge, Geschichte und Plakate*, 214–35, 293–95. Behrend, *The History of Wagons-Lits*, 12.

38 Behrend, *The History of Wagons-Lits*.

39 On East European intellectuals' travel experiences, see: Bracewell, "Travels Through the Slav World," 162; Parusheva, "Orient Express and Everyday Life in the Balkans"; Schivelbusch, "Railroad Space and Railroad Time." See also: Stilgoe, *Metropolitan Corridor*. On health dangers, see: Argenbright, "Lethal Mobilities"; Huber, "International Sanitary Conferences on Cholera"; Essner, "Cholera der Mekkapilger und internationale Sanitätspolitik in Ägypten."

40 Nikles, *Soziale Hilfe am Bahnhof*; Kirchhof, *Das Dienstfraulein auf dem Bahnhof*; Gerodetti and Bieri, "Train Stations as Gateways." See also: http://www. aletta.nu/aletta/nl/vitrine/stationswerk, accessed September 15, 2012.

41 Brinkmann, "The Impact of American Immigration Policies."

42 Ibid. Weindling, *Epidemics and Genocide in Eastern Europe*, 65; Steiner, *On the Trail of the Immigrant*.

43 Brinkmann, "The Impact of American Immigration Policies"; Torpey's important study does not focus on the opposite movement of the growing passportization and surveillance for immigrants, laborers, and undesirables, Torpey, *The Invention of the Passport*. Cf. Shearer, "Passportization in the Stalinist State, 1932–1952."

44 Weindling, *Epidemics and Genocide in Eastern Europe*; Brinkmann, "From Oświęcim to Ellis Island."

45 Weindling, *Epidemics and Genocide in Eastern Europe*, 56–70; Zevenbergen, *Toen zij uit Rotterdam vertrokken*, 98–105; Brinkmann, "The Impact of American Immigration Policies."

46 Weindling, *Epidemics and Genocide in Eastern Europe*, 60.

47 Brinkmann, "The Impact of American Immigration Policies"; Zevenbergen, *Toen zij uit Rotterdam vertrokken*; Weindling, *Epidemics and Genocide in Eastern Europe*, 62.

48 Weindling, *Epidemics and Genocide in Eastern Europe*, 56–61; Brinkmann, "The Impact of American Immigration Policies," 478; Brinkmann, "From Oświęcim to Ellis Island"; Fairchild, *Science at the Borders*.

49 Ibid.

50 Fairchild, *Science at the Borders*.

51 Brinkmann, "The Impact of American Immigration Policies," 472.

52 Zevenbergen, *Toen zij uit Rotterdam vertrokken*. Brinkmann, "From Oświęcim to Ellis Island."

53 Weindling, *Epidemics and Genocide in Eastern Europe*, 65. See also: Ballin-Stadt Emigration Museum Hamburg. Lubbers, *Lloydhotel*, 1–69. See also: http:// playproxyemu.vpcore.snkn.nl/play_proxy/mmc/37113/AVN_14.wmv, accessed January 15, 2010. We thank Suzanne Oxenaar of the Lloyd Hotel for her generosity in sharing her archives and in helping trace the steps of emigrants in this building.

54 Steiner, *On the Trail of the Immigrant*, quote on 41 and chapter 3. A U.S. government study confirmed these conditions: Government, *Abstracts of Reports of the Immigration Commission*, 296–303. Zevenbergen, *Toen zij uit Rotterdam vertrokken*, 49.

55 Steiner, *On the Trail of the Immigrant*, 72.

56 Brinkmann, "The Impact of American Immigration Policies," 478–82.

57 Garton Ash, "Mitteleuropa?"; Van Laak, *Imperiale Infrastrukturen*, section 3; Headrick, *The Tools of Empire*; Divall, "Railway Imperialisms."

58 Macmillan, *Paris 1919*; Behrend, *The History of Wagons-Lits*, 12; Sherwood, *Venice Simplon Orient-Express*, chapter 1 and 30–32.

59 Anastasiadou, *Constructing Iron Europe*, chapter 2; Buiter and Anastasiadou, "Regulations and Flows in Transnational Rail Traffic," 9.

60 Brendon, *Thomas Cook*, 258; Behrend, *The History of Wagons-Lits*, 14; Lloyd, *Battlefield Tourism*. We thank Alec Badenoch for the insight on war tourism. Garton Ash, "Mitteleuropa?"; Köhler, *Die Geschichte der MITROPA*; Behrend,

The History of Wagons-Lits, 12; Sherwood, *Venice Simplon Orient-Express*, chapters 1–2.

61 Behrend, *The History of Wagons-Lits*, 14–15; Köhler, *Die Geschichte der MITROPA*, 20.

62 Wilkerson, *The Warmth of Other Suns*.

4 Bicycling & Driving Europe

1 Helge Berglund, "Vilka praktiska lärdomar kan vi i Sverige draga ur stads- och trafikplaneringen i U.S.A.?" in *Bilstaden* (Stockholm: Royal Institute of Technology 1960), 9–16, as quoted in Emanuel, *Trafikslag på undantag*, 274; Sven Lundberg, "Att lära om trafik I USA," *Kommunalteknisk tidskrift* 22 (1956): 145–54, as quoted in Lundin, *Bilsamhället*, 29; Sven Tynelius, "Amerikanska parkeringsproblem," *Kommunalteknisk tidskrift* 22 (1956): 175–80 (italics in the original), as quoted in Lundin, *Bilsamhället*, 30; Horn, "Zur Geschichte der Städlichen Radverkehrsplanung," 121–26, 134–70. Lundin, "Mediators of Modernity."

2 Elwell, *Cycling in Europe*, 3.

3 A.M. Thompson, "King of the Road," *Clarion Press* (October 1897), as quoted in Rubinstein, "Cycling in the 1890s," 51.

4 *Irish Society*, May 15, 1897, as quoted in Griffin, "Cycling and Gender in Victorian Ireland," 236.

5 Thompson, "Bicycling, Class, and Politics in France," 133–34.

6 *Bulletin PVU* (July 1907): 1. We thank João Machado for helping to locate sources on the Portuguese history of cycling for this chapter.

7 Garrison, *Wheeling through Europe*, 9, 11; Burri, "Das Fahrrad."

8 Pennell, *Over the Alps on a Bicycle*; Pennell and Pennell, "Over the Alps on a Bicycle," 849; Herlihy, *The Lost Cyclist*; Thompson, "Bicycling, Class, and Politics in France."

9 Pinch and Bijker, "The Social Construction of Facts and Artifacts."

10 Herlihy, *Bicycle*, 210–16; Thompson, "Bicycling, Class, and Politics in France"; "Gossip of the Cyclers. Royal Personages Who Ride Wheels"; Bijker, *Of Bicycles, Bakelites, and Bulbs*, 92.

11 Hounshell, *From the American System to Mass Production*.

12 Concerning the concept "tourist gaze," see: Urry, *The Tourist Gaze*. See also: Sachs, *For Love of the Automobile*.

13 For more details on the relationship between innovations and governance, see: Kaiser and Schot, *Writing the Rules for Europe*.

14 See, for example, works focusing on the European history of consumption and transportation, like the special issue *Fra Frihed til Fritid – Det europæiske fritidsmenneske mellem forbrug og fornøjelse* of the Danish journal *Den jyske historiker*, nos. 127–28 (2012). Characteristically, the leading *Journal of Transport History* devoted a special issue to the history of cycling only in 2012.

15 Fraser, "Cycling over the Caucasus Mountains"; Fraser, *Round the World*, 46. The book was a bestseller and was immediately translated into Swedish, for example.

16 Elwell, *The Elwell European Bicycle Tours*; Stevens, *Around the World on a Bicycle I*; Stevens, *Around the World on a Bicycle II*; Jefferson, *Across Siberia on a Bicycle*.

For a detailed, well-researched story on one of many round-the-world cyclists, see: Herlihy, *The Lost Cyclist*.

17 "Gossip of the Cyclers. Wheeling About Cairo."

18 Schueler, *Materialising Identity*. On the tropes of anti-tourism travels, see: Pennell, "The Most Picturesque Place"; Buzard, *The Beaten Track*, chapter 2; Emanuel, "Europe as Perceived from the Bicycle Saddle"; Bosch, "Parallelle levens op de fiets"; Rubinstein, "Cycling in the 1890s," 60–61.

19 "La fiesta del pedal." See also: Plans Bosch, "Bodas de oro de la fiesta del Pedal." We thank Santiago Gorostiza for directing our attention to this fascinating, but forgotten history of Barcelona cycling traditions. See, for other examples, the collection on www.Europeana.eu and Bertho Lavenir, *La roue et le stylo*, chapter 6.

20 Quote from *Bulletin PVU* (July 1907): 1; *O Sport* (January 21, 1904), as quoted in Costa, "O Desporto e a Sociedade," 70.

21 *Le vélo* (January 10, 1895), as quoted in Holt, "The Bicycle and Discovery of Rural France," 132.

22 Tobin, "The Bicycle Boom of the 1890's," 842; Holt, "The Bicycle and Discovery of Rural France," 131; Rubinstein, "Cycling in the 1890s," 55; Herlihy, *The Lost Cyclist*.

23 Helphand, "The Bicycle Kodak," 30.

24 Ibid. Holt, "The Bicycle and Discovery of Rural France," 130–35. For a contemporary observation: Pennell and Pennell, "Over the Alps on a Bicycle." For the best analysis on the close relationship between bicycle tourism, the camera, and writing, see: Bertho Lavenir, *La roue et le stylo*, chapters 4–6.

25 Henry, "Technical Trials."

26 Burr, "National Cycle Organizations." Stevens, *Around the World on a Bicycle I*; Michel, "Les cyclistes en Algérie"; Mason, "Good-Roads Movement"; Milroy, "Rough Road Cyclists Display Political Power"; Veraart, "Fietspaden," 22; Briese, "Bicycle Path Construction in Germany." Bosworth, "Touring Club Italiano."

27 Because railroad companies let passengers bring their bicycles for free, the French lobby efforts concentrated only on careful handling rather than on reduced rates, while German and British cyclists bitterly complained over their railroads' lack of liability for damaged bikes. "Rijwielmishandeling op Duitsche spoorwegen." Burr, "National Cycle Organizations"; Pennell, "How to Cycle in Europe"; Elwell, *Cycling in Europe*.

28 "Verband- en reparatiekisten." *La boîte de secours* or *hulpdoos* needs a serious historian, but see: Zutphen, "Sociale geschiedenis van het fietsen te Leuven," 218–20. "Sjonge, sjonge"; "De Rijwiel- en Automobieltentoonstelling."; "De Tentoonstelling te Leeuwarden."

29 Rubinstein, "Cycling in the 1890s," 50; Bertho Lavenir, *La roue et le stylo*, 106; B.H., "Het Internationaal Touristen-Congres," 810 ; "L'algemeene Nederlandsche Wielrijdersbond." Bruhèze and Veraart, *Fietsverkeer*, 44; Tobin, "The Bicycle Boom of the 1890's"; Burr, "National Cycle Organizations," 37–39; Veraart, "Fietspaden." Cf. Schipper credits motorist clubs for these inventions. Schipper, *Driving Europe*.

30 Redationeel, "Waarom?," 758.

31 "Wet op de veiligheid van het verkeer."

32 Fraser, *Round the World*, 12.

33 "Het Congres van den Wereldbond op 15 en 16 augustus 1902," 67; Redation-eel, "Waarom?"; B.H., "Het Internationaal Touristen-Congres"; J.C.R[edele], "Rondom het Internationaal Congrès," 550; Ewart, "Agenda no. 100–104 August 1897 International Touring Congress."

34 Cyclists' Touring Club, "Report of the Foreign Touring Arrangements Committee"; Schipper, *Driving Europe*, chapter 2.

35 Bertz, *Philosophie des Fahrrads*, 10.

36 As quoted in Griffin, "Cycling and Gender in Victorian Ireland," 239.

37 Thompson, "Bicycling, Class, and Politics in France"; Bertho Lavenir, *La roue et le stylo*, chapter 4; Pivato, "The Bicycle as a Political Symbol"; Bosworth, "Touring Club Italiano"; Ebert, "Cycling towards the Nation"; Mom, Staal, and Schot, "Civilizing Moterized Adventure"; Taylor, "Cycling and Politics in the 1890s." Links to British liberal capitalism are notable: two founders of the Dutch cycling association were British expats; the first cycling newspaper in Berlin was started by a Brit: Ebert, "Cycling towards the Nation," 350; Italian founder Johnson was allied to British banking interests through his father's family connections. Bosworth, "Touring Club Italiano," 374. The Algerian club in the French colony of Oran was also established by a British citizen M. Coll, accessed January 15, 2011, http://ancien.oraniecycliste.net/1885velocecluboran.html and http://ancien.oraniecycliste.net/1894uva.html.

38 On prohibition of Jews in Germany, see: "Sport en politiek." On Zola, see: "Buitenlandsch nieuws" and "A Week's Paris Gossip." On women: Bleck-mann, *Die Anfänge des Frauenradfahrens in Deutschland*; Mackintosh and Norc-liffe, "Men, Women and the Bicycle"; Männistö-Funk, "Gendered Practices in Finnish Cycling"; Simpson, "New Zealand Nineteenth-Century 'New Women'"; Thompson, "Bicycling, Class, and Politics in France"; Steendijk-Kuypers, "Freedom on the Bicycle"; Griffin, "Cycling and Gender in Victorian Ireland"; Lapinskiene, "Discourses on the Cycling of Woman." On national-ism and internationalism, see: Ebert, "Cycling towards the Nation."

39 Touring Club Italia, *Attraverso l'Italia: illustrazioine delle regioni italiane*, Vol. 1 *Piemonte* (Milan 1930), as quoted in Bosworth, "Touring Club Italiano," 383.

40 *Rivista mensile del Touring* 21 (January 1915), as quoted in ibid., 388.

41 On nationalism and bicyclism: Thompson, "Bicycling, Class, and Politics in France," 137; Holt, "The Bicycle and Discovery of Rural France"; Ebert, *Radelnde Nationen*.

42 B.H., "Het Internationaal Touristen-Congres"; Thompson, "Bicycling, Class, and Politics in France," 136–37; Rubinstein, "Cycling in the 1890s," 50; Holt, "The Bicycle and Discovery of Rural France," 128; Bertho Lavenir, *La roue et le stylo*, 90, 106.

43 Bijker, *Of Bicycles, Bakelites, and Bulbs*, 92. See for contemporary assessments, for example: Redationeel, "Rijwielen en automobielen op de Wereldtentoon-stelling te Parijs," 690.

44 The most comprehensive study on urban cycling is: Bruhèze and Veraart, *Fietsverkeer*. See also: *Stadt- und Landesplanung Bremen, 1926–1930*, 347–48; Zutphen, "Sociale geschiedenis van het fietsen te Leuven," 107–15; Caracciolo, "Bicicleta, circulación vial." Growing number of studies present the bicycle as the first vehicle of individual mass transport towards the "ride to modernity," see: Stoffers, Oosterhuis, and Cox, "Bicycle History as Transport History."

How the automobile became the dominant mode of transportation entrenched in a political-industrial-technical-cultural complex, see: Patterson, *Automobile Politics*. On utility bicycles for commuting, see: Veraart, "Geschiedenis van de fiets"; Taylor, "Cycling and Politics in the 1890s." On working-class and rural leisure cycling, see: Holt, "The Bicycle and Discovery of Rural France," 131–37; Männistö-Funk, "Vernacular Bicycles in Rural Finland."After 1922, cheap and stripped German bicycles flooded the Dutch market. Veraart, "Geschiedenis van de fiets," 83–84.

45 *Travel* (June 1934), as quoted in Herlihy, *Bicycle*, 328. In the 1930s British parliamentary debates, government officials claimed at least 9,000,000 cyclists. We thank Peter Cox for the reference.

46 The most comprehensive study remains Bruhèze and Veraart, *Fietsverkeer*. See also: Hodges, "Did the Emergence of the Automobile End the Bicycle Boom?"; Schacht, *Der Radwegebau in Deutschland*, 8; Herlihy, *Bicycle*, 328; Ebert, "Cycling towards the Nation"; Männistö-Funk, "Gendered Practices in Finnish Cycling."

47 Bardou, *The Automobile Revolution*; Barker, *Spread of Motor Vehicles*; Georgano, *The American Automobile*; Gemeente Amsterdam, "Het Verkeer te Amsterdam," Bijlage A, 10.

48 Mom, "Roads without Rails"; Schipper, *Driving Europe*, 66–67.

49 Oldenziel and Bruhèze, "Contested Spaces"; Bruhèze and Veraart, "Fietsgebruik in negen West-Europese steden," 186–89; Bruhèze and Veraart, *Fietsverkeer*.

50 Caracciolo, "Bicicleta, circulación vial," 24–25, 30. Similar developments have been observed for other European countries by Bruhèze and Veraart, *Fietsverkeer*.

51 Magdeburger Verein für Radfahrwege e.V, as quoted in Briese, "Bicycle Path Construction in Germany," 125; Peters, *Time, Innovation and Mobilities*, 131. Kreuzer, "Historische Verkehrsutopien," 261–64, 280; Veraart, "Geschiedenis van de fiets," 88–90; Ebert, *Radelnde Nationen*, 378–90.

52 Briese, "Bicycle Path Construction in Germany," 124. German cyclists financed the first lanes in Bremen, Hamburg, Hannover, Lübeck, and Magdeburg. In Bremen, the two-way middle-of-the-road sections were paved with coal cinders or copper dross, but transferred to road sides in the 1910s.

53 Briese, "Radwegebau vor dem Zweiten Weltkrieg"; Horn, "Zur Geschichte der Städlichen Radverkehrsplanung," 79, 100; Möser, "Motorization of German Societies in East and West"; Zezina, "Motor Vehicles on a Mass Scale in the USSR," 43–48; Caracciolo, "Bicicleta, circulación vial," 26; Schipper, *Driving Europe*, 68–71.

54 Piero Gambarotta, "La velocità dell'automobile ha veramente bisogno di vittime?" *Avvenire d'Italia* [Bolona], September 13, 1928, as quoted in Caracciolo, "Bicicleta, circulación vial," 26.

55 "Il crepúsculo della bicicletta," *La bicicletta* (January 1928): 30–31, as quoted in ibid., 26.

56 Guido Giardini, "La Gioia della bicicletta," *Lo sport fascista* (November 1928): 50 and Mario Museila, "Significato dello sport," *Lo sport fascista* (January 1930): 8, as quoted in ibid., 28; Peters, *Time, Innovation and Mobilities*, 130–31. Kreuzer, "Historische Verkehrsutopien," 260–80. Bonham and Cox, "The Disruptive Traveller?," 44–45; Wagenaar, *Stedebouw en burgerlijke vrijheid*; Mom, "Roads without Rails." See also: Edgerton, *The Shock of the Old*.

57 "Staten-Generaal. Tweede Kamer. Vergadering van Donderdag 16 October…Tariefwet"; "Staten Generaal. Eerste Kamer. Vergadering 23 dec….Belastingonderwerpen." Reitsma, "De rijwielbelasting als bestemmingsheffing"; Grapperhaus, *Nederlandse rijwielbelasting*, 20.

58 Gulick Jr., "Vienna Taxes since 1918," 552.

59 Italo Bonardi, "Ciclo-moto-turismo. Le bachine ciclistiche," *Le vie d'Italia* (January 1938): 9–13, as quoted in Caracciolo, "Bicicleta, circulación vial," 30.

60 W. Banfield, Accidents (pedal cyclists), House of Commons debates February, 28 1935, vol. 298: 1281–2. We thank Peter Cox for this and the following reference.

61 Sir W. Sugden, Road Traffic Acts, House of Commons debates May 24, 1935, vol. 302: 703–39.

62 Pye, *The National Clarion Cycling Club*; Prynn, "Clarion Clubs"; Rabenstein, *Radsport und Gesellschaft*; Briese, "Radwegebau vor dem Zweiten Weltkrieg"; Briese, "Bicycle Path Construction in Germany," 125; Hoffmann, "Arbejderindrættens forhold"; Pivato, "The Bicycle as a Political Symbol."

63 *Regionplan för Stockholm med omnejd, huvudsakligen anseende förortsområdet* (Stockholm: Stockholmsförorternas regionplaneförbund, 1936), 149, as quoted in Emanuel, "Constructing the Cyclist," 71.

64 "Il Parère degli altri. Il traffico di via d'Azeglio," *Resto del Canino*, Bolonia, June 27, 1934, as quoted in Caracciolo, "Bicicleta, circulación vial," 32.

65 Oldenziel and Bruhèze, "Contested Spaces."

66 Nils Lidvall, "Cykeltrafiken," *Kommunalteknisk tidskrift* 1 (1935): 2–10 and "Vägtrafikstadgans svaga punkter," *Kommunalteknisk tidskrift* 1 (1937): 4–10, as quoted in Emanuel, "Constructing the Cyclist," 80.

67 C. D. [Cesare Dorici], "Disciplina stradale," *Il Politécnico* (September 1934): 317, as quoted by Caracciolo, "Bicicleta, circulación vial," 27; Bruhèze and Veraart, *Fietsverkeer*; Papon, "The Evolution of Bicycle Mobility in France."

68 As quoted from Caracciolo, "Bicicleta, circulación vial," 27. Bruhèze and Veraart, *Fietsverkeer*, 186–89; Papon, "The Evolution of Bicycle Mobility in France."

69 Bruhèze and Veraart, *Fietsverkeer*, 120; Caracciolo, "Bicicleta, circulación vial," 32; Briese, "Bicycle Path Construction in Germany," 124.

70 Quarta conferenza per l'unificazíone delle norme e dei segnali per la circulazione urbana. Ordini del giorno approvati della Conferenza, Genova June 26–29 (s.l., s.n.), as quoted in Caracciolo, "Bicicleta, circulación vial," 29.

71 Bruhèze and Veraart, *Fietsverkeer*, chapter 9.

72 National Cycling Archive, University of Warwick: CTC Minutes, National Committee on Cycling, August 28, 1938; *Memorandum Stating the National Committee's view on the T.A.C. Report*; and National Cycling Archive, "An Amazing Memorandum," *Bicycling News*, April 15, 1937. See also: Oakley, *Winged Wheel*. Bruhèze and Veraart, *Fietsverkeer*, 131–33.

73 Bruhèze and Veraart, *Fietsverkeer*, 133–35; Caracciolo, "Bicicleta, circulación vial," 29.

74 Oldenziel and Bruhèze, "Contested Spaces"; Veraart, "Geschiedenis van de fiets"; Briese, "Bicycle Path Construction in Germany," 123–28; Ebert, "Cycling towards the Nation"; Ebert, *Radelnde Nationen*, 128; Mom, Staal, and Schot, "Civilizing Moterized Adventure"; Mom and Filarski, *Van Transport naar Mobiliteit*. The great exception was interwar Sweden. In 1937, the Swedish

government decided to build cycle trails in the rural counties of Norrbotten and Västerbotten in the North, to create fire breaks and improve the lives of government officials, forestry managers, loggers, and workers, whose jobs were considered at the heart of the state's economy. By then, about 1,200 km of narrow and 112 km of wide cycle trails had been built in these sparsely populated provinces. The cycle trail system was a unique Swedish invention to integrate the northern area, where forestry and mining rapidly developed, with the south. Rautio and Östlund, "'Starvation Strings'." On bicycle taxes, see: Grapperhaus, *Nederlandse rijwielbelasting*, 34–38.

75 Henneking, *Der Radfahverkehr*; Schacht, *Der Radwegebau in Deutschland*; Horn, "Zur Geschichte der Städlichen Radverkehrsplanung," 57–63, 87–109; Briese, "Bicycle Path Construction in Germany," 126; Briese, "Radwegebau vor dem Zweiten Weltkrieg." Briese, "Opium für Radfahrer"; Caracciolo, "Bicicleta, circulación vial," 35.

76 We thank Martin Kohlraush for helping us to understand the significance of Amsterdam's urban plans within architectural circles and for urban planning experts. See also Trischler and Kohlrausch, *Building Europe on Expertise*; Herlihy, *Bicycle*, 328; Eesteren, *De functionele stad*. Rossem, *Algemeene Uitbreidingsplan*; Mumford, *The CIAM Discourse*; Somer, *The Functional City*, chapters 3 and 4; Mumford, *Defining Urban Design*, chapter 3; Buiter, "Constructing Dutch Streets," 158.

77 "Oltre 140.000 ciclisti ferraresi beneficeranno della abolizione della tassa," *Resto del Carlino*, Bolonia, December 4, 1938, as quoted in Caracciolo, "Bicicleta, circulación vial," 36.

78 Ibid., 38.

79 "Ciclo raduno del Sebino," *Giorrude Luce* C0166, Istituto Luce, July 31, 1941, as quoted in ibid., 38.

80 Oldenziel and Bruhèze, "Bicycle Taxes as Tools of Policy"; Bertho Lavenir, "Scarcity, Poverty, Exclusion."

81 Madras 1937 police report in 1937, as quoted in Arnold and DeWald, "Cycles of Empowerment?", 992.

82 *Report on the Administration of the Police of the Madras Presidency*, 1937, 58, as quoted in ibid., 992.

83 Ibid., 974. Lundin, "Mediators of Modernity." Buiter, *Riool, rails en asfalt*; Buiter, "Constructing Dutch Streets"; Misa, "Appropriating the International Style"; Schot and Lagendijk, "Technocratic Internationalism."

84 Uno Åhrén, "Förord," *Bilstaden* (1960), as quoted in Emanuel, "Constructing the Cyclist," 78. The author effectively uses the term "proactive city traffic regime" to describe the phenomenon in more detail. See also: Emanuel, *Trafikslag på undantag*.

85 We thank Santiago Gorostiza for pointing us to this great Spanish movie classic. On 1970s activism, see: Furness, *One Less Car*.

5 Eating around the Continent

1 *Die Neuen Zürcher Nachrichten*, April 3, 1948 as quoted in Brändli, *Der Supermarkt im Kopf*, 49.

2 Ibid., 14.

3 Elsa Gasser, "Europe's Mrs. Consumer takes to Self-Service," reprint from *Annual Boston Conference of Distribution: A National Forum for Problems of Distribution* (Boston: Greater Boston Chamber of Commerce, 1954), 26, as quoted in ibid., 197–99.

4 P. Bützenberger et al., "USA-Reise 1961," *Manuscript* (Zurich 1961), as quoted in ibid., 216.

5 Ibid., 108, 112. On similar Dutch study tour, see: Inklaar, *Van Amerika geleerd.* For study tours the area of traffic planning, see, for example, Lundin, "Mediators of Modernity."

6 Emil Matter et al., "Vier Wochen in den USA," *Manuscript* (Basel 1956), as quoted in Brändli, *Der Supermarkt im Kopf*, 215–16.

7 On European reception of American consumer goods, see: Pells, *Not Like US*;

8 Gasser, "Europe's Mrs. Consumer takes to Self-Service," as quoted in Brändli, *Der Supermarkt im Kopf*, 198. De Grazia, "Changing Consumption Regimes"; Scarpellini, "Shopping American-Style."

9 Sluijter, "Kijken is grijpen"; Du Gay, "Self-Service," 158.

10 Gerhard, *Selbstbedienung und Selbstwahl*, 17–18. We thank Simon Bihr for supplying us with this information.

11 Chatriot, Chessel, and Hilton, eds., *The Expert Consumer*, chapter 4.

12 Otterloo, "Fast food and Slow Food," 262; König, *Geschichte der Konsumgesellschaft*, 148–49; Wagner, *Konserven und Konservenindustrie*, 95–101; Spiekermann, "Twentieth-Century Product Innovations," 306.

13 Hård, *Machines are Frozen Spirit.*

14 Hierholzer, *Nahrung nach Norm*, 65; Spiekermann, "Twentieth-Century Product Innovations," 306–10; Wagner, *Konserven und Konservenindustrie*, 9.

15 König, *Geschichte der Konsumgesellschaft*, 140–41; Wiedemann, *Herrin im Hause*, 67; Hietala and Lepistő, "Food Preservation and Refrigeration in Finland"; Spiekermann, "Twentieth-Century Product Innovations," 311.

16 Bruegel, "Appertising and Food Preservation in Rural France."

17 Wiedemann, *Herrin im Hause*, 35, 68.

18 Commission des recherches collectives, "Enquête 3 sur l'alimentation populaire quotidienne en 1936," microfilm, Musée d'arts et traditions populaires, Paris, as quoted in Bruegel, "Appertising and Food Preservation in Rural France," 216.

19 Renée Raymond, *Les conserves à la maison* Vol 1 (Paris: Hachette, 1913), 6, as quoted in ibid., 209.

20 Madeleine Maraval, *Ma pratique des conserves de fruits et légumes* (Paris: Libraire de la maison rustique, 1911), as quoted in ibid., 213.

21 Petrick, "'Purity as Life'"; Cowan, "The 'Industrial Revolution' in the Home," 8; Raetsch, *Köstliche Küche*, 75; Spiekermann, "Twentieth-Century Product Innovations," 310.

22 Fenton, "The Scottish Women's Rural Institutes," 66. Oldenziel et al., *Huishoudtechnologie*, chapter 5.

23 Fenton, "The Scottish Women's Rural Institutes," 60; Atkins, "The Empire Marketing Board"; Trentmann, *Free Trade Nation*, 231.

24 Trentmann, *Free Trade Nation*, 228–40.

25 *Home and Politics*, May 25, 1925, as quoted in ibid., 232.

26 Atkins, "The Empire Marketing Board," 252–53.

27 Merki, *Zucker gegen Saccharin*, chapter 1; König, *Geschichte der Konsumgesellschaft*, 155–56; Burnett, "Time, Place and Content," 118.

28 Magdalen S. Reeves, *Round About a Pound a Week* (London: Bell, 1913), as quoted in Burnett, "Time, Place and Content," 121.

29 Hierholzer, *Nahrung nach Norm*, 46; Merki, *Zucker gegen Saccharin*, 38.

30 Teuteberg and Weigelmann, *Unsere tägliche Kost*, 148–49; Burnett, "Time, Place and Content," 118–9.

31 Kisbán, "Coffee in Hungary," 74; Tascon, "The Coffee Innovation Culture versus the Chocolate Tradition"; Rischbieter, "Globalisierungsprozesse vor Ort," 36–37; Burnett, "Time, Place and Content," 122.

32 Scholliers, "Chocolate in Belgium," 131–32; Moss and Badenoch, *Chocolate*, 34.

33 Chiapparino, "Milk Chocolate and the Swiss Chocolate," 332.

34 Ibid., 334.

35 Hans J. Berckum, *Die Bedeutung des Kakaos und der Kakaoprodukte vom Standpunkte der Volksernährung* (Bern: Drechsel, 1916), 21, as quoted in Rossfeld, *Schweizer Schokolade*, 103.

36 Ibid., 88.

37 Tanner, *Fabrikmahlzeit*, 109.

38 Rossfeld, *Schweizer Schokolade*, 103.

39 Ibid., 431–35. Moss and Badenoch, *Chocolate*, 79.

40 Rossfeld, *Schweizer Schokolade*, 439–40, 443, 447.

41 Orland, "Milky Ways"; Orland, "Turbo-Cows," 170; Teuteberg and Weigelmann, *Unsere tägliche Kost*, 164; Bringéus, *Man, Food, and Milieu*.

42 Teuteberg and Weigelmann, *Unsere tägliche Kost*, 177–8; Fenton, "Milk and Milk Products in Scotland," 92; Bruhèze and Otterloo, "The Milky Way," 254.

43 Fenton, "Milk and Milk Products in Scotland," 97; Kjærnes, "Milk in Norway."

44 From the magazine *Liv og Sundhet: Norsk blad for riktig levesett, legemets og sinnets helse* 3 (1936), as quoted in Lyngø, "Symbols in the Rhetoric on Diet and Health," 159.

45 *Liv og Sundhet: Norsk blad for riktig levesett, legemets og sinnets helse* 4 (1937), as quoted in ibid., 160–62, 167.

46 *Liv og Sundhet: Norsk blad for riktig levesett, legemets og sinnets helse* 3 (1936), as quoted in ibid., 163–64.

47 Furlough and Strikwerda, "Introduction." Christiansen, "Between Farmers and Workers."

48 For the following, we thank Slawomir Lotysz, University of Zielona Gora, for his research in relevant Polish sources.

49 Baranowski, *Polska Karczma*, 84. Żerkowski, "Spółdzielczość Spożywców," 13–16. Małolepszy, "Między starym a nowym brzegiem Świdra," 39. See also: *Polska Gospodarcza* 11, no. 29 (1930): 9; ibid., 15, no. 28 (1933): 5.

50 Żerkowski, "Spółdzielczość Spożywców"; "Zadania Polskiej Ligi Nabiałowej," 13–16; "Miejskie zakłady mleczarskie 'Agril.'"

51 Brzostek, *PRL na widelcu*, 268.

52 Metchnikoff, *The Prolongation of Life*, 90. Stoilova, "Bulgarian Yoghurt," chapter 1.

53 M. Chemin, as quoted in Metchnikoff, *The Prolongation of Life*, 90. Stoilova, "Bulgarian Yoghurt," chapter 1.

54 Metchnikoff, *The Prolongation of Life*, xvi.

55 Stoilova, "Bulgarian Yoghurt," chapter 2.

56 Spiekermann, "Twentieth-Century Product Innovations," 297–304; Stoilova, "Bulgarian Yoghurt," chapter 2.

57 Stoilova, "Bulgarian Yoghurt," chapter 2.

58 Ibid., chapter 2. Koese, "Nestlé in the Ottoman Empire."

59 Stoilova, "Bulgarian Yoghurt," chapter 3.

60 Ibid., chapter 4.

61 Ibid., chapter 5.

62 Ibid., chapter 5.

63 Zimmerman, *The Super Market*. Bihr, "The Super-Market Empire"; Bowlby, *Carried Away*; Zimmerman, *Los supermercados*; De Grazia, *Irresistible Empire*, 389.

64 Letter by R.W. Boogaart to W.B. Dixon Stoud, February 27, 1957, Rockefeller Family Archives, as quoted in Scarpellini, "Shopping American-Style," 635.

65 De Grazia, *Irresistible Empire*, 389; Scarpellini, "Shopping American-Style."

66 Hamilton, "Supermarket USA."

67 Ibid., 146, 153; Scarpellini, "Shopping American-Style."

68 Ibid., 147. Scarpellini, "Shopping American-Style."

6 Living in State-Sponsored Europe

1 Nolan, "Consuming America"; Holder, "The Nation State or the United States?"; Castillo, "The American 'Fat Kitchen'"; Zachmann, "Managing Choice." See also: Winner, *The Whale and the Reactor*.

2 For one of the first historical but modernist-inspired accounts on kitchens, see: Giedion, *Mechanization Takes Command*, pt. 6. See also: Marling, *As Seen on TV*, 262–63.

3 Carbone, "Staging the Kitchen Debate"; Harrison E. Salisbury, *The New York Times*, 25 July, 1959, 1 as quoted in Marling, *As Seen on TV*, 49; Castillo, "The American 'Fat Kitchen'."

4 Rosenberg, "Consuming Women"; Hilton, "Kitchen in a Global Context"; Jakabovics, "Displaying American Abundance Abroad"; Haddow, *Pavilions of Plenty*, chapters 6 and 8; Hixson, *Parting the Curtain*, chapter 6 and 7; Marling, *As Seen on TV*, 242–83; May, *Homeward Bound*, 18–19; De Grazia, *Irresistible Empire*, 454–56; Henthorn, *From Submarines to Suburbs*, 1–3; Nickles, "Preserving Women."

5 Annie M.G. Schmidt, "Keukens," in *In Holland staat mijn huis* (Amsterdam: Querido, 1955), 17, as quoted in Oldenziel, "Exporting the American Cold War Kitchen," 329.

6 Ibid.

7 Saarikangas, "What's New?"

8 Castillo, "The American 'Fat Kitchen'."

9 Hixson, *Parting the Curtain*, chapter 6; Haddow, *Pavilions of Plenty*.

10 Carbone, "Staging the Kitchen Debate"; Marling, *As Seen on TV*, chapter 7; Jakabovics, "Displaying American Abundance Abroad."

11 Reid, "'Our Kitchen Is Just as Good'"; Crew, ed., *Consuming Germany in the Cold War*; Gerchuk, "Everyday Life in the Khrushchev Thaw."

12 For example, see: Inklaar, *Van Amerika geleerd.*

13 De Grazia, *Irresistible Empire*, 345–50; Maier, "The Politics of Productivity"; Theien, "Shopping for the 'People's Home'."

14 Landström, "National Strategies"; Hellman, "The Other American Kitchen."

15 Hayden, *Grand Domestic Revolution*, 77–79; 183–88.

16 For example, see: Nickles, "Preserving Women"; Lintsen and Oldenziel, *Het technisch paradijs.*

17 Bellamy, "A Vital Domestic Problem."

18 Hayden, *Grand Domestic Revolution*, appendix Table A2, 352–55.

19 Atwater, "For the Homemaker," 420; Andrews, *Economics of the Household*, 284; Oldenziel et al., *Huishoudtechnologie*, 31–33.

20 Braun, *Frauenarbeit und Hauswirtschaft.*

21 Uhlig, *Kollektivmodell 'Einküchenhaus'*; Terlinden and von Oertzen, *Die Wohnungsfrage ist Frauensache*; Oldenziel et al., *Huishoudtechnologie*, 40–41. For the transnational and transnational exchange: Rodgers, *Atlantic Crossings.*

22 Lenin, "A Great Beginning," 429.

23 Lenin, "The Tasks of the Working Women's Movement."

24 Stieber, *Housing Design and Society in Amsterdam*, 142–44.

25 Sklar, *Catharine Beecher*; Hayden, *Grand Domestic Revolution*, 58–60.

26 Sklar, *Catharine Beecher*; Forty, *Objects of Desire*; Caigny, "Bouwen aan een nieuwe thuis," 193.

27 Schot and Bruhèze, "Mediated Design"; Oldenziel, Bruhèze, and de Wit, "Europe's Mediation Junction"; Oldenziel and Bruhèze, "Theorizing the Mediation Junction"; Stage and Vincenti, eds., *Rethinking Home Economics.*

28 Erna Meyer's book became an international bestseller that went through at least 27 German editions between 1926 and 1928. Meyer, *Der neue Haushalt.* It was translated into Dutch and Czech as respectively: Meyer, *De nieuwe huishouding*; Meyerová, *Moderní domácnost.* The publications of a French home economist also found their way into Germany: respectively, Bernège, *De la méthode ménagère*; Bernège, *Die methodische Führung des Haushaltes.* See also: Stage and Vincenti, eds., *Rethinking Home Economics*, 2, 7; Akcan, "Civilizing Housewives," 188–89; Jørgensen, "Tidens krav-Framveksten"; Oldenziel et al., *Huishoudtechnologie*, 17–19, 50–51, 69–72; Rodgers, *Atlantic Crossings*; Bender, *A Nation Among Nations*, chapter 5; Caigny, "Bouwen aan een nieuwe thuis," 94–98, 108, 35–41, 257; Hayden, *Grand Domestic Revolution*, 285; Bervoets, "Consultation Required!"; Rudolf, "Domestic Politics," 488; Saarikangas, "What's New?"; Hessler, "The Frankfurt Kitchen," 167; Nolan, *Visions of Modernity*; Betts and Eghigian, eds., *Pain and Prosperity.*

29 Pinsolle, "Gas Consumers' Strikes"; Thoms, "Changes in the Kitchen Range"; Oldenziel et al., *Huishoudtechnologie*, chapter 1.

30 Clenneding, "'Deft Fingers'"; Clenneding, "Gas and Water Feminism."

31 Oldenziel et al., *Huishoudtechnologie*; Goldstein, "From Service to Sales: Home Economics in Light and Power, 1920–1940."

32 Haslett, *The Electrical Handbook for Women*; Pursell, "Domesticating Modernity"; Oldenziel et al., *Huishoudtechnologie*, 73, 80–85; Robinson, "Safeguarded by Your Refrigerator"; Goldstein, "Part of the Package"; Blaszczyk, "Where

Mrs. Homemaker Is Never Forgotten." Cf. Bix, "Equipped for Life"; Kline, "Agents of Modernity."

33 Saarikangas, "What's New?"; Kramer in Oedekoven-Gerischer, *Frauen im Design*; Henderson, "A Revolution in the Woman's Sphere"; Hessler, "The Frankfurt Kitchen."

34 Ryan, "The Apartment Question," 179–80; Bervoets, "Consultation Required!"; Akcan, "Civilizing Housewives."

35 Saarikangas, "What's New?" 290–91; Oldenziel et al., *Huishoudtechnologie*, 64–66; Goldstein, "Educating Consumers, Representing Consumers."

36 Nolan, "'Housework Made Easy'"; Reagin, *Sweeping the German Nation*; Oldenziel et al., *Huishoudtechnologie*, chapter 4; Lövgren, "Hemarbete som politik"; Saarikangas, "What's New?" Cf. Theien, "Shopping for the People's Home," 145; Zachmann, "Managing Choice."

37 Caudenberg and Heynen, "The Rational Kitchen," 548; Caigny, "Bouwen aan een nieuwe thuis," 145–49; Holder, "The Nation State of the United States?" 250–53; Saarikangas, "What's New?"; Bervoets, "Consultation Required!"; Bijker and Bijsterveld, "Women Walking through Plans"; Rudolph, "Architects versus Housewives"; Zachmann, "Managing Choice"; Akcan, "Civilizing Housewives."

38 Caigny, "Bouwen aan een nieuwe thuis," 199.

39 Rudolph, "Architects versus Housewives," 549.

40 Architect Georges-Henri Pingusson, cited by Chombart de Lauwe, *Famille et habitation* Vol. 1 (Paris: C.N.R.S., 1959), 178, as quoted in ibid., 551.

41 Ibid., 555.

42 Ibid., 549.

43 L. Veillon-Duverneuil, "Les Cloisons de l'appartement de demain seront mobiles," *Combat*, February 22, 1960, 8, as quoted in ibid., 549.

44 André Wogenscky interviewed by Chombart de Lauwe, *Famille et habitation*, 191, as quoted in ibid., 557.

45 Ionel Schein, letter to Jeanne Picard, February 16, 1958, as quoted in ibid., 550.

46 Ibid., 556.

47 Bijker and Bijsterveld, "Women Walking through Plans," 498; Nolan, "Housework Made Easy," 563; Rudolph, "Architects versus Housewives"; Nickles, "More is Better"; Veenis, "Dromen van dingen."

48 Bervoets, "Consultation Required!"; Rudolph, "Architects versus Housewives"; Bijker and Bijsterveld, "Women Walking through Plans"; Caigny, "Bouwen aan een nieuwe thuis," 43; Gartman, "Why Modern Architecture Emerged in Europe, not America"; Bervoets, "From 'Normalized Dwellings to Normalized Dwellers'?"

49 *Magyar Építőművészet*, Vol. 1 (1962), 3; *Magyar Építőművészet*, Vol. 2 (1962) 6, and L. Zoltan, "A delosi deklaráció" (The Delos Decree), *Magyar Építőművészet*, Vol. 1 (1964), 61 respectively, as quoted in Molnar, "In Search of the Ideal Socialist Home," 71.

50 J. Kenedi, *Do It Yourself! Hungary's Hidden Economy* (London: Pluto Press, 1981), as quoted from ibid., 76.

51 Pingusson in Chombart de Lauwe, *Famille et habitation*, 178 as cited in Rudolph, "Architects versus Housewives," 551.

52 Henthorn, *From Submarines to Suburbs*, 202–206; Marling, *As Seen on TV*, 253. See also: Olney, *Buy Now, Pay Later*; Kilgannon, "Change Blurs Memories

in a Famous Suburb"; Whyte, *The Organization Man*; Friedan, *The Feminine Mystique*.

53 Caigny, "Bouwen aan een nieuwe thuis"; Kilgannon, "Change Blurs Memories in a Famous Suburb."

54 Caigny, "Bouwen aan een nieuwe thuis"; De Vos, *Hoe zouden we graag wonen?*; Saarikangas, "What's New?," 295–301.

55 Griffith, "The Selling of America"; Holder, "The Nation State of the United States?"

56 Inklaar, *Van Amerika geleerd*, 275–90; Hanchett, "The Other 'Subsidized Housing'"; Hessler, "The Frankfurt Kitchen"; Betts and Eghigian, eds., *Pain and Prosperity*.

57 Oldenziel et al., *Huishoudtechnologie*, chapter 4.

58 Cohen, *A Consumer's Republic*. For a focused counter narrative from the point of view of working-class families, see: Nickles, "More is Better"; Zieger, "Paradox of Plenty"; Whitfield, *The Culture of the Cold War*, chapter 1; Marling, *As Seen on TV*, 268.

59 Eric Sevareid, "Super-Dupering the War: A Report on the Battle of the Adjectives," *Saturday Review of Literature*, February 12, 1944, 9, as quoted in Henthorn, *From Submarines to Suburbs*, 175.

60 Raymond Loewy, "What of the Promised Post-War World," *The New York Times Magazine*, 26 September 1943, 14, 27, as quoted in ibid., 176.

61 Miles L. Colean, "The Miracle House Myth," *House Beautiful* (December 1944): 78–79, 115–17 as quoted in ibid., 184–85.

62 Oldenziel et al., *Huishoudtechnologie*, 107–10; Halliday, *The Making of the Second Cold War*. See also: Wiesen, "Miracles for Sale"; Inklaar, *Van Amerika geleerd*.

63 Holliday, "Kitchen Technologies," 104.

64 "That Famous Debate in Close-Up Pictures," *Life* (August 3, 1959): 28, as quoted in Marling, *As Seen on TV*, 276.

65 "Whirlpool Corporation in the 1950s: Domestic Expansion and the Miracle Kitchen," accessed 20 March 2006, http://www.whirlpoolcorp.com/about/history/1950s.asp. Carbone, "Staging the Kitchen Debate." Marling, *As Seen on TV*. Haddow, *Pavilions of Plenty*. Hixson, *Parting the Curtain*.

66 *Libelle* 1964, as quoted in Groot and Kunz, *Libelle 50*, 123.

67 Speech, Edit Baumann, head of the Department of Women's Affairs of the Central Committee of the SED, January 1957, as quoted in Zachmann, "Managing Choice," 270.

68 Saarikangas, "What's New?"

69 Holliday, "Kitchen Technologies," 92–97.

70 Oldenziel et al., *Huishoudtechnologie*; Saarikangas, "What's New?"; De Vos, "The American Kitchen in Belgium?"; Rudolph, "Architects versus Housewives." On dish racks, for example, see: Phillips, *The Servantless House*, 58–69; Meyer, *Der neue Haushalt*.

71 Oldenziel, "Exporting the American Cold War Kitchen," 329–33; See also: Nickles, "Preserving Women"; Holder, "The Nation State of the United States?"

72 Oldenziel, "Exporting the American Cold War Kitchen," 329–33; Overbeeke, *Kachels, geisers en fornuizen*.

73 Bervoets, "Consultation Required!"; Holder, "The Nation State of the United States?"; Caigny, "Bouwen aan een nieuwe thuis," 248"; Oldenziel et al.,

Huishoudtechnologie, 57–61, 108–15; Rudolph, "Architects versus Housewives"; De Vos, "The American Kitchen in Belgium?"

74   Otterloo, *Eten en eetlust*, 175.
75   Badenoch, "Cooling the Cold War." Montijn, *Aan tafel!*, 83. In 1984, only 49 percent had a freezer. Finstad, "Cool Alliances"; Nickles, "Preserving Women."
76   Cieraad, "The Radiant American Kitchen"; Jakabovics, "Displaying American Abundance Abroad"; Sørensen, "Domestication"; Haddon, "Using the Domestication Framework."
77   RIX Charlotte, "Dans le cadre du Marché Commun. La première Cuisine Européenne," *La Maison* (February 2, 1961): 63, as quoted in De Vos, "The American Kitchen in Belgium?", 20.
78   Hilton, "Kitchen in a Global Context."

7   Saving the Nation, Saving the Earth

1   We are grateful to Heike Weber for helping to co-organize the workshop "Re/Cycling Histories: Users and the Paths to Sustainability in Everyday Life" at the Rachel Carson Center, May 25–27, 2011, and co-editing a special issue on recycling in *Contemporary European History*, 22, no. 3 (2013), as well as to Milena Veenis in providing the case study of the recycling container in the Netherlands. Jan-Hendrik Meyer has been so kind to read the chapter in depth, providing us with critical readings in environmental history.
2   Letters to the Interior Ministry: Dipl. Kfm. Karl B., Heepen, January 22, 1972; Volker B., Braunschweig May 21, 1971; Werner V., Dipl. Ing. Architekt, Fellbach, November 29, 1971, as respectively quoted in Westermann, "When Consumer Citizens Spoke Up," 487, 487, and 493.
3   Meadows and Rome, *The Limits to Growth*.
4   Pfister, "The '1950s Syndrome'"; Strasser, *Waste and Want*, chapter 5.
5   Federation, "Good Practices in Collection and Closed-Loop Glass Recycling in Europe."
6   Oldenziel and Veenis, "The Glass Recycling Container." The situation was vastly different in socialist countries after the Second World War, when efforts continued to separate and re-use waste as much as possible. See: Gille, *From the Culture of Waste to the Trash Heap of History*, 41–104.
7   Cooper, "War on Waste," 60–61.
8   Jørgensen, *The Green Machine*. See also: Jørgensen, "Green Citizenship at the Recycling Junction."
9   Jørgensen, *The Green Machine*, chapter 5.
10 Meyer, "Recycled Glass."
11 Multinational beverage companies like Coca Cola Global pushed for disposable containers in many countries against the locally based return-bottle systems because it prohibited their global expansion. At the same, the glass-deposit system's decline in 1970 brought the glass stream to a virtual standstill; shipping empty bottles back to the right manufacturers was the transport company Maltha's important source of income; the company was pleasantly surprised to find frugal housewives on their side when they sought to restore the glass chain. On the latter point: Agerbeek, "De glasbak."

12 Oldenziel and Veenis, "The Glass Recycling Container." See for the generational divide: Inglehart, "The Silent Revolution in Europe"; Inglehart, "Changing Values." We thank Jan-Hendrik Meyer for pointing out the irony of the baby-boomer generation's post-materialist attitude to the environment.

13 For a pointed critique about the sustainability of recycling see: MacBride, *Recycling Reconsidered*.

14 Oldenziel and Veenis, "The Glass Recycling Container." On the U.S. see: Rome, ""Give Earth a Chance'," 534–40.

15 Meyer, "Challenging the Atomic Community." For the role of Protestant churches in the early environmental movement, see: Michael Schüring's research project, accessed September 23, 2012, http://www.carsoncenter.uni-muenchen.de/fellows/alumni/visiting_fellows/michael_schuering/index. html.

16 W. de Ru-Schouten, "Interview Elizabeth Aiking-van Wageningen," *Hervormd Nederland*, April 4, 1970, 4, as quoted in Oldenziel and Veenis, "The Glass Recycling Container," 457.

17 "Kadercursus Milieuverontreiniging: bewijs van deelneming," May 1972, as quoted in ibid., 458.

18 Ibid.

19 Interview Milena Veenis with B. Riemens-Jagerman, December 16, 1999, as quoted in ibid., 460.

20 Dr. Ir.F. de Soet, "Wat betekent een geïntegreerd milieubeheer voor individu en samenleving," Lecture at kadercursus Milieuzorg, May 10, 1972, as quoted in ibid., 460.

21 Ibid. On similar middle-class women's activism, see: Rome, "'Give Earth a Chance'"; Evans, *Personal Politics*.

22 Meyer, "Saving Migrants."

23 Denton, "Récupérez!" ; Thorsheim, "Salvage and Destruction"; Weber, "Towards 'Total' Recycling"; Oldenziel et al., *Huishoudtechnologie*, chapter 3; Klemann, "Dutch Industrial Companies," 6.

24 Marie-Elisabeth Lüders, *Das unbekannte Heer. Frauen kämpfen für Deutschland, 1914–1918* (Berlin: E.S. Mittler und Sohn, 1937 [1936]), 74, 75–76, as quoted in Weber, "'Towards 'Total' Recycling."

25 Ibid.

26 "Klugheit und Einsicht der deutschen Frau" "Verantwortung," "Pflichten," "Bedenken Sie, dass 80 % des deutschen Volkseinkommen durch die Hände der Hausfrauen gehen!", as quoted in Köstering, "'Millionen im Müll?'" 148.

27 From a leaflet issued by the *Reich* Commissioner of Scrap Utilization, as quoted in ibid., 143.

28 Petzina, *Autarkiepolitik*.

29 Ibid.

30 Köstering, "'Pionere der Rohstoffbeschaffung'," 47, 55.

31 The trade journal *Der Rohproduktenhandel* (1934), as quoted in ibid., 49.

32 Petzold, *Die Altstoffe und die Rohstoffversorgung Deutschlands*, 92; Weber, "Towards 'Total' Recycling."

33 *Anordnungen und Richtlinien der Geschäftsgruppe Rohstoffverteilung und des Reichskommissars für Altmaterialverwertung in der Zeit vom November 1936 bis*

Februar 1940 (Berlin 1940), as quoted in Hutching, "Abfallwirtschaft im Dritten Reich," 264. See also 58–59.

34　*Produkten-Markt* 44 (1934): 49, as quoted in Köstering, "'Pionere der Rohstoffbeschaffung'," 51.

35　*Das Rohproduktengewerbe* 2, no. 33 (1936): 2.

36　Köstering, "'Pionere der Rohstoffbeschaffung'," 56; Hutching, "Abfallwirtschaft im Dritten Reich," 263.

37　Hutching, "Abfallwirtschaft im Dritten Reich," 266, 268. Kunitz, "'Recycling fürs Reich'"; Frilling and Mischer, *Pütt un Pann'n*, 137–39.

38　From the *Hamburger Anzeiger*, February 5, 1943, as quoted in Frilling and Mischer, *Pütt un Pann'n*, 146; Hutching, "Abfallwirtschaft im Dritten Reich," 264, 267.

39　Frilling and Mischer, *Pütt un Pann'n*, 136; Poster of 1938, as reprinted in Kunitz, "'Recycling fürs Reich'."

40　"Giornata della Fede." Significantly, *fede* means both "wedding ring" and "faith". The the propaganda effectively made use of this double meaning. Terhoeven, *Liebespfand*, 173–77.

41　*The Times*, December 19, 1935, as quoted in in Thorsheim, "Salvage and Destruction."

42　*The Times*, May 6, 1942, as quoted in in ibid., 445.

43　*The Times*, July 27, 1940, as quoted in ibid., 440.

44　*The Times*, November 7, 1941, quoted in ibid., 444.

45　Thorsheim, "Salvage and Destruction"; Weber, "Towards 'Total' Recycling."

46　Kunitz, "'Recycling fürs Reich'." On "philosophy of prey," see: Petzina, *Autarkiepolitik*, 141.

47　Weber, "Towards 'Total' Recycling"; Denton, "Récupérez!"; Thorsheim, "Salvage and Destruction"; Oldenziel and Veenis, "The Glass Recycling Container."

48　Schein, *Organisation und Technik des deutschen Rohproduktenhandels*, 122–26. *Das Rohproduktengewerbe* 2, no. 33 (1936): 1.

49　Denton, "Récupérez!"; Terhoeven, *Liebespfand*, 427.

50　Denton, "Récupérez!"

51　Edgerton, *Britain's War Machine*. Henthorn, *From Submarines to Suburbs*.

52　Helvenston and Bubolz, "Home Economics and Home Sewing"; Oldenziel et al., *Huishoudtechnologie*, chapter 4.

53　Letter by the Right Honourable Hugh Dalton M.P., UK Public Record Office, as quoted in Reynolds, "'Your Clothes are Materials of War'," 327 and 330.

54　Ibid., 328. Oldenziel et al., *Huishoudtechnologie*, chapter 4; Green, *Ready-to-Wear*, 95–99; Judt, *Postwar*, 14. Sudrow, *Der Schuh im Nationalsozialismus*. On sewing classes: Reynolds, "'Your Clothes are Materials of War'," 332–33; Helvenston and Bubolz, "Home Economics and Home Sewing."

55　Reynolds, "'Your Clothes are Materials of War'," 331; Oldenziel et al., *Huishoudtechnologie*, chapter 4.

56　Oldenziel et al., *Huishoudtechnologie*, chapter 4 and 89–90; Hofstede, Hoitsma, and Jong, *Kleding op de bon*.

57　Okawa, "Licensing Practices at Maison Christian Dior"; Gelber, "'Do-it-yourself'"; Goldstein, *Do it Yourself*; Wezel, "Doe-het-zelf"; Atkinson, "Do It Yourself"; Maines, *Hedonizing Technologies*, 91–98; Molnar, "In Search of the Ideal Socialist Home."

58　Gurova, "The Life Span of Things in Soviet Society."

59 Annelies Glaner, "Das Kleid dort auf der Stange...," *BZ am Abend*, May 15, 1968, as quoted in Stitziel, *Fashioning Socialism*, 151.

60 Unless indicated the case study and its details come from ibid., 148–52. Stitziel, "Shopping, Sewing." See also: Medvedev, "Hungarian Women Piece Together a New Communist Fashion."

61 Gurova, "The Life Span of Things in Soviet Society," 55.

62 Ibid., 52. Gordon, *Make It Yourself.*

63 Stitziel, *Fashioning Socialism*, 52–53.

64 Ward, *Spaceship Earth*. Höhler, "Spaceship Earth"; Cosgrove, "Contested Global Visions"; Chelcea, "The Culture of Shortage during State-Socialism"; Gurova, "The Life Span of Things in Soviet Society."

65 Oldenziel et al., *Huishoudtechnologie*, chapter 5; Zweiniger-Bargielowska, *Austerity in Britain.*

66 Cooper, "War on Waste"; Westermann, "When Consumer Citizens Spoke Up"; Oldenziel and Veenis, "The Glass Recycling Container."

67 Rome, "'Give Earth a Chance'"; Rome, "The Genius of Earth Day."

68 Mahrane, "De la nature à la biosphère"; Hamblin, "Environmentalism for the Atlantic Alliance"; Schulz, "Transatlantic Environmental Security"; Selin and Linnér, "The Quest for Global Sustainability"; Vickery, "Conservative Politics"; Macekura, "The Limits of the Global Community"; Hünemörder, "Vom Experten-netzwerk zur Umweltpolitik"; Meyer, "Appropriating the Environment"; Meyer, "Green Activism."

69 Patterson, "A Decade of Friendship."

70 Vogel, *Trading Up.*

71 Jamison, *The Making of Green Knowledge*; Jamison, Eyerman, and Cramer, *The Making of the New Environmental Consciousness*, 20.

72 Jørgensen, "Green Citizenship at the Recycling Junction."

73 Garrett De Bell, "Recycling," in J. Barr, ed., *Environmental Handbook* (London: Friends of the Earth, 1970), 217, as quoted in Cooper, "War on Waste," 62. The *Consumers' Guide* was published in 1971. Jamison, *The Making of Green Knowlege*, chapter 4.

74 Turner, *From Counterculture to Cyberculture.*

75 Brand, "Purpose."

76 Turner, *From Counterculture to Cyberculture.*

77 Orig: *Nogle oplysninger om den jord, vi sammen lever på* (NOAH: Copenhagen, 1970).

78 See also: http://www.smilingsun.org/index.html, accessed October 11, 2012.

79 Est, *Winds of Change*, 243. See also the website tracking the historical development, accessed October 11, 2012, http://windsofchange.dk/.

80 *Vedvarende Energikilder* was first published in 1975 and substantially expanded in 1977. *Solenergi—vindkraft* in 1976.

81 Kamp, "The Role of Policy in Inverse Developments"; Est, *Winds of Change*, 69–81; 238–47. Jørgensen and Karnøe, "The Danish Wind Turbine Story"; Dörner, *Drei Welten*. See also: Heymann, "Signs of Hubris."

82 Davis and Warring, "Living Utopia." The remarkable international success of the Danish cargo bike still needs a serious historian. See also: Haugbøll, *Svajerne*; Popplow, "Vom Lastenrad zum Cargobike"; Dambeck, "Renaissance des Cargobikes."

83 Edgerton, *The Shock of the Old.*

84 Vogel, *Trading Up*, 83–97; Meyer, "Appropriating the Environment."

8 Toying with America, Toying with Europe

1 We are grateful to Gerard Alberts for helping us understand the ludological aspect of technological change. Schleinzer, "Die Freude der Kinder."

2 De Vries, *De jongens van de Hobbyclub*. Unless otherwise indicated, the following is based on the research and broadcast by Jansen Henderiks, *De Hobbyclub*. See also: van Lente, "Romance of Technology in the Cold War Era." Scott Smith, *Networks of Empire*, 222, fn 90. W.H.G. Stuiver, "Dank u meneer Leonard de Vries," June 1992, accessed December 24, 2011, http://www.radio-wereld. nl/pg000019.htm. For some time, the Dutch electronics firm Philips hosted the club's secretariat.

3 De Vries, *De Hobbyclub breekt baan*.

4 De Vries, *De jeugdgemeenschap Hobbyclub. Een handleiding voor het oprichten van een hobby club*. Scott Smith, *Networks of Empire*, 222–25. On the response and counter response from Catholic organizations, see: "Het Vrije Volkje HOBBY CLUB!"; "Hobby-club te Kerkrade"; "Katholieke Hobby-clubs opgericht"; "Oude school werd jeugdgebouw"; "Vlijtige handen kan een succes worden."

5 De Vries, *De Hobbyclub*.

6 Huizinga, *Homo Ludens*.

7 "The Meccano Guild: A Fellowship of Meccano Boys," *Meccano Magazine* 1, no. 10 (September–October, 1919): 2–3, here 2. Most issues are available online: accessed February 22, 2013, http://www.nzmeccano.com/MMviewer.php. For a similar guild, founded by General Motors, see: Oldenziel, "Boys and Their Toys."After the war, GM's European subsidiaries launched their own contest for a few years. In 1965, the Adam Opel Company in Rüsselsheim, Germany, founded the "Model Builder Guild" (*Modellbauer-Gilde*); courtesy of Mr Ernst-Peter Berresheim, Opel Heritage & Institutional Communications, Rüsselsheim, Germany.

8 "The Meccano Guild: A Fellowship of Meccano Boys," *Meccano Magazine* 1, no. 10 (September–October 1919): 2–3, here 2, accessed February 22, 2013, http://www.nzmeccano.com/MMviewer.php.

9 Ibid.

10 "The Meccano Guild: A Fellowship of Meccano Boys," *Meccano Magazine*, 1, no. 11 (November–December, 1919): 2–3, here: 2, accessed February 22, 2013, http://www.nzmeccano.com/MMviewer.php.

11 Ibid.

12 *Meccano Magazine*, 1, no. 5 (November–December 1917): 1, accessed February 22, 2013, http://www.nzmeccano.com/MMviewer.php.

13 Myrdal, *En Meccano-pojke berättar*, 31.

14 *Miniaturbahnen* 1, no. 11 (1948/49): 14–15 refers to international gatherings, documenting contacts among German, Swiss, French, American, and Spanish model railroad buffs. We thank Burkhard Breiing for supplying us with this source.

15 Myrdal, *En Meccano-pojke berättar*, 47–52.

16 Advertisement by the *Franckh'sche Verlagshandlung, Abt. Kosmos-Lehrmittel*, March 1955, as quoted in Öxler, *Vom tragbaren Labor zum Chemiebaukasten*, 213. See also: ibid., chapters 8 and 9.

17 We are grateful to Giel van Hooff for locating relevant documents and to the archivist at the Philips Company Archives, Eindhoven, for permission to

view the material. See also: "Serie CL Ubersicht," accessed December 23, 2011, http://norbert.old.no/kits/cl/.

18 VEB Laborchemie Apolda, *Kleiner Experimentator: 135 Versuche aus dem Wunderland der Chemie* (Apolda 1960) 4, as quoted in Öxler, *Vom tragbaren Labor zum Chemiebaukasten*, 245.

19 Based on interviews, as quoted in ibid., 237. See discussion in ibid., chapter 10.

20 Myrdal, *En Meccano-pojke berättar*, 47–52.

21 "Om klantenbinding Barbie-fans." For similar sentiments see: Jerman, "Sex and the Single Doll"; Buchwald, "Barbie needs... Ken."

22 For French fashion titles from1776 to 1937, see: Ghering-van Ierlant, *Vrouwen-mode in Prent*, 227–29; Odin, *Lilas*, 8. See further: "L'Europe de 25," Musée de la Poupée, exhibition, 2006; Musée de la Poupée, "Temporary Exhibits," accessed January 5, 2011, http://www.museedelapoupeeparis.com/tempo/tempoen-sybarite.html.

23 Gerber, *Barbie and Ruth*, 5; Handler and Shannon, *Dream Doll*, 91.

24 Cartoons as quoted in "German Products History Series: The Bild Lilli Doll Story," accessed January 15, 2011 http://donauschwaben-usa.org/pdf%20 forms/TCT%20pdf%20files/German%20Products%20History%20Series%20 Bild%20Lilli%20Story.pdf.

25 Handler and Shannon, *Dream Doll*, 7, 12; Knaak, *Deutsche modepuppen*, 70–165; Lord, *Forever Barbie*; Anne Zielinski-Old interview Inge Astor-Kaiser "Bild Lilli: a History Catches Up," reprint from *Fashion Doll Quarter*, accessed December 7, 2011, http://www.annezielinski-old.com/html/lilli2.html; Peers, *The Fashion Doll*, 143–49.

26 Handler and Shannon, *Dream Doll*, 11; Gerber, *Barbie and Ruth*, 137.

27 Miller, *Toy Wars*, 69; Handler and Shannon, *Dream Doll*, 88; Oppenheimer, *Toy Monster*, 25, 32. By the 1980s, the company diversified and lowered prices to beat the steep decline in consumer spending. While mobilizing history, tradition, and crafts, the company never again achieved the eye for haute-couture detail it had in the 1960s. Spigel, *Welcome to the Dreamhouse*, 323–35.

28 Winkelman, *Hier is Barbie*.

29 The subject awaits serious historical research. An Ebay search generates thousands of examples of such Barbie paper patterns.

30 *Neveda* patroon, no. 3156, 3157 in Amsterdam Museum Archives, Collection Maria Over, KA 21004. Neveda advertised in local newspapers for example, *Friese koerier*, February 25, 1966 and *Limburgs dagblad*, March 4, 1966. Ruth Oldenziel, "Mijn Barbie, mijn oma," voordracht ter gelegenheid van Internationale Vrouwendag en Barbie's 50ste verjaardag bij de opening van collectie Miep Over," March 8, 2009, Amsterdam Museum, Archives, Collection Maria Over, inv. no. KA 21004 1/120. For examples of other Barbie collections, see: Nuremberg Toy Museum, the Prague Toy Museum, the Paris Musée de la Poupée, and the Swedish Nordiska museet.

31 Scott, "California Casual"; Kramer, "Blauhemd und Bluejeans."

32 Kramer, "Blauhemd und Bluejeans," 132, 202; Scheuring, "Heavy Duty Denim," 226–30; Polhemus, *Streetstyle*, 26–32. For the relationships in fashion and popular culture, see: Wilson, *Adorned in Dreams*, chapter 8.

33 Scheuring, "Heavy Duty Denim," 226–30.

34 Ross, *Clothing*, 149–51.

35 See: the well-researched case for Hungary in Hammer, "The Real One," 133; Iakovleva-Zakharova, "S'habiller à la Sovietique," 276, 282–84.

36 Pelka, *Jugendmode und Politik*; Kramer, "Blauhemd und Bluejeans."

37 Pelka, *Jugendmode und Politik*, 49–52. We thank Dagmara Jajesniak-Quast for helping to understand Hoff's importance.

38 Story from H. Istvan, as quoted in Hammer, "Sartorial Manoeuvres," 58.

39 Story from K.T. (b. 1950), as quoted in Hammer, "The Real One," 46.

40 Ibid., 134. See also the intriguing news item from Italy: La moda dei blue jeans tra I giovani, La Attimana Incom, 1959, accessed September 15, http://euscreen.eu/play.jsp?id=EUS_857DD60EC88A4FBE996454BCCD5B6FD4, 2012. Flint, "Unzipping the USSR," 22.

41 J. Kádár, *Hazafiságés internacianalizmuz* (Budapest, 1968), 188, as quoted in Hammer, "Sartorial Manoeuvres," 65.

42 Reuters, "Girls, Watchers of the World, Arise," September 30, 1966, as quoted in Flint, "Unzipping the USSR," 4.

43 M. György (b. 1952), as quoted in Hammer, "Sartorial Manoeuvres," 62–65; Kramer, "Blauhemd und Bluejeans," 129.

44 Pelka, *Jugendmode und Politik*, passim and 51; Flint, "Unzipping the USSR," 20.

45 "Soviet Jeans a Parody, Says Soviet Magazine," August 14, 1976, as quoted in Flint, "Unzipping the USSR," 24–28.

46 Ibid., 29–30.

47 Ibid., 33–42, 52, 62.

48 "IDC-Eurocast, Personal Computers Market. Reference Book Western Europe," London, September 1983, "Markets and Products Reports," 55, Box 16, Charles Babbage Institute Minneapolis, MN. We are grateful to Frank Veraart for sharing this and the next document with us.

49 Theodore Spinoulas *RAM* (1994), as quoted in Lekkas, "Legal Pirates Ltd"

50 From an internal document it becomes clear that IBM had completely lost its rudder and found Europe a thoroughly confusing and fragmented place where it could not enter in a single sweep. Frost & Sullivan, "The IBM Market in Europe" Vol. II, 1983, Market & Products Reports 55, Box 4. Charles Babbage Institute Minneapolis, MN. See also: Veraart, "Vormgevers van Persoonlijk Computergebruik," 125–40.

51 Heide, *Punched-Card Systems*; Friedewald, *Der Computer als Werkzeug und Medium*. Lubar, *Infoculture*, 331–52.

52 Lean, "The Making and Remaking of Sinclair Personal Computing"; Veraart, "Transnational (dis)connection"; Kaiserfeld, "Computerizing the Swedish Welfare State."

53 Lean, "The Making and Remaking of Sinclair Personal Computing."

54 Veraart, "Transnational (dis)connection"; Veraart, "Vormgevers van Persoonlijk Computergebruik."

55 Kaiserfeld, "Computerizing the Swedish Welfare State."

56 Jakić, "Galaxy and the New Wave"; Wasiak, "Playing and Copying."

57 Marek Jędrzejewski, "Gdy komputer jest bożkiem," *Argumenty*, September 27, 1987, 12, as quoted in Wasiak, "Playing and Copying."

58 Ibid.

59 Lean, "The Making and Remaking of Sinclair Personal Computing"; Kaiserfeld, "Computerizing the Swedish Welfare State"; Jakić, "Galaxy and the New Wave"; Veraart, "Transnational (dis)connection"; Wasiak, "Playing and Copying." After a decade of experimentation to create its own standards

and encourage national manufacturers, the Dutch Ministry of Education succumbed to the corporate dance. The government signed a license for Microsoft Windows 3.0 to be installed in schools as the standard software on IBM machines—the first country in the world to do so. Veraart, "Transnational (dis)connection."

60 Tympas, Tsaglioti, and Lekkas, "Universal Machines vs. National Languages."

61 Lekkas, "Legal Pirates Ltd"; Wasiak, "Playing and Copying"; Saarikoski and Suominen, "Computer Hobbyists in Finland."

62 Veraart, "Transnational (dis)connection"; Veraart, "Basicode"; Lean, "The Making and Remaking of Sinclair Personal Computing."

63 Lean, "The Making and Remaking of Sinclair Personal Computing."

64 Ibid; Lekkas, "Legal Pirates Ltd"; Jakić, "Galaxy and the New Wave"; Wasiak, "Playing and Copying."

65 Saarikoski and Suominen, "Computer Hobbyists in Finland."

66 Marc A. Keck-Szajbel, PhD candidate at the University of California is writing a dissertation on trade tourism and the transnational flow of consumer goods in socialist countries during the Cold War.

67 Wasiak, "Playing and Copying."

68 Jakić, "Galaxy and the New Wave"; Veraart, "Basicode."

69 Jakić, "Galaxy and the New Wave."

70 Denker, "Heroes yet Criminals."

71 Castells, *The Internet Galaxy*, 146, as quoted in Nevejan and Badenoch, "How Amsterdam Invented the Internet."

72 Nevejan and Badenoch, "How Amsterdam Invented the Internet."

73 Ibid. Haar, "Projectverslag," accessed January 15, 2011; Martz, "Waag Society's first fifteen years"; Turner, *From Counterculture to Cyberculture*. Despite the impressive number of first-time users, only 9 percent were women. Oldenziel, "Feminist Narratives of Technology"; Rommes, "Gender Scripts and the Internet."

74 Nevejan and Badenoch, "How Amsterdam Invented the Internet."

75 Ibid. Denker, "Heroes yet Criminals"; Haar, "Projectverslag."

76 Söderberg, "Users in the Dark."

77 Ibid.

78 David Jolly, "A New Question of Internet Freedom," *International Herald Tribune* (January 5, 2012).

79 Ibid.

80 Stewart, "Copying and Copyrighting Haute Couture."

81 Ibid.

Conclusion

1 The issue of large technical systems will be taken up in great detail in Kaijser, Högselius, and van der Vleuten, *Europe's Infrastructure Transition*.

2 Fickers and Griset, *Communicating Europe*.

3 See also: Schivelbusch, *Railway Journey*.

4 On "relational space," see: Löw, *Raumsoziologie*.

5 The issue of the relation between Europe and other countries in a global context will be treated head on in Diogo and Van Laak, *Europe Globalizing*.

Bibliography

Ackermann, Ch. *Répertoire de jurisprudence en materière de transports: Voyageurs, bagages ou le consiller-guide du voyeur*. Geneva: Georg & Co, 1936.

Agerbeek, Marjan. "De glasbak." *Trouw*, 16 April 2003.

Akcan, Esra. "Civilizing Housewives versus Participatory Users: Margarete Schütte-Lihotsky in the Employ of the Turkish Nation State." In *Cold War Kitchen: Americanization, Technology, and European Users*, edited by Ruth Oldenziel and Karin Zachmann, 185–207. Cambridge, MA: MIT Press, 2009.

Alberts, Gerard, and Ruth Oldenziel, eds. *Hacking Europe: From Computer Cultures to Demoscenes*. New York: Springer, 2013.

Aliberti, Giovanni. *L' economia domestica italiana da Giolitti a De Gasperi: 1900–1960*. Roma: 50 & più, 1992.

Ames, Kenneth L. "Meaning in Artifacts: Hall Furnishings in Victorian America." *Journal of Interdisciplinary History* 9, no. 1 (1978): 19–46.

Anastasiadou, Irene. *Constructing Iron Europe: Transnationalism and Railways in the Interbellum*. Amsterdam: Amsterdam University Press, 2012.

Andrews, Benjamin R. *Economics of the Household: Its Administration and Finance*. New York: Macmillan, 1924.

Argenbright, Robert. "Lethal Mobilities: Bodies and Lice on Soviet Railroads, 1918–1922." *Journal of Transport History* 29, no. 2 (2008): 259–76.

Armstrong, John, and David M. Williams. "The Steamboat and Popular Tourism." *Journal of Transport History* 26 (2005): 61–77.

Arnold, David, and Erich DeWald. "Cycles of Empowerment? The Bicycle and Everyday Technology in Colonial India and Vietnam." *Comparative Studies in Society and History* 53, no. 4 (2011): 971–96.

Arons, Ger et al. *Kollektivering van huishoudelijke arbeid: Verslag van een onderzoek*. Amsterdam: Universiteit van Amsterdam, 1980.

Arrighi, Giovanni. *The Long Twentieth Century: Money, Power, and the Origins of Our Times*. London: Verso, 1994.

Asplund, Gunnar. *acceptera*. Stockholm: Tiden, 1931.

Atkins, Peter J. "The Empire Marketing Board." In *The Diffusion of Food Culture in Europe from the Late Eighteenth Century to the Present Day*, edited by Derek J. Oddy and Lydia Petráňová, 248–55. Prague: Academia, 2005.

Atkinson, Paul. "Do It Yourself: Democracy and Design." *Journal of Design History* 19, no. 1 (2006): 1–10.

Atwater, Helen W. "For the Homemaker: The Woman's Committee Survey of Agencies for the Sale of Cooked Food." *Journal of Home Economics* 10, no. 9 (1918): 419–24.

B.H., Jan. "Het Internationaal Touristen-Congres." *De Kampioen*, 30 July 1897, 809–18.

Badenoch, Alexander. "Cooling the Cold War: The American Fridge as a Symbol," Inventing Europe, http://www.inventingeurope.eu/story/cooling-the-cold-war-the-american-fridge-as-a-symbol, as accessed July 7, 2013.

Baranowski, Bohdan. *Polska Karczma: Restauracja, Kawiarnia*. Wrocław: Ossolineum 1979.

Bardou, Jean-Pierre. *The Automobile Revolution: The Impact of an Industry.* Chapell Hill, NC: University of North Carolina Press, 1982.

Barker, Mary L. "Traditional Landscape and Mass Tourism in the Alps." *Geographical Review* 72, no. 4 (1982): 395–415.

Barker, Theo. *The Economic and Social Effects of the Spread of Motor Vehicles: An International Centenary Tribute.* Basingstoke: Macmillan, 1987.

Barlow, Tani E., Madeleine Yue Dong, Uta G. Poiger, Priti Ramamurthy, Lynn M. Tomas, and Alys Eve Weinbaum. "Modern Girl Around the World: A Research Agenda and Preliminary Findings." *Gender & History* 17, no. 2 (2005): 245–94.

Barnes, David S. *The Great Stink of Paris and the Nineteenth-Century Struggle against Filth and Germs.* Baltimore, MD: Johns Hopkins University Press, 2006.

Baron, Ava, and Susan Klepp. "'If I Didn't Have My Sewing Machine...': Women and Sewing Machine Technology." In *A Needle, a Bobbin, a Strike: Women Needleworkers in America,* edited by Joan M. Jensen and Sue Davidson, 20–59. Philadelphia: Temple University Press, 1984.

Barton, Susan. *Working-Class Organisations and Popular Tourism, 1840–1970.* Edited by Jeffrey Richard, Studies in Popular Culture. Manchester: Manchester University Press, 2005.

Behrend, George. *The History of Wagons-Lits 1875–1955.* London: Modern Transport Publishing Co., 1959.

Bellamy, Edward. "A Vital Domestic Problem: Household Service Reform." *Good Housekeeping* 10, no. 4 (1889): 74–77.

Bender, Thomas. *A Nation Among Nations: America's Place in World History.* New York: Hill and Wang, 2006.

Berg, Maxine. *The Machinery Question and the Making of Political Economy.* Cambridge: Cambridge University Press, 1980.

———, and Elizabeth Eger, eds. *Luxury in the Eighteenth Century: Debates, Desires and Deletable Goods.* New York: Palgrave, 2003.

Berghout, Saskia, ed. *Wij wonen: 100 jaar woningwet.* Amsterdam: Stedelijke Woningdienst Amsterdam, 2001.

Bernège, Paulette. *De la méthode ménagére.* Paris: Mon chez moi, Dunod, 1928.

———. *Die methodische Führung des Haushaltes.* Translated by Marga Bestmann. Gotha: Schmidt & Thelow, 1928.

Berner, Boel. "The Meaning of Cleaning: The Creation of Harmony and Hygiene in the Home." *History and Technology* 14, no. 4 (1998): 313–52.

Bertho Lavenir, Catherine. *La roue et le stylo: Comment nous sommes devenus touristes.* Paris: Éditions Odille Jacob, 1999.

———. "Scarcity, Poverty, Exclusion: Negative Associations of the Bicycle's Uses and Cultural History in France." In *Re/Cycling Histories: Paths to Sustainability,* edited by Ruth Oldenziel and Helmuth Trischler. London: Berghahn, Submitted.

Bertz, Eduard. *Philosophie des Fahrrads.* Dresden and Leipzig: Carl Reißer, 1900.

Bervoets, Liesbeth. "'Consultation Required!' Women Coproducing the Modern Kitchen in the Netherlands, 1920 to 1970." In *Cold War Kitchen: Americanization, Technology, and European Users,* edited by Ruth Oldenziel and Karin Zachmann, 209–32. Cambridge, MA: MIT Press, 2009.

Bervoets, Liesbeth. *Telt zij wel of telt zij niet: Een onderzoek naar de beweging voor de rationalisatie van huishoudelijke arbeid in de jaren twintig.* Amsterdam: Sociologisch Instituut Universiteit van Amsterdam, 1982.

———. "A Utopia in Stone: Mediating in the Name of Working-Class Collectivism." In *Manufacturing Technology, Manufacturing Consumers: The Making of Dutch Consumer Society*, edited by Adri A. de la Bruhèze and Ruth Oldenziel, 95–114. Amsterdam: Aksant, 2009.

———. "From 'Normalized Dwellings to Normalized Dwellers?' The Interpretative Flexibility of Modern Housing in Europe." Paper presented at the workshop "The European: An Invention at the Interface between Technology and Consumption, " Munich, 4–5 October, 2007.

Betts, Paul, and Greg Eghigian, eds. *Pain and Prosperity: Reconsidering Twentieth-Century German History*. Stanford: Stanford University Press, 2003.

Bihr, Simon. "The Super-Market Empire: Self Service in the United States and Western Europe, 1915–1960." Paper presented at the workshop "Techno-Topologies: Spatial Perspectives—Spatial Practices, Darmstadt, 3–5 March, 2011.

Bijker, Wiebe E. *Of Bicycles, Bakelites, and Bulbs: Toward a Theory of Sociotechnical Change*. Cambridge, MA: MIT Press, 1995.

———, and Karin Bijsterveld. "Women Walking through Plans: Technology, Democracy, and Gender Identity." *Technology and Culture* 41, no. 3 (2000): 485–515.

Bix, Amy. "Equipped for Life: Gendered Technical Training and Consumerism in Home Economics, 1920–1980." *Technology and Culture* 43, no. 4 (2002): 728–54.

Blaszczyk, Regina L. "'Where Mrs. Homemaker Is Never Forgotten': Lucy Maltby and Home Economics at Corning Glass Works, 1929–1965." In *Rethinking Home Economics: Women and the History of a Profession*, edited by Sarah Stage and Virginia B. Vincenti, 163–80. Ithaca, NY: Cornell University Press, 1997.

Bleckmann, D. *Wehe wenn sie losgelassen! Über die Anfänge des Frauenradfahrens in Deutschland*. Gera-Leipzig: Maxime-Verlag, 1999.

Bonham, Jennifer, and Peter Cox. "The Disruptive Traveller? A Foucauldian Analysis of Cycleways." *Road & Transport Research* 19, no. 2 (2010): 42–53.

Bonzon, Alfred. "Convention internationale sur le transport des voyageurs et des bagages par chemins de fer." PhD diss., Université de Lausanne, 1897.

Bosch, Mineke. "Parallelle levens op de fiets: Het (wielrijders) verbond van Aletta Jacobs en Carel Gerritsen." In *Jaarboek voor Vrouwengeschiedenis* 18 (1998), edited by Corrie van Eijl, 59–78. Amsterdam, 1998.

Bosworth, Richard J.B. "The Touring Club Italiano and the Nationalization of the Italian Bourgeoisie." *European History Quarterly* 27, no. 3 (1997): 371–410.

Bowlby, Rachel. *Carried Away: The Invention of Modern Shopping*. New York: Columbia University Press, 2001.

Bracewell, Wendy. "Travels Through the Slav World." In *Under Eastern Eyes: A Comparative Introduction to East European Travel Writings on Europe*, edited by Wendy Bracewell and Alec Drace-Francis,, 147–94. Budapest and New York: Central European University Press, 2008.

Brand, Stuart, "Purpose." *Whole Earth Catalog: Access to Tools*, Fall 1969, 3.

Brändli, Sibylle. *Der Supermarkt im Kopf: Konsumkultur und Wohlstand in der Schweiz nach 1945*. Wien: Böhlau, 1997.

Braun, Lily. *Frauenarbeit und Hauswirtschaft*. Berlin: Expedition der Buchhandlung Vorwärts, 1901.

Brendon, Piers. *Thomas Cook: 150 Years of Popular Tourism*. London: Secker and Warburg, 1991.

Breuss, Susanne. "Die Stadt, der Staub und die Hausfrau: Zum Verhältnis von schmutziger Stadt und sauberem Heim." In *Urbane Welten: Referate der Österre-*

ichischen Volkskundetagung 1998 in Linz, edited by Andrea Euler, 353–76. Vienna: Selbstverlag des Vereins für Volkskunde, 1999.

Breward, Christopher. "Patterns of Respectability: Publishing, Home Sewing and the Dynamics of Class and Gender, 1870–1914." In *The Culture of Sewing: Gender, Consumption, and Home Dressmaking,* edited by Barbara Burman, 21–32. Oxford: Berg, 1999.

Brewer, John, and Roy Porter, eds. *Consumption and the World of Goods.* New York: Routledge, 1993.

Brewer, Priscilla J. *From Fireplace to Cookstove: Technology and the Domestic Ideal in America.* Syracuse, NY: Syracuse University Press, 2000.

Briese, Volker. "From Cycling Lanes to Compulsory Bike Path: Bicycle Path Construction in Germany, 1897–1940." *Cycle History* 5 (1994): 123–8.

———. "Opium für Radfahrer." *Radfahren* 1 (1994): 36–42.

———. "Radwegebau vor dem Zweiten Weltkrieg: Zurück in die Zukunft." *RadMarkt* 5 (1993): 50–62.

Bringéus, Nils-Arne. *Man, Food, and Milieu: A Swedish Approach to Food Ethnology.* East Lothian: Tuckwell Press, 2001.

Brinkmann, Tobias. "From Oświęcim to Ellis Island: Jewish and other Transmigrants and the Evolution of Border Controls along Germany's Eastern Border, 1885–1914." In *Between the Old and the New World: Studies in Overseas Migrations,* edited by Agnieszka Małek and Dorota Praszałowicz, 109–24. Frankfurt am Main: Peter Lang, 2012.

———. "'Travelling with Ballin': The Impact of American Immigration Policies on Jewish Transmigration within Central Europe, 1880–1913." *International Review of Social History* 53 (2008): 459–84.

Bruegel, Martin. "From the Shop Floor to the Home: Appertising and Food Preservation in Households in Rural France, 1810–1930." In *Food and Material Culture: Proceedings of the Fourth Symposium of the International Commission for Research into European Food History,* edited by Martin R. Schärer and Alexander Fenton, 203–28. East Linton: Tuckwell Press, 1998.

Bruhèze, Adri A. de la, and Anneke H. van Otterloo. "The Milky Way: Infrastructures and the Shaping of Milk Chains." *History and Technology* 20, no. 3 (2004): 249–69.

———, and Frank Veraart. "Fietsen en verkeersbeleid: Het fietsgebruik in negen Westeuropese steden in de twintigste eeuw." *NEHA-jaarboek* 62 (1999): 138–70.

———, and Frank C.A. Veraart. *Fietsverkeer in praktijk en beleid in de twintigste eeuw: Overeenkomsten en verschillen in fietsgebruik in Amsterdam, Eindhoven, Enschede, Zuidoost Limburg, Antwerpen, Manchester, Kopenhagen, Hannover en Basel.* Eindhoven: Stichting Historie der Techniek, 1999.

Brzostek, Błażej. *PRL na widelcu.* Warsaw: Baobab, 2010.

Buchwald, Art. "Barbie needs… Ken." In *I Chose Capitol Punishment,* edited by Art Buchwald. Cleveland: World Publishing, 1963.

Buckley, Cheryl. "On the Margins: Theorizing the History and Significance of Making and Designing Clothes at Home." *Journal of Design History* 11, no. 2 (1998): 157–71.

"Buitenlandsch nieuws." *Nieuws van de dag,* 28 March 1899, 3.

Buiter, Hans. "Constructing Dutch Streets: A Melting Pot of European Technologies." In *Urban Machinery: Inside Modern European Cities,* edited by Mikael Hård and Thomas J. Misa, 141–62. Cambridge, MA: MIT Press, 2008.

———. *Riool, rails en asfalt: 80 jaar straatrumoer in vier Nederlandse steden.* Zutphen: Walburg Pers, 2005.

———, and Irene Anastasiadou. "Regulations and Flows in Transnational Rail Traffic: The Relationship between the Develpment of Governance of International Railway Traffic and the Flows across Dutch and Belgian Cross-Border Rail Links, 1850–2000." Paper presented at the Tensions of Europe Conference, Sofia, 17–20 June, 2010.

Burman, Barbara. "Home Sewing and *'Fashions for All'* 1908–1937." *Costume* 28 (1994): 71–80.

Burnett, John. *A Social History of Housing, 1815–1970.* Newton Abbot: David & Charles, 1978.

———. "Time, Place and Content: The Changing Structure of Meals in Britain in the Nineteenth and Twentieth Centuries." In *Food and Material Culture: Proceedings of the Fourth Symposium of the International Commission for Research into European Food History*, edited by Martin R. Schärer and Alexander Fenton, 116–32. East Linton: Tuckwell, 1998.

Burr, Thomas. "National Cycle Organizations in Britain, France, and the United States, 1875–1905." *Cycle History* 18 (2007): 34–42.

Burri, Monika. "Das Fahrrad: Wegbereiter oder überrolltes Leitbild? Eine Fussnote zur Technikgeschichte des Automobils." *Preprints zur Kulturgeschichte der Technik*, no. 5 (1998).

Buzard, James. *The Beaten Track: European Tourism, Literature, and the Ways to 'Culture,' 1800–1918.* Oxford: Clarendon Press, 1993.

Buzinkay, Géza. "Das Wohnideal des Mittelstandes." In *Bürgerliche Wohnkultur des Fin de siècle in Ungarn*, edited by Péter Hanák, 25–43. Vienna: Böhlau, 1994.

Caigny, Sofie De. "Bouwen aan een nieuwe thuis: Wooncultuur in Vlaanderen tijdens het Interbellum." PhD diss., KU Leuven, 2007.

Caracciolo, Carlos Héctor. "Bicicleta, circulación vial y espacio público en la Italia fascista." *Historia Critica* 39 (2009): 20–42.

Carbone, Cristina. "Staging the Kitchen Debate: How Splitnik Got Normalized in the United States." In *Cold War Kitchen: Americanization, Technology, and European Users*, edited by Ruth Oldenziel and Karin Zachmann, 59–81. Cambridge, MA: MIT Press, 2009.

Castells, Manuel. *The Internet Galaxy: Reflections on the Internet, Business, and Society.* Oxford and New York: Oxford University Press, 2001.

Castillo, Greg. "The American 'Fat Kitchen' in Europe: Postwar Domestic Modernity and Marshall Plan Strategies of Enchantment." In *Cold War Kitchen: Americanization, Technology, and European Users*, edited by Ruth Oldenziel and Karin Zachmann, 33–58. Cambridge, MA: MIT Press, 2009.

Caudenberg, Anke van, and Hilde Heynen. "The Rational Kitchen in the Interwar Period in Belgium: Discourses and Realities." *Home Cultures* 1, no. 1 (2004): 23–50.

Cepl-Kaufmann, Gertrude, and Antje Johanning. *Mythos Rhein: Zur Kulturgeschichte eines Stromes.* Darmstadt: Wissenschaftliche Buchgesellschaft, 2003.

Chadwick, Edwin. *Report on the Sanitary Condition of the Labouring Population of Gt. Britain.* Edinburgh: Edinburgh University Press, 1965 [1842].

Chatriot, Alain, Marie-Emmanuelle Chessel, and Matthew Hilton, eds. *The Expert Consumer: Associations and Professionals in Consumer Society.* Aldershot: Ashgate, 2006.

Chekhov, Anton. "Peasants." In *The Lady with the Little Dog and Other Stories, 1896–1904*, 22–60. London: Penguin, 2002.

Chelcea, Liviu. "The Culture of Shortage during State-Socialism: Consumption Practices in a Romanian Village in the 1980s." *Cultural Studies* 16, no. 1 (2002): 16–43.

Chessel, Marie-Emmanuelle. "Consumers' Leagues in France: A Transatlantic Perspective." In *The Expert Consumer: Associations and Professionals in Consumer Society*, edited by Alain Chatriot, Marie-Emmanuelle Chessel, and Matthew Hilton, 53–69. Aldershot: Ashgate, 2006.

Chiapparino, Francesco. "Milk and Fondant Chocolate and the Emergence of the Swiss Chocolate Industry at the Turn of the Twentieth Century." In *Food and Material Culture: Proceedings of the Fourth Symposium of the International Commission for Research into European Food History*, edited by Martin R. Schärer and Alexander Fenton, 328–44. East Linton: Tuckwell Press, 1998.

Christiansen, Niels Finn. "Between Farmers and Workers: Consumer Cooperation in Denmark, 1850–1940." In *Consumers Against Capitalism? Consumer Cooperation in Europe, North America, and Japan, 1840–1990*, edited by Ellen Furlough and Carl Strikwerda, 221–40. Lanham, MD: Rowman & Littlefield, 1999.

"Chronicle December." In *The Annual Register, or a View of the History and Politics of the Year 1860*, 181–84. London, 1861.

Cieraad, Irene. "The Radiant American Kitchen: Domesticating Dutch Nuclear Energy." In *Cold War Kitchen: Americanization, Technology, and European Users*, edited by Ruth Oldenziel and Karin Zachmann, 113–36. Cambridge, MA: MIT Press, 2009.

Clendinning, Anne. "'Deft Fingers' and 'Persuasive Eloquence': The 'Lady Demons' of the English Gas Industry, 1888–1918." *Women's History Review* 9, no. 3 (2000): 501–37.

———. "Gas and Water Feminism: Maud Adeline Brereton and Edwardian Domestic Technology." *Canadian Journal of History/Annales canadiennes d'histoire* 33 (1998): 1–24.

Coffin, Judith G. "Credit, Consumption, and Images of Women's Desires: Selling the Sewing Machine in Late Nineteenth-Century France." *French Historical Studies* 18, no. 3 (1994): 749–83.

Cohen, Elizabeth. *A Consumer's Republic: The Politics of Mass Consumption in Post-War America*. New York: Alfred Knopf, 2003.

"Comité vrije-tijdsbesteding." *Limburgsch Dagblad*, June 10, 1952.

Cooper, Timothy. "War on Waste? The Politics of Waste and Recycling in Post-War Britain, 1950–1975." *Capitalism, Nature, Socialism* 20, no. 4 (2009): 53–72.

Corn, Jospeh. *User Unfriendly: Consumer Struggles with Personal Technologies, from Clocks and Sewing Machines to Cars and Computers*. Baltimore, MD: Johns Hopkins University Press, 2011.

Cosgrove, Denis. "Contested Global Visions: One-World, Whole-Earth, and the Apollo Space Photographs." *Annals of the Association of American Geographers* 84, no. 2 (1994): 270–94.

Cosseta, Katrin. *Ragione e sentimento dell'abitare: la casa e l'architettura nel pensiero femminile tra le due guerre*. Milano: F. Angeli, 2000.

Costa, Vítor. "O Desporto e a Sociedade em Portugal—Fins do século XIX- princípios do século XX." Tese de mestrado em História Social Contemporânea, ISCTE, 1999.

Cowan, Ruth Schwartz. "The 'Industrial Revolution' in the Home: Household Technology and Social Change in the 20th Century." *Technology and Culture* 17, no. 1 (1976): 1–23.

———. *More Work for Mother: The Ironies of Household Technology from the Open Hearth to the Microwave*. New York: Basic Books, 1983.

Crane, Diana. *Fashion and its Social Agendas: Class, Gender, and Social Identity in Clothing*. Chicago: University of Chicago Press, 2000.

Crew, David F., ed. *Consuming Germany in the Cold War*. Oxford: Berg, 2003.

Cyclists' Touring Club. "Report of the Foreign Touring Arrangements Committee." The Council of the Cyclists' Touring Club, 1897.

Dambeck, Holger. "Renaissance des Cargobikes: Lastesel zum Treten." *Der Spiegel*, 7 November 2012.

Dapples, Ernest. *Le matériel roulant des chemins de fer au point de vue du confort et de la sécurité des voyageurs: Moyens d'intercommunication proposés pour permettre aux voyageurs de tous les compartiments et aux conducteurs des trains de communiquer entr'eux pendant la marche*. Lausanne: Société vaudoise de typographie, 1866.

Davies, Robert B. "'Peacefully Working to Conquer the World': The Singer Manufacturing Company in Foreign Markets, 1854–1889." *Business History Review* 43, no. 3 (1969): 299–325.

Davis, John, and Anette Warring. "Living Utopia: Communal Living in Denmark and Britain." *Cultural and Social History* 8, no. 4 (2011): 513–30.

de Grazia, Victoria. "Changing Consumption Regimes in Europe 1930–1970: Comparative Perspectives on the Distribution Problem." In *Getting and Spending: European and American Consumer Societies in the Twentieth Century*, edited by Susan Strasser, Charles McGovern and Matthias Judt, 59–83. Cambridge: Cambridge University Press, 1998.

———. *Irresistible Empire: America's Advance Through Twentieth-Century Europe*. Cambridge, MA: Belknap Press of Harvard University Press, 2005.

"De Rijwiel- en Automobieltentoonstelling." *Algmeen Handelsblad*, March 19, 1899.

"De Tentoonstelling te Leeuwarden." *Nieuws van den Dag*, August 26, 1902.

De Vos, Els. "The American Kitchen in Belgium? A Story of Countering, Reversing, Selective Appropriation and Sidelining." *Tensions of Europe Working Paper* no. 18 (2009).

———. *Hoe zouden we graag wonen? Woonvertogen in Vlaanderen tijdens de jaren zestig en zeventig*. Leuven: Leuven University Pers, 2012.

De Vries, Jan. *The Industrious Revolution: Consumer Behavior and the Household Economy, 1650 to the Present*. Cambridge and New York: Cambridge University Press, 2008.

de Vries, Leonard. *De Hobbyclub*. Amsterdam: Arbeiderspers, 1966.

———. *De Hobbyclub breekt baan*. Amsterdam: De Bezige Bij, 1953.

———. *De jeugdgemeenschap Hobbyclub: Een handleiding voor het oprichten van een hobby club*, 1947.

———. *De jongens van de Hobbyclub: Technische romans voor jonge mensen*. Amsterdam: De Bezige Bij, 1947.

Deleuze, Gilles, and Félix Guattari. *A Thousand Plateaus: Capitalism and Schizophrenia*. Vol. 2. Minneapolis: University of Minnesota Press, 1987 [1980].

Denker, Kai. "Heroes yet Criminals of the German Computer Revolution." In *Hacking Europe: From Computer Cultures to Demoscenes*, edited by Gerard Alberts and Ruth Oldenziel. New York: Springer, 2013.

Denton, Chad. "Récupérez! – The German Origins of French Wartime Salvage Drives, 1939–1945." *Contemporary European History* 22, no. 3 (2013): 399–430.

"Der Mord auf der französische Ostbahn." *Eisenbergisches Nachrichtsblatt*, December 18, 1860, 1.

Dinçkal, Noyan. "Arenas of Experimentation: Modernizing Istanbul in the Late Ottoman Empire." In *Urban Machinery: Inside Modern European Cities*, edited by Mikael Hård and Thomas J. Misa, 49–69. Cambridge, MA: MIT Press, 2008.

———. "Reluctant Modernization: The Cultural Dynamics of Water Supply in Istanbul, 1885–1950." *Technology and Culture* 49 (2008): 675–700.

Diogo, Maria Paula, and Dirk Van Laak. *Europe Globalizing: Mapping, Exploiting, Exchanging*. London: Palgrave Macmillan, forthcoming.

Disco, Cornelis. "Taming the Rhine: Economic Connection and Urban Competition." In *Urban Machinery: Inside Modern European Cities*, edited by Mikael Hård and Thomas J. Misa, 23–47. Cambridge, MA: MIT Press, 2008.

Divall, Colin. "Railway Imperialisms, Railway Nationalisms." In *Die Internationalität der Eisenbahn 1850–1970*, edited by Monika Burri, Kilian T. Elsasser, and David Gugerli, 195–209. Zurich: Chronos, 2003.

Dörner, Heiner. *Drei Welten – ein Leben, Prof. Dr. Ulrich Hütter, Hochschullehrer – Konstrukteur – Künstler*. Heilbronn: H. Dörner, 1995.

Douglas, Diane M. "The Machine in the Parlor: A Dialectical Analysis of the Sewing Machine." *Journal of American Culture* 5, no. 1 (1982): 20-29.

du Gay, Paul. "Self-Service: Retail, Shopping and Personhood." *Consumption, Markets and Culture* 7 (2004): 149–63.

Ebert, Anne-Katrin. "Cycling towards the Nation: The Use of the Bicycle in Germany and the Netherlands, 1880–1940." *European Review of History—Revue européenne d'histoire* 11, no. 3 (2004): 347–64.

———. *Radelnde Nationen: Die Geschichte des Fahrrads in Deutschland und den Niederlanden bis 1940*. Frankfurt a.Main: Campus Verlag 2010.

Edelman, Robert. "Everybody's Got to be Someplace: Organizing Space in the Russian Peasant House, 1880 to 1930." In *Russian Housing in the Modern Age: Design and Social History*, edited by William Craft Brumfield and Blair A. Ruble, 7–24. New York: Cambridge University Press, 1993.

Edgerton, David. *Britain's War Machine: Weapons, Resources, and Experts in the Second World War*. London: Allen Lane, 2011.

———. *The Shock of the Old: Technology and Global History since 1900*. London: Profile Books, 2006.

Eesteren, Cornelis van. *De functionele stad*. Amsterdam: De 8 en opbouw, 1935.

Elias, Norbert. *The Civilizing Process: The History of Manners*. Oxford: Basil Blackwell, 1978 [1939].

Elwell, F[rank] A. *Cycling in Europe: An Illustrated Hand-Book of Information for the Use of Touring Cyclists*. Boston: The League of American Wheelmen, 1899.

———. *The Elwell European Bicycle Tours, Tour # 1 (June 8–August 7)—England, France, Switzerland, Germany and Holland. Tour # 2 (August 3–September 30), England (London and Southern England), France (Paris, the Morvan, Auvergne, Gorges of the Tarn, the Cevennes Mountains, Monpellier le Vieux, Provence)*. Portland, ME: F. A. Elwell, 1895.

Emanuel, Martin. "Constructing the Cyclist: Ideology and Representations in Urban Traffic Planning in Stockholm, 1930–70." *The Journal of Transport History* 33, no. 1 (2012): 67–91.

———. "Europe as Perceived from the Bicycle Saddle in the Interwar Period." Paper presented at the Tensions of Europe conference, Sofia, 17–20 June, 2010.

———. *Trafikslag på undantag: Cykeltrafiken i Stockholm 1930–1980*. Stockholm: Stockholmia förlag, 2012.

Emery, Joy Spanabel. "Dreams on Paper: A Story of the Commercial Pattern Industry." In *The Culture of Sewing: Gender, Consumption, and Home Dressmaking*, edited by Barbara Burman, 235–53. Oxford: Berg, 1999.

Engels, Friedrich. *The Condition of the Working Class in England*. London: Penguin, 1987 [1844].

Esbester, Mike. "Designing Time: The Design and Use of Nineteenth-Century Transport Timetables." *Journal of Design History* 22, no. 2 (2009): 91–113.

Essner, Cornelia. "Cholera der Mekkapilger und internationale Sanitätspolitik in Ägypten (1866–1938)." *Die Welt des Islams* 32, no. 1 (1992): 41–82.

Est, Quirinus Cornelis van. *Winds of Change: A Comparative Study of the Politics of Wind Energy Innovation in California and Denmark*. Utrecht: International Books, 1999.

Etzemüller, Thomas. "Die Romantik des Reißbretts: Social engineering und demokratische Volksgemeinschaft in Schweden: Das Beispiel Alva und Gunnar Myrdal (1930–1960)." *Geschichte und Gesellschaft* 32 (2006): 445–66.

Evans, Richard J. *Death in Hamburg: Society and Politics in the Cholera Years, 1830–1910*. Oxford: Clarendon Press, 1987.

Evans, Sara M. *Personal Politics: The Roots of Women's Liberation in the Civil Rights Movement and the New Left*. New York: Knopf: distributed by Random House, 1979.

Ewart, W. "Agenda no. 100–104 August 1897 International Touring Congress." The Council of the Cyclists' Touring Club, 1897.

Fairchild, Amy L. *Science at the Borders: Immigrant Medical Inspection and the Shaping of the Modern Industrial Labor Force*. Baltimore, MD: Johns Hopkins University Press, 2003.

Federation, The European Container Glass. "Good Practices in Collection and Closed-Loop Glass Recycling in Europe." Brussels: Report prepared by the Association of Cities and Regions for Recycling and sustainable Resource management (ACR+) in partnership with the European Container Glass Federation (FEVE), 2012.

Fenton, Alexander. "Milk and Milk Products in Scotland: The Role of the Milk Marketing Boards." In *Food Technology, Science and Marketing: European Diet in the Twentieth Century*, edited by Adel P. den Hartog, 89–102. East Linton: Tuckwell Press 1995.

———. "The Scottish Women's Rural Institutes: A Case Study of Their Impact on Twentieth-Century Eating Habits." In *The Diffusion of Food Culture in Europe from the Late Eighteenth Century to the Present Day*, edited by Derek J. Oddy and Lydia Petráňová, 58–69. Prague: Academia, 2005.

Fickers, Andreas, and Pascal Griset. *Communicating Europe: Technologies, Information, Events*. London: Palgrave Macmillan, forthcoming.

Finnane, Antonia. "Yangzhou's 'Mondernity': Fashion and Consumption in the Early Nineteenth Century." *Position* 11, no. 1 (2011): 395–425.

Finstad, Terje. "Cool Alliances: Freezers, Frozen Fish and the Shaping of Industry–Rretail Relations in Norway, 1950–1960." In *Transformations of Retailing in Europe after 1945*, edited by Ralph Jessen and Lydia Langer, 195–210. Farnham: Ashgate, 2012.

Flint, Larisa. "Unzipping the USSR: Jeans as a Symbol of the Struggle between Consumerism and Consumption in the Brezhnev Era." Master's Thesis, Central European University, 1997.

Forêt, Philippe. "Railroad Literature on Suitable Places: How the Japanese Government Railways Forged an 'Old China' Travel Culture." In *Die Internationalität der Eisenbahn 1850–1970*, edited by Monika Burri, Kilian T. Elsasser, and David Gugerli, 309–26. Zurich: Chronos, 2003.

Forty, Adrian. *Objects of Desire: Design and Society, 1780–1980*. London: Thames and Hudson, 1986.

Fox, Warren. "Murder in Daily Installments: The Newspapers and the Case of Franz Müller (1864)." *Victorian Periodicals Review* 31, no. 3 (1998): 271–98.

Fraser, John Foster. "Cycling over the Caucasus Mountains." *Cassell's Family Magazine*, November 1897, 612–18.

———. *Round the World on a Wheel: Being the Narrative of a Bicycle Ride of Nineteen Thousand Two Hundred and Thirty-Seven Miles through Seventeen Countries and across Three Continents by John Foster Fraser, S. Edward Lunn, and F.H. Lowe*. London: Chatto and Windus, 1982 [1899].

Frei, Alfred Georg. *Rotes Wien: Austromarxismus und Arbeiterkultur: Sozialdemokratische Wohnungs- und Kommunalpolitik 1919–1934*. Berlin (West): DVK-Verlag, 1984.

Frevert, Ute. "'Fürsorgliche Belagerung': Hygienebewegung und Arbeiterfrauen im 19. und frühen 20. Jahrhundert." *Geschichte und Gesellschaft* 11, no. 4 (1985): 420–46.

Frey, Manuel. *Der reinliche Bürger: Entstehung und Verbreitung bürgerlicher Tugenden in Deutschland 1760–1860*. Göttingen: Vandenhoek & Ruprecht, 1997.

Friedan, Betty. *The Feminine Mystique*. New York: Norton, 1963.

Friedewald, Michael. *Der Computer als Werkzeug und Medium: Die geistigen und technischen Wurzeln des Personal Computers*. Berlin: Verlag für Geschichte der Naturwissenschaften und der Technik, 1999.

Frilling, Hildegard, and Olaf Mischer. *Pütt un Pann'n: Geschichte der Hamburger Hausmüllbeseitigung*. Hamburg: Ergebnisse Verlag, 1994.

Frykman, Jonas. *Modärna tider: Vision och vardag i folkhemmet*. Malmö: Liber, 1985.

———, and Orvar Löfgren. *Culture Builders: A Historical Anthropology of Middle-Class Life*. New Brunswick, NJ: Rutgers University Press, 1987 [1979].

Furlough, Ellen, and Carl Strikwerda. "Economics, Consumer Culture, and Gender: An Introduction to the Politics of Consumer Cooperation." In *Consumers Against Capitalism? Consumer Cooperation in Europe, North America, and Japan, 1840–1990*, edited by Ellen Furlough and Carl Strikwerda, 1–66. Lanham, MD: Rowman & Littlefield, 1999.

Furness, Zachary Mooradian. *One Less Car: Bicycling and the Politics of Automobility*. Philadelphia: Temple University Press, 2010.

Gamber, Wendy. *The Female Economy: The Millinery and Dressmaking Trades, 1860–1930*, Women in American history. Urbana: University of Illinois Press, 1997.

———. "'Reduced to Science': Gender, Technology, and Power in the American Dressmaking Trade, 1860–1910." *Technology and Culture* 36, no. 3 (1995): 455–82.

Garrison, Winfred Ernest. *Wheeling through Europe*. St. Louis: Christian Publishing Company, 1900.

Gartman, David. "Why Modern Architecture Emerged in Europe, not America: The New Class and the Aesthetics of Technocracy." *Theory, Culture and Society* 17, no. 5 (2000): 75–96.

Gecser, Ottó, and Dávid Kitzinger. "Fairy Sales: The Budapest International Fairs as Virtual Shopping Tours." *Cultural Studies* 16, no. 1 (2002): 145–64.

Gelber, Steven M. "Do-it-yourself: Constructing, Repairing and Maintaining Domestic Masculinity." *American Quarterly* 49, no. 1 (1997): 66–112.

Gemeente Amsterdam. "Het Verkeer te Amsterdam volgens de uitkomsten van de openbare verkeerstelling 1930." Amsterdam: Dienst Publieke Werken, 1934.

Gemeentemuseum, Den Haag. *Haute Couture: Voici Paris!* Zwolle: Waanders, 2010.

Georgano, Nick. *The American Automobile: A Centenary 1893–1993*. London: Prion, 1992.

Georgitsogianni, E.N. *Panagis Charokopos, 1835–1911: His Life and Work*. Athens: Livani, 2000 [Γεωργιτσογιάννη, *Παναγής Χαροκόπος, 1835–1911. Η ζωή και το έργο του*. Αθήνα: Λιβάνη, 2000].

Gerber, Robin. *Barbie and Ruth: The Story of the World's Most Famous Doll and the Woman who Created Her*. New York: Collins Business, 2009.

Gerchuk, Iurii. "The Aesthetics of Everyday Life in the Khrushchev Thaw in the USSR (1954–64)." In *Style and Socialism: Modernity and Material Culture in Post-War Eastern Europe*, edited by Susan E. Reid and David Crowley, 81–99. Oxford: Berg, 2000.

Gerhard, Herbert. *Selbstbedienung und Selbstwahl: Ein ausführlicher Leitfaden für das neue Verkaufs-System—Einrichtungs-Probleme, Kosten, Werbung, Erfahrungen aus dem In- und Ausland*. Zurich: Verlag Organisator, 1956.

Gerodetti, Natalia, and Sabin Bieri. "(Female Hetero)Sexualities in Transition: Train Stations as Gateways." *Feminist Theory* 7, no. 1 (2006): 69–87.

Ghering-van Ierlant, M.A. *Mode in Prent, 1550–1914*. Den Haag: Nederlands Kostuum Museum, 1988.

———. *Vrouwenmode in Prent: Modeprenten 1780–1930*. Utrecht: Ghering Books, 2007.

Giedion, Siegfried. *Mechanization Takes Command: A Contribution to Anonymous History*. New York: Oxford University Press, 1948.

Gille, Zsuzsa. *From the Culture of Waste to the Trash Heap of History: The Politics of Waste in Socialist and Postsocialist Hungary*. Bloomington, IN: University of Indiana Press, 2007.

Gläntzer, Volker. "Nord-Süd-Unterschiede städtischen Wohnens um 1800 im Spiegel der zeitgenössischen Literatur." In *Nord-Süd-Unterschiede in der städtischen und ländlichen Kultur Mitteleuropas*, edited by Günter Wiegelmann, 73–88. Münster: F. Copperath, 1985.

Gloag, John. *Victorian Comfort: A Social History of Design from 1830–1900*. 2nd ed. Newton Abbot: David & Charles, 1973.

Gnoli, Sofia. *Un secolo di moda italiana*. Roma: Meltemi, 2005.

Godley, Andrew. "Homeworking and the Sewing Machine in the British Clothing Industry, 1850–1905." In *The Culture of Sewing: Gender, Consumption, and Home Dressmaking*, edited by Barbara Burman, 255–68. Oxford: Berg, 1999.

———. "Selling the Sewing Machine Around the World: Singer's International Marketing Strategies, 1850–1920." *Enterprise & Society* 7, no. 2 (2006): 266–314.

Goldstein, Carolyn M. *Do it Yourself: Home Improvement in 20th-Century America*. Washington, DC and New York: National Building Museum; Princeton Architectural Press, 1998.

———. "Educating Consumers, Representing Consumers: Reforming the Marketplace through Scientific Expertise at the Bureau of Home Economics, United

States Department of Agriculture, 1923–1940." In *The Expert Consumer: Associations and Professionals in Consumer Society*, edited by Alain Chatriot, Marie-Emmanuelle Chessel, and Matthew Hilton, 73–88. Aldershot: Ashgate, 2006.

———. "From Service to Sales: Home Economics in Light and Power, 1920–1940." *Technology and Culture* 38, no. 1 (1997): 121–52.

———. "Part of the Package: Home Economists in the Consumer Products Industries, 1920–1940." In *Rethinking Home Economics: Women and the History of a Profession*, edited by Sarah Stage and Virginia B. Vincenti, 271–96. Ithaca, NY: Cornell University Press, 1997.

Gordon, Andrew. "Selling the American Way: The Singer Sales System in Japan, 1900–1938." *Business History Review* 82, no. 4 (2008): 671–99.

Gorton Ash, Timothy. "Mitteleuropa?" *Daedalus* 119, no. 1 (1990): 1–21.

Gordon, Sarah A. "'Any Desired Length': Negotiating Gender through Sports Clothing, 1870–1925." In *Beauty and Business: Commerce, Gender, and Culture in Modern America*, edited by Philip Scranton, 24–51. New York and London: Routledge, 2001.

———. "'Boundless Possibilities': Home Sewing and the Meanings of Women's Domestic Work in the United States, 1890–1930," *Journal of Women's History* 16, no. 2 (2004): 68–91.

———. *'Make It Yourself': Home Sewing, Gender, and Culture, 1890–1930*. New York: Columbia University Press, 2009.

"Gossip of the Cyclers: Royal Personages Who Ride Wheels." *The New York Times*, July 19, 1896.

"Gossip of the Cyclers: Wheeling About Cairo." *The New York Times*, April 18, 1897.

Government, U.S. *Abstracts of Reports of the Immigration Commission, with Conclusions and Recommendations and Views of the Minority*, Vol. II. Washington: Government Printing Office, 1911.

Grapperhaus, Ferdinand H.M. *Over de loden last van het koperen fietsplaatje: De Nederlandse rijwielbelasting 1924–1941*. Deventer: Kluwer, 2005.

Green, Nancy. *Ready-to-Wear, Ready-to-Work: A Century of Industry and Immigrants in Paris and New York*. Durham, NC: Duke University Press, 1997.

Griffin, Brian. "Cycling and Gender in Victorian Ireland." *Éire-Ireland* 41, nos. 1&2 (2006): 213–41.

Griffith, Robert. "The Selling of America: The Advertising Council and American Politics, 1942–1960." *Business History Review* 57, no. 3 (1983): 388–412.

Groot, Marianne, and Trudy Kunz. *Libelle 50: 50 jaar dagelijks leven in Nederland*. Utrecht: het Spectrum, 1984.

Guerrand, Roger-Henri. *Une Europe en construction: Deux siècles d'habitat social en Europe*. Paris: Éditions la découverte, 1992.

Gulick Jr., Charles A. "Vienna Taxes since 1918." *Political Science Quarterly* 53, no. 4 (1938): 533–83.

Gurova, Ol'ga. "The Life Span of Things in Soviet Society." *Russian Studies in History* 48, no. 1 (2009): 46–57.

Gyáni, Gábor. *Parlor and Kitchen: Housing and Domestic Culture in Budapest, 1870–1940*. Budapest: Central European University Press, 2002 [1998].

Haar, Rob van der. "Projectverslag De Digitale Stad. Voorlopige Versie." http://www.dds.nl/html/dds/jarig/3.0project/, last accessed 22 July, 2013.

Haddon, Leslie. "Empirical Studies Using the Domestication Framework." In *Domestication of Media and Technology*, edited by Thomas Berker, Maren Hartmann, Yves Punie, and Katie J. Ward, 103–22. Maidenhead: Open University Press, 2006.

Haddow, Robert H. *Pavilions of Plenty: Exhibiting American Culture Abroad in the 1950s*. Washington, DC: Smithsonian Institution Press, 1997.

Halliday, Fred. *The Making of the Second Cold War*. London: Verso, 1986.

Hamblin, Jacob Darwin. "Environmentalism for the Atlantic Alliance: NATO's Experiment with the 'Challenges of Modern Society'." *Environmental History* 15, no. 1 (2010): 54–75.

Hamilton, Jill. *Thomas Cook: The Holiday-Maker*. Stroud: Sutton Pub., 2005.

Hamilton, Shane. "Supermarket USA Confronts State Socialism: Airlifting the Technopolitics of Industrial Food Distribution into Cold War Yugoslavia." In *Cold War Kitchen: Americanization, Technology, and European Users*, edited by Ruth Oldenziel and Karin Zachmann, 137–59. Cambridge, MA: MIT Press, 2009.

Hamlin, Christopher. "Edwin Chadwick and the Engineers, 1842–1854: Systems and Antisystems in the Pipe-and-Brick Sewers War." *Technology and Culture* 33, no. 4 (1992): 680–709.

Hammer, Ferenc. "A Gasoline Scented Sindbad: The Truck Driver as a Popular Hero in Socialist Hungary." *Cultural Studies* 16, no. 1 (2002): 80–126.

———. "The Real One: Western Brands and Competing Notions of Authenticity in Socialist Hungary." In *Cultures of Commodity Branding*, edited by Andrew Bevan and David Wengrow, 131–54. Walnut Creek, CA: Left Coast Press, 2010.

———. "Sartorial Manoeuvres in the Dusk: Blue Jeans in Socialist Hungary." In *Citizenship and Consumption*, edited by Kate Soper and Frank Trentmann, 51–68. Basingstoke: Palgrave Macmillan, 2008.

Hanchett, Thomas W. "The Other 'Subsidized Housing': Federal Aid to Suburbanization, 1940s-1960s." In *From Tenements to the Taylor Homes: In Search of an Urban Housing Policy in Twentieth-Century America*, edited by John F. Bauman, Roger Roger Biles, and Kristin M. Szylvian, 163–79. University Park, PA: Pennsylvania State University Press, 2003 [2000].

Handler, Ruth, and Jacqueline Shannon. *Dream Doll: The Ruth Handler Story*. Stamford, CT: Longmeadow Press, 1994.

Hård, Mikael. *Machines are Frozen Spirit: The Scientification of Refrigeration and Brewing in the 19th Century—A Weberian Interpretation*. Frankfurt a.Main and Boulder, CO: Campus & Westview, 1994.

Hardy, Anne I. *Ärzte, Ingenieure und städtische Gesundheit: Medizinische Theorien in der Hygienebewegung des 19. Jarhhunderts*. Frankfurt a.Main: Campus, 2005.

Haslett, Caroline. *The Electrical Handbook for Women: Edited for the Electrical Association for Women*. London: Hodder and Stoughton, 1934.

Haugbøll, Charles. *Svajerne: De københavnske cykelbude og mælkedrenges liv og virke belyst gennem samtaler*. Copenhagen: Foreningen Danmarks Folkeminder, 1979.

Hausen, Karin. "Technical Progress and Women's Labour in the Nineteenth Century: The Social History of the Sewing Machine." In *The Social History of Politics: Critical Perspectives in West German Historical Writing since 1945*, edited by Georg Iggers, 259–81. New York: St. Martin's Press, 1985.

Hayden, Dolores. *The Grand Domestic Revolution: A History of Feminist Designs for American Homes, Neighborhoods and Cities*. Cambridge, MA: MIT Press, 1981.

Hazbun, Waleed. "The East as an Exhibit: Thomas Cook & Son and the Origins of the International Tourism Industry in Egypt." In *The Business of Tourism: Place, Faith, and History*, edited by Philip Scranton and Janet F. Davidson, 3–33. Philadelphia: University of Pennsylvania Press, 2007.

Headrick, Daniel R. *The Tools of Empire: Technology and European Imperialism in the Nineteenth Century.* Oxford: Oxford University Press, 1981.

Heafford, Michael. "Between Grand Tour and Tourism: British Travellers to Switzerland in a Period of Transition, 1814–1860." *Journal of Transport History* 27, no. 1 (2006): 25–47.

Heide, Lars. *Punched-Card Systems and the Early Information Explosion, 1880–1945*, Studies in Industry and Society. Baltimore, MD: Johns Hopkins University Press, 2009.

Hellman, Caroline. "The Other American Kitchen: Alternative Domesticity in 1950s Design, Politics, and Fiction." *Americana* (2004), http://www.american-popularculture.com/journal/articles/fall_2004/hellman.htm, last accessed 23 July, 2013.

Helphand, Kenneth I. "The Bicycle Kodak." *Environmental Review* 4, no. 3 (1980): 24–33.

Helvenston, Sally I., and Margaret M. Bubolz. "Home Economics and Home Sewing in the United States, 1870–1940." In *The Culture of Sewing: Gender, Consumption, and Home Dressmaking*, edited by Barbara Burman, 303–26. Oxford: Berg, 1999.

Henderson, Susan R. "A Revolution in the Woman's Sphere: Grete Lihotzky and the Frankfurt Kitchen." In *Architecture and Feminism*, edited by Debra L. Coleman, Elizabeth Ann Danze, and Carol Jane Henderson, 221–53. New York: Princeton Architectural Press, 1996.

Henneking, Carl. *Der Radfahrverkehr. Seine Wirtschafliche Bedeutung und die Anlage van Radfahrwegen.* Magdeburg: Verein Deutscher Fahrrad-Industrieller e.V., 1927.

Henry, Raymond. "Touring Bicycle Technical Trials in France, 1901–1950." *Cycle History* 5 (1994): 79–86.

Henthorn, Cynthia Lee. *From Submarines to Suburbs: Selling a Better America, 1939–1959.* Athens, OH: Ohio University Press, 2006.

Herlihy, David V. *Bicycle: The History.* New Haven and London: Yale University Press, 2004.

———. *The Lost Cyclist: The Epic Tale of an American Adventurer and his Mysterious Disappearance.* Boston: Houghton Mifflin Harcourt, 2010.

Hess, Heather. "The Wiener Werkstätte and the Reform Impulse." In *Producing Fashion: Commerce, Culture and Consumers*, edited by Regina L. Blaszczyk, 111–29. Philadelphia: University of Pennsylvania Press, 2008.

Hessler, Martina. "The Frankfurt Kitchen: The Model of Modernity and the 'Madness' of Traditional Users, 1926 to 1933." In *Cold War Kitchen: Americanization, Technology, and European Users*, edited by Ruth Oldenziel and Karin Zachmann, 163–84. Cambridge, MA: MIT Press, 2009.

———. *'Mrs. Modern Woman': Zur Sozial- und Kulturgeschichte der Haushaltstechnisierung.* Frankfurt a.Main and New York: Campus, 2001.

"Het Congres van den Wereldbond op 15 en 16 augustus 1902." *De Kampioen*, August 29, 1902, 645–53.

"Het Vrije Volkje HOBBY CLUB! Een pracht gelegenheid om je liefhebberijen te bedrijven Want wat jij alleen niet kan, lukt je samen met anderen." *Het Vrije Volk*, December 9, 1947.

Heymann, Matthias. "Signs of Hubris: The Shaping of Wind Technology Styles in Germany, Denmark, and the United States, 1940–1990." *Technology and Culture* 39, no. 4 (1998): 641–70.

Heynen, Hilde. *Architecture and Modernity: A Critique*. Cambridge, MA: MIT Press, 1999.

———. "Modernity and Domesticity: Tensions and Contradictions." In *Negotiating Domesticity: Spatial Productions of Gender in Modern Architecture*, edited by Hilde Heynen and Gülsüm Baydar, 1–29. London and New York: Routledge, 2005.

Hierholzer, Vera. *Nahrung nach Norm: Regulierung von Nahrungsmittelqualität in der Industrialisierung 1871–1914*. Göttingen: Vandenhoeck & Ruprecht, 2010.

Hietala, Marjatta, and Vuokko Lepistö. "Arctic Finland and the New Technology of Food Preservation and Refrigeration, 1850–1990." In *Food and Material Culture: Proceedings of the Fourth Symposium of the International Commission for Research into European Food History*, edited by Martin R. Schärer and Alexander Fenton, 310–29. East Linton: Tuckwell Press, 1998.

Hilton, Matthew. "The Cold War and the Kitchen in a Global Context: The Debate over the United Nations Guidelines on Consumer Protection." In *Cold War Kitchen: Americanization, Technology, and European Users*, edited by Ruth Oldenziel and Karin Zachmann, 341–62. Cambridge, MA: MIT Press, 2009.

Hirdman, Yvonne. "Utopia in the Home: An Essay." *International Journal of Political Economy* 22, no. 2 (1992): 5–99.

Hixson, Walter L. *Parting the Curtain: Propaganda, Culture, and the Cold War, 1945–1961*. New York: St. Martin's Press, 1997.

"Hobby-club te Kerkrade: Nuttige aanwinst." *Limburgsch dagblad*, 13 December 1949, 2.

Hodges, Karl. "Did the Emergence of the Automobile End the Bicycle Boom?" In *Cycle History* 4 (1993): 39–42.

Hoffmann, Aage. "Arbejderidrættens forhold til Socialdemokratiet ca. 1880 – ca. 1925." *Arbejderhistorie* 1 (2008): 96–115.

Hofstede, Ellen ter, Sjouk Hoitsma, and Maydy de Jong. *Kleding op de bon: kleding- en textielschaarste in Nederland 1939–1949*. Assen: Drents Museum, 1995.

Hoganson, Kristin L. *Consumers' Imperium: Global Production of American Domesticity, 1865–1920*. Chapel Hill, NC: University of North Carolina Press, 2007.

Högselius, Per, Arne Kaiser, and Erik van der Vleuten. *Europe's Infrastructure Transition: Economy, War, Nature*. London: Palgrave Macmillan, forthcoming.

Höhler, Sabine. "'Spaceship Earth': Envisioning Human Habitats in the Environmental Age." *GH Bulletin* 42 (Spring 2008): 65–85.

Holder, Julian. "The Nation State or the United States? The Irresistible Kitchen of the British Ministry of Works, 1944 to 1951." In *Cold War Kitchen: Americanization, Technology, and European Users*, edited by Ruth Oldenziel and Karin Zachmann, 235–58. Cambridge, MA: MIT Press, 2009.

Holliday, Laura Scott. "Kitchen Technologies: Promises and Alibis, 1944–1966." *Camera Obscura* 16, no. 2 (2001): 79–131.

Holm, Lennart. *HSB*. Stockholm: HSB, 1954.

Holt, Richard. "The Bicycle, the Bourgeoisie and the Discovery of Rural France, 1880–1914." *British Journal of Sports History* 2, no. 2 (1985): 127–39.

Horn, Burkhard. "Vom Niedergang eines Massenverkehrsmittels: Zur Geschichte der Städlichen Radverkehrsplanung." Diplomarbeit Gesamthochschule Kassel, 1990.

"Horrible Murder in Railway Carriage." *The New York Times*, December 21, 1860.

Hounshell, David A. *From the American System to Mass Production, 1800–1932: The Development of Manufacturing Technology in the United States*, Studies in Industry and Society. Baltimore, MD: Johns Hopkins University Press, 1984.

Hoy, Suellen. *Chasing Dirt: The American Pursuit of Cleanliness*. New York: Oxford University Press, 1995.

Huber, Valeska. "The Unification of the Globe by Disease? The International Sanitary Conferences on Cholera, 1851–1894." *Historical Journal* 49, no. 2 (2006): 453–76.

Huisman, Jaap, Irene Cieraad, Karin Gaillard, and Rob van Engelsdrop Gastelaars, eds. *Honderd jaar wonen in Nederland, 1900–2000*. Rotterdam: 010, 2000.

Huizinga, Johan. *Homo Ludens: A Study of the Play-Element in Culture*. London: Routledge & Kegan Paul, 1949 [1938].

Hünemörder, Kai F. "Vom Expertennetzwerk zur Umweltpolitik: Frühe Umweltkonferenzen und die Ausweitung der öffentlichen Aufmerksamkeit für Umweltfragen in Europa (1959–1972)." *Archiv für Sozialgeschichte* 43 (2003): 275–96.

Hunter, Robert F. "Tourism and Empire: The Thomas Cook & Son Enterprise on the Nile, 1868–1914." *Middle Eastern Studies* 40, no. 5 (2004): 28–54.

Hutching, Friedrich. "Abfallwirtschaft im Dritten Reich." *Technikgeschichte* 48 (1981): 252–73.

Hyldtoft, Ole. "Arbeiterwohnungen in Kopenhagen 1840–1914." In *Homo habitans: Zur Sozialgeschichte des ländlichen und städtischen Wohnens in der Neuzeit*, edited by Hans Jürgen Teuteberg, 199–221. Münster: F. Coppenrath, 1985.

Iakovleva-Zakharova, Larissa. "S'habiller à la soviétique: La mode sous Khrouchtchev: transfers, production, consommation." Paris: Ecole des Hautes Etudes en Sciences Sociales, 2006.

Inglehart, Ronald. "Changing Values among Western Publics from 1970 to 2006." *West European Politics* 31, nos. 1–2 (2008): 130–46.

———. "The Silent Revolution in Europe: Intergenerational Change in Post-Industrial Societies." *American Political Science Review* 65, no. 4 (1971): 991–1017.

Inklaar, Frank. *Van Amerika geleerd. Marshallhulp en kennisimport in Nederland*. Den Haag: Sdu, 1997.

J.C.R[edele]. "Rondom het Internationaal Congrès." *De Kampioen*, July 10, 1903, 549–606.

Jakabovics, Barrie Robyn. "Displaying American Abundance Abroad: The Misinterpretation of the 1959 American National Exhibition in Moscow." 2007.

Jakić, Bruno. "Galaxy and the New Wave: Yugoslav Computer Culture in 1980s." In *Hacking Europe: From Computer Cultures to Demoscenes*, edited by Gerard Alberts and Ruth Oldenziel. New York: Springer, 2013.

Jamison, Andrew. *The Making of Green Knowledge: Environmental Politics and Cultural Transformation*. Cambridge: Cambridge University Press, 2001.

———, Ron Eyerman, and Jacqueline Cramer. *The Making of the New Environmental Consciousness: A Comparative Study of the Environmental Movements in Sweden, Denmark, and the Netherlands*, Environment, Politics, and Society series. Edinburgh: Edinburgh University Press, 1990.

Jansen Henderiks, Gerda. *De Hobbyclub*. Hilversum: VPRO Andere Tijden, Radio broadcast, 2000.

Jefferson, Robert Louis. *Across Siberia on a Bicycle*. London: The Cycle Press, 1897.

Jerman, Betty. "Sex and the Single Doll." *The Guardian*, April 23, 1968, 7.

Jørgensen, Finn Arne. "Green Citizenship at the Recycling Junction: Consumers and Infrastructures for the Recycling of Packaging in Twentieth-Century Norway." *Contemporary European History* 22, no. 3 (2013): 499–516.

────. *The Green Machine: The Infrastructure of Beverage Container Recycling*. New Brunswick: Rutgers University Press, 2011.

────. "Tidens krav: Framveksten av det vitenskapelige husstellet i Norge, 1900–1940." *STS Report* no 52, NTNU, Trondheim, 2002.

Jørgensen, Ulrik, and Peter Karnøe. "The Danish Wind Turbine Story: Technical Solutions to Political Visions?" In *Managing Technology in Society: The Approach of Constructive Technology Assessment*, edited by Arie Rip, 57–82. London: Pinter, 1995.

Judt, Tony. "The Glory of the Rails." *The New York Review of Books*, December 23, 2010.

────. *Postwar: A History of Europe Since 1945*. London: Pimlico, 2007 [2005].

Kaiser, Wolfram, and Johan Schot. *Writing the Rules for Europe: Experts, Cartels, and International Organizations*. London: Palgrave Macmillan, forthcoming.

Kaiserfeld, Thomas. "Computerizing the Swedish Welfare State: The Middle Way of Technological Success and Failure." *Technology and Culture* 37, no. 2 (1996): 249–79.

Kamp, Linda M. "The Role of Policy in Inverse Developments: Comparing Dutch and Danish Wind Energy." In *Inverse Infrastructures: Disrupting Networks from Below*, edited by Tineke M. Egyedi and Donna C. Mehos, 125–40. Cheltenham: Edward Elgar, 2012.

"Katholieke Hobby-clubs opgericht: Nieuwe tak van Katholieke Jeugdbeweging." *Nieuwe Limburger Koerier*, 5 January 1951.

Kidwell, Claudia B., and Margaret C. Christman. *Suiting Everyone: The Democratization of Clothing in America*. Washington, DC: Smithsonian Institution Press for the National Museum of History and Technology, 1974.

Kilgannon, Corey. "Change Blurs Memories in a Famous Suburb." *The New York Times*, October 13, 2007, 1.

Kirchhof, Astrid Mignon. *Das Dienstfräulein auf dem Bahnhof: Frauen im öffentlichen Raum im Blick der Berliner Bahnhofsmission 1894–1939*. Stuttgart: Franz Steiner Verlag, 2011.

Kisbán, Eszter. "Coffee in Hungary: Its Advent and Integration into the Hierarchy of Meals." In *Kaffee im Spiegel europäischer Trinksitten: Coffee in the Context of European Drinking Habits*, edited by Daniela U. Ball, 69–82. Zurich: Johann Jacobs Museum, 1991.

Kjærnes, Unni. "Milk: Nutritional Science and Agricultural Development in Norway, 1890–1990." In *Food Technology, Science and Marketing: European Diet in the Twentieth Century*, edited by Adel P. den Hartog, 103–16. East Linton: Tuckwell Press, 1995.

Klemann, Hein A.M. "Dutch Industrial Companies and the German Occupation, 1940–1945." *Vierteljahrschrift für Sozial- und Wirtschaftsgeschichte* 93, no. 1 (2006): 1–22.

Kline, Ronald. "Agents of Modernity: Home Economists and Rural Electrification, 1925–1950." In *Rethinking Home Economics: Women and the History of a Profession*, edited by Sarah Stage and Virginia B. Vincenti, 237–52. Ithaca, NY: Cornell University Press, 1997.

Kloek, Els. *Vrouw des huizes: Een cultuurgeschiedenis van de Hollandse huisvrouw*. Amsterdam: Balans, 2009.

Knaak, Silke. *Deutsche Modepuppen der 50er und 60er*. Seevetal: Privately published, 2004.

Koese, Yavuz. "Nestlé in the Ottoman Empire: Global Marketing with Local Flavor 1870–1927." *Enterprise & Society* 9, no. 4 (2008): 724–61.

Köhler, Thilo. *Sie werden pläziert! Die Geschichte der MITROPA.* Berlin: TRANSIT Buchverlag, 2002.

König, Wolfgang. *Geschichte der Konsumgesellschaft.* Stuttgart: Franz Steiner, 2000.

Köstering, Susanne. "'Millionen im Müll?' Altmaterialverwertung nach dem Vierjahresplan." In *Müll von gestern? Eine umweltgeschichtliche Erkundung in Berlin und Brandenburg*, edited by Susanne Köstering and Renate Rüb, 139–49. Münster: Waxmann, 2003.

———. "'Pionere der Rohstoffbeschaffung': Lumpensammler im Nationalsozialismus, 1934–1939." *WerkstattGeschichte* 17 (1997): 45–65.

Kramer, Karen Ruoff. "Blauhemd und Bluejeans in Filmen der DEFA." In *Jeans, Rock & Vietnam: Amerikanische Kultur in der DDR*, edited by Therese Hörnigk and Alexander Stephan, 129–51. Berlin: Theater der Zeit, 2002.

Kreuzer, Bernd. "Historische Verkehrsutopien für die Stadt der Zukunft: Von der Utopie zur Realität." In *Stadt: Strom-Strasse-Schiene: Die Bedeutung des Verkehrs für die Genese der mitteleuropäischen Städtelandschaft*, edited by Alois Niederstätter, 257–305. Linz: Österreichischen Arbeidskreis für Stadtgeschichtsforschung, 2001.

Kunitz, Sylvia. "'Recycling fürs Reich': Secondary Raw Materials in Nazi Germany 1933–1945." Paper presented at the Rachel Carson Center Workshop "Re/Cycling Histories: Users and the Paths to Sustainability in Everyday Life". Munich, May 27–29, 2011.

"L'algemeene Nederlandsche Wielrijdersbond." *Cycliste belge illustré*, July 22, 1897, 1.

"La fiesta del pedal." *La Vanguardia* (1913): 4.

"Laatste berigten: Paris 6 december." *Nieuw Amsterdamsch Handels en Effectenblad*, December 8, 1860.

Landström, Catharina. "National Strategies: The Gendered Appropriation of Household Technology." In *The Intellectual Appropriation of Technology: Discourses on Modernity, 1900–1939*, edited by Mikael Hård and Andrew Jamison, 163–88. Cambridge, MA: MIT Press, 1998.

Lapinskiene, Laura. "From 'Hygienic Saddles' to the 'Vehicle of Beauty': Discourses on the Cycling Woman from the Turn of the Century to the 1930s in Lithuania." Budapest: Central European University, 2008.

Lean, Thomas. "'Inside a Day You'll be Talking to it Like an Old Friend': The Making and Remaking of Sinclair Personal Computing in 1980s Britain." In *Hacking Europe: From Computer Cultures to Demoscenes*, edited by Gerard Alberts and Ruth Oldenziel. New York: Springer, 2013.

Leeuw, Denis de. "Le transport international des voyageurs et des bagages par chemins de fer: Matériaux pour une convention à conclure." PhD diss., University of Amsterdam, 1908.

Lekkas, Theodore. "Legal Pirates Ltd: Home Computing Cultures in Early 1980s Greece." In *Hacking Europe: From Computer Cultures to Demoscenes*, edited by Gerard Alberts and Ruth Oldenziel. New York: Springer, 2013.

Lenin, V.I. "A Great Beginning: Heroism of The Workers in the Rear: 'Communist Subbotniks'". In *Collected Works*, 29: 409–34. Moscow: Progress, 1965.

———. "The Tasks of the Working Women's Movement in The Soviet Republic: Speech Delivered at the Fourth Moscow City Conference of Non-Party Working Women September 23, 1919." In *Collected Works*, 30: 40-46. Moscow: Progress, 1965.

Levitt, Sarah. "Cheap Mass-Produced Men's Clothing in the Nineteenth and Early Twentieth Centuries." *Textile History* 22, no. 2 (1991): 179–92.

Lewis, Su Lin. "Cosmopolitanism and the Modern Girl: A Cross-Cultural Discourse in 1930s Penang." *Modern Asian Studies* 43, no. 6 (2009): 1385–419.

Lintsen, Harry, and Ruth Oldenziel. "Het technisch paradijs: Honderd jaar elektriciteit in huis." Haarlem: Teylers Museum, 2000.

Lloyd, David. *Battlefield Tourism: Pilgrimage and the Commemoration of the Great War in Britain, Australia and Canada, 1919–1939.* London: Berg, 1998.

Löfgren, Orvar. "The Sweetness of Home: Class, Culture and Family Life in Sweden." In *The Anthropology of Space and Place: Locating Culture*, edited by Setha M. Low and Denise Lawrence-Zúñiga, 142–59. Malden, MA: Blackwell, 2003.

Lord, M.G. *Forever Barbie: The Unauthorized Biography of a Real Doll.* New York: Morrow and Co., 1994.

Lövgren, Britta. *Hemarbete som politik: Diskussioner om hemarbete, Sverige 1930–40-talen, och tillkomsten av Hemmens Forskningsinstitut.* Stockholm: Almqvist & Wiksell International, 1993.

Löw, Martina. *Raumsoziologie.* Frankfurt a.Main: Suhrkamp, 2001.

Lubar, Steven. *Infoculture: The Smithsonian Book of Information Age Inventions.* Boston and New York: Houghton Mifflin Company, 1993.

Lubbers, Annette. *Lloydhotel.* Amsterdam: Uitgeverij Bas Lubberhuizen, 2004.

Lundin, Per. *Bilsamhället, Ideologi, expertis och regelskapande i efterkrigstidens Sverige.* Stockholm: Stockholmia förlag, 2008.

———. "Mediators of Modernity: Planning Experts and the Making of the 'Car-Friendly' City in Europe." In *Urban Machinery: Inside Modern European Cities*, edited by Mikael Hård and Thomas J. Misa, 257–79. Cambridge, MA: MIT Press, 2008.

Lupo, Salvatore. *Storia della mafia: Dalle origini ai giorni nostri.* Roma: Donzelli, 1996 [1993].

Lyngø, Inger Johanne. "Symbols in the Rhetoric on Diet and Health—Norway 1930: The Relation between Science and the Performance of Daily Chores." In *Order and Disorder: The Health Implications of Eating and Drinking in the Nineteenth and Twentieth Centuries*, edited by Alexander Fenton, 155–69. East Linton: Tuckwell Press, 2000.

MacBride, Samantha. *Recycling Reconsidered: The Present Failure and Future Promise of Environmental Action in the United States.* Cambridge, MA: MIT Press, 2012.

MacCannell, Dean. *The Tourist: A New Theory of the Leisure Class.* New York: Schocken Books, 1976.

Macekura, Stephen. "The Limits of the Global Community: The Nixon Administration and Global Environmental Politics." *Cold War History* 11, no. 4 (2011): 489–518.

Mackintosh, Phillip Gordon, and Glen Norcliffe. "Men, Women and the Bicycle: Gender and Social Geography of Cycling in the Late Nineteenth Century." In *Cycling and Society*, edited by Dave Horton, Peter Cox, and Paul Rosen, 153–77. Aldershot: Ashgate, 2007.

Macmillan, Margaret Owen. *Paris 1919: Six Months that Changed the World.* New York: Random House, 2002.

MacNamara, Walter Henry. *A Digest of the Law of Carriers of Goods and Passengers by Land and Internal Navigation.* London: Stevens and Sons, 1888.

Maher, Vanessa. *Tenere le fila: Sarte, sartine a Torino e cambiamento sociale 1860–1960.* Turin: Rosenberg & Sellier, 2007.

Mahrane, Yannik et al. "De la nature à la biosphere: L'invention politique de l'environnement global 1945–1972." *Vingtième Siècle: Revue d'histoire* 113, no. 1 (2012): 127–41.

Maier, Charles S. "The Politics of Productivity: Foundations of American International Economic Policy after World War II." *International Organization* 31, no. 4 (1977): 607–33.

Maines, Rachel P. *Hedonizing Technologies: Paths to Pleasure in Hobbies and Leisure.* Baltimore, MD: Johns Hopkins University Press, 2009.

Małolepszy, Wojciech. "Między starym a nowym brzegiem Świdra." *Aspiracje* (2010–2011): 38–45.

Männistö-Funk, Tiina. "The Crossroads of Technology and Tradition: Vernacular Bicycles in Rural Finland, 1880–1910." *Technology and Culture* 52, no. 4 (2011): 733–56.

———. "Gendered Practices in Finnish Cycling, 1890–1939." *Icon* 16 (2010): 53–73.

Marcus, Sharon. *Apartment Stories: City and Home in Nineteenth-Century Paris and London.* Berkeley, CA: University of California Press, 1999.

Marling, Karal Ann. *As Seen on TV: The Visual Culture of Everyday Life in the 1950s.* Cambridge, MA: Harvard University Press, 1994.

Martz, Laura. "Don't Take the World as It Is: Waag Society's First Fifteen Years." *Waag Society Magazine* 17 (2009).

Mason, Philip P. "The League of American Wheelmen and the Good-Roads Movement, 1880–1905." PhD Diss., University of Michigan, 1957.

May, Elaine Tyler. *Homeward Bound: American Families in the Cold War Era.* New York: Basic Books, 1988.

May, Jan Andreas. "Die Villa als Wohnkultur und Lebensform: Der Grunewald vor dem Ersten Weltkrieg." In *Berliner Villenleben: Die Inszenierung bürgerlicher Wohnwelten am grünen Rand der Stadt um 1900*, edited by Heinz Reif, 285–308. Berlin: Gebr. Mann Verlag, 2008.

Meadows, Donella H., and The Club of Rome. *The Limits to Growth: A Report for the Club of Rome's Project on the Predicament of Mankind.* New York: Universe Books, 1972.

Medvedev, Katalin. "Ripping Up the Uniform Approach: Hungarian Women Piece Together a New Communist Fashion." In *Producing Fashion: Commerce, Culture, and Consumers*, edited by Regina L. Blaszczyk, 250–72. Philadelphia: University of Pennsylvania Press, 2008.

Merki, Christoph Maria. *Zucker gegen Saccharin: Zur Geschichte der künstlichen Süsstoffe.* Frankfurt a.Main: Campus, 1993.

Merlo, Elisabetta. *Moda italiana: Storia di un'industria dall'Ottocento ad oggi.* Venezia: Marsilio 2003.

———, and Francesca Polese. "Accessorizing, Italian Style: Creating a Market for Milan's Fashion Merchandise." In *Producing Fashion: Commerce, Culture and Consumers*, edited by Regina L. Blaszczyk, 42–61. Philadelphia: University of Pennsylvania Press, 2008.

Metchnikoff, Elie. *The Prolongation of Life: Optimistic Studies.* London: Heinemann, 1907.

Meyer, Christian. "Recycled Glass – From Waste Material to Valuable Resource," Paper presented at the Symposium on Recycling and Reuse of Glass Cullet, Dublin 19–20 March, 2001.

Meyer, Erna. *De nieuwe huishouding*. Translated by R. Lotgering-Hillebrand. Edited by R. Lotgering-Hillebrand. Amsterdam: Van Holkema, 1928.

———. *Der neue Haushalt: ein Wegweiser zu wirtschaftlicher Hausfürung*. Stuttgart: Frankh'sche Verlagshandlung, 1926.

Meyer, Jan-Henrik. "Appropriating the Environment: How the European Institutions Received the Novel Idea of the Environment and Made it Their Own Research College 'The Transformative Power of Europe'." *KFG Working Paper 31*, Berlin: Free University Berlin, 2011.

———. "Green Activism: The European Parliament's Environmental Committee Promoting a European Environmental Policy in the 1970s." *Journal of European Integration History* 17, no. 1 (2011): 73–85.

———. "Saving Migrants: A Transnational Network Supporting Supranational Bird Protection Policy." In *Transnational Networks in Regional Integration: Governing Europe, 1945–83*, edited by Wolfram Kaiser, Brigitte Leucht, and Michael Gehler, 176–98. Basingstoke: Palgrave Macmillan, 2010.

———. "Challenging the Atomic Community: The European Environmental Bureau and the Europeanization of Anti-Nuclear Protest." In *Societal Actors in European Integration: Polity-Building and Policy-Making 1958–1992*, edited by Wolfram Kaiser and Jan-Henrik Meyer. Basingstoke: Palgrave Macmillan, forthcoming 2013.

Meyer, Sibylle. *Das Theater mit der Hausarbeit: Bürgerliche Repräsentation in der Familie der wilhelminischen Zeit*. Frankfurt a.Main: Campus, 1982.

Meyerová, Erna. *Moderní domácnost: Rádce usporného vedení domácnosti*. Translated by Anna Hrubá. Edited by Jarmila Kraftová. Praha: B. Kočí, 1928.

Michel, Eugène. "Les cyclistes en Algérie: Le Congrès de l'union vélocipédique algérienne." 1894.

"Miejskie zakłady mleczarskie 'Agril'." *Codzienna Gazeta Handlowa*, October 23, 1934, 4.

Mihm, Andrea. *Packend…Eine Kulturgeschichte des Reisekoffers*. Marburg: Jonas Verlag, 2001.

Miller, G. Wayne. *Toy Wars: The Epic Struggle between G.I. Joe, Barbie, and the Companies that Make them*. New York: Times Books, 1998.

Miller, Michael B. *The Bon Marché: Bourgeois Culture and the Department Store, 1869–1920*. Princeton: Princeton University Press, 1981.

Milroy, Patrick D. "Rough Road Cyclists Display Political Power: The League of American Wheelmen in the Good Roads Movement as Reported by *The New York Times*, 1880–1900." *North American Society for Sport History: Proceedings and Newsletter* (1989): 39–41.

Misa, Thomas J. "Appropriating the International Style: Modernism in East and West." In *Urban Machinery: Inside Modern European Cities*, edited by Mikael Hård and Thomas J. Misa, 71–95. Cambridge, MA: MIT Press, 2008.

Molnar, Virag. "In Search of the Ideal Socialist Home in Post-Stalinist Hungary: Prefabricated Mass Housing or Do-It-Yourself Family Home?" *Journal of Design History* 23, no. 1 (2010): 61–81.

Mom, Gijs. "Roads without Rails: European Highway-Network Building and the Desire for Long-Range Motorized Mobility." *Technology and Culture* 46, no. 4 (2005): 745–72.

Mom, Gijs P.A., and Ruud Filarski. *Van Transport naar Mobiliteit: De Mobiliteitsexplosie (1895–2005)*. Zutphen: Walburgpers, 2008.

Mom, Gijs, Peter Staal, and Johan W. Schot. "Civilizing Motorized Adventure: Automotive Technology, User Culture and the Dutch Touring Club as Mediator." In *Manufacturing Technology, Manufacturing Consumers: The Making of Dutch Consumer Society*, edited by Adri A. de la Bruhèze and Ruth Oldenziel, 139–58. Amsterdam: Aksant, 2009.

Montijn, Ileen. *Aan tafel! Vijftig jaar eten in Nederland*. Amsterdam: Uitgeverij Kosmos, 1991.

Möser, Kurt. "Motorization of German Societies in East and West." In *Towards Mobility: Varieties of Automobilism in East and West*, edited by Corinna Kuhr-Korolev and Dirk Schlinkert, 55–72. Wolfsburg: Volkswagen AG, 2009.

Moser, Oskar. "Anmerkungen zum Nord-Süd-Vergleich in Hausbau, Wohnung und Gerät." In *Nord-Süd-Unterschiede in der städtischen und ländlichen Kultur Mitteleuropas*, edited by Günter Wiegelmann, 63–72. Münster: F. Copperath, 1985.

Moss, Sarah, and Alexander Badenoch. *Chocolate: A Global History*. London: Reaktion Books, 2009.

Mühl, Albert, and Jürgen Klein. *125 Jahre Internationale Schlafwagen-Gesellschaft: die Luxuszüge, Geschichte und Plakate; 125 ans Compagnie Internationale des Wagons-Lits: les trains de luxe, historique et affiches; 125 years International Sleeping Car Company: Trains de luxe, History and Posters; 125 ans Compagnie Internationale des Wagons-Lits 125 years International Sleeping Car Company*. Freiburg: EK-Verlag, 1998.

Mullen, Richard, and James Munson. *The Smell of the Continent: The British Discover Europe*. London: Pan Books, 2009.

Mumford, Eric. *The CIAM Discourse on Urbanism, 1928–1960*. Cambridge, MA: MIT Press, 2000.

———. *Defining Urban Design: CIAM Architects and the Formation of a Discipline, 1937–1969*. New Haven, CT: Yale University Press, 2009.

Myrdal, Alva. *Nation and Family: The Swedish Experiment in Democratic Family and Population Policy*. London: Kegan Paul, 1945 [1941].

———, and Gunnar Myrdal. *Kris i befolkningsfrågan*. Stockholm: Albert Bonniers, 1934.

Myrdal, Jan. *En Meccano-pojke berättar, eller anteckningar om ett borgerligt förnuft* Höganäs: Wiken, 1989.

Nevejan, Caroline, and Alexander Badenoch. "How Amsterdam Invented the Internet: European Networks of Significance, 1980–1995." In *Hacking Europe: From Computer Cultures to Demoscenes*, edited by Gerard Alberts and Ruth Oldenziel. New York: Springer, 2013.

Nickles, Shelley. "More is Better: Mass Consumption, Gender, and Class Identity in Postwar America." *American Quarterly* 54, no. 4 (2002): 581–622.

———. "'Preserving Women': Refrigerator Design as Social Process in the 1930s." *Technology and Culture* 43, no. 4 (2002): 693–727.

Nikles, Bruno W. *Soziale Hilfe am Bahnhof. Zur Geschichte der Bahnhofsmission in Deutschland (1894–1960)*. Freiburg im Breisgau: Lambertus Verlag, 1994.

Nolan, Mary. "Consuming America, Producing Gender." In *The American Century in Europe*, edited by R. Laurence Moore and Maurizio Vaudagna, 243–61. Ithaca, NY: Cornell University Press, 2003.

———. "'Housework Made Easy': The Taylorized Housewife in Weimar Germany's Rationalized Economy." *Feminist Studies* 16, no. 3 (1990): 549–77.

———. *Visions of Modernity: American Business and the Modernization of Germany*. New York: Oxford University Press, 1995.

Nordström, Ludvig. *Lort-Sverige*. Stockholm: Kooperativa förbundets bokförlag, 1938.

Oakley, William. *Winged Wheel: The History of the First Hundred Years of the Cyclists' Touring Club*. Godalming: CTC, 1977.

Oddy, Nicholas. "A Beautiful Ornament in the Parlour or Boudoir: The Domestication of the Sewing Machine." In *The Culture of Sewing: Gender, Consumption, and Home Dressmaking*, edited by Barbara Burman, 285–302. Oxford: Berg, 1999.

Odin, Samy. *Lilas: The Exemplary Life of a Fashion Doll under Napoleon III*. Paris: Musée Poupée, 2010.

Oedekoven-Gerischer, Angela. *Frauen im Design: Berufsbilder und Lebenswege seit 1900*. 2 vols. Stuttgart: Design Center Stuttgart, 1989.

Offen, Karen. "'Powered by a Woman's Foot': A Documentary Introduction to the Sexual Politics of the Sewing Machine in Nineteenth-Century France." *Women's Studies International Forum* 11, no. 2 (1988): 93–101.

Okawa, Tomoko. "Licensing Practices at Maison Christian Dior." In *Producing Fashion: Commerce, Culture and Consumers*, edited by Regina L. Blaszczyk, 82–107. Philadelphia: University of Pennsylvania Press, 2008.

Oldenziel, Ruth. "Boys and Their Toys: The Fisher Body Craftsman's Guild, 1930–1968, and the Making of a Male Technical Domain." *Technology and Culture* 38, no. 1 (1997): 60–96.

———. "Exporting the American Cold War Kitchen: Challenging Americanization, Technological Transfer, and Domestication." In *Cold War Kitchen: Americanization, Technology, and European Users*, edited by Ruth Oldenziel and Karin Zachmann, 315–39. Cambridge, MA: MIT Press, 2009.

———. "Of Old and New *Cyborgs*: Feminist Narratives of Technology." *Letterature d'America* 14, no. 55 (1994): 95–111.

———, and Adri A. de la Bruhèze. "Bicycle Taxes as Tools of Policy, 1890–2012." In *Re/Cycling Histories: Paths to Sustainability*, edited by Ruth Oldenziel and Helmuth Trischler. London: Berghahn, Submitted.

———, and Adri A. de la Bruhèze. "Contested Spaces: Bicycle Lanes in Urban Europe, 1900–1995." *Transfers* 1, no. 2 (2011): 31–49.

———, and Adri A. de la Bruhèze. "Theorizing the Mediation Junction for Technology and Consumption." In *Manufacturing Technology, Manufacturing Consumers: The Making of Dutch Consumer Society*, edited by Adri A. de la Bruhèze and Ruth Oldenziel, 9–40. Amsterdam: Aksant, 2009.

———, Adri A. de la Bruhèze, and Onno de Wit. "Europe's Mediation Junction: Technology and Consumer Society in the 20th Century." *History and Technology* 21, no. 1 (2005): 107–39.

———, et al. *Huishoudtechnologie en medische techniek*. Vol. 4, Techniek in Nederland in de twintigste eeuw. Zutphen: Walburg Pers, 1998.

———, and Milena Veenis. "The Glass Recycling Container in the Netherlands: Symbol in Times of Scarcity and Abundance, 1939–1978." *Contemporary European History* 22, no. 3 (2013): 453–76.

Olney, Martha L. *Buy Now, Pay Later: Advertising, Credit, and Consumer Durables in the 1920s*. Chapel Hill, NC: University of North Carolina Press, 1991.

"Om klantenbinding Barbie-fans." *De Tijd*, March 12, 1966, 6.

Oppenheimer, Jerry. *Toy Monster: The Big, Bad World of Mattel*. Hoboken, NJ: Wiley, 2009.

Orland, Barbara. "Milky Ways: Dairy, Landscape and Nation Building until 1930." In *Land, Shops and Kitchens: Technology and the Food Chain in Twentieth-Century Europe*, edited by Carmen Sarasúa, Peter Scholliers, and Leen Van Molle, 212–54. Turnhout: Brepols, 2005.

———. "Turbo-Cows: Producing a Competitive Animal in the Nineteenth and Early Twentieth Centuries." In *Industrializing Organisms: Introducing Evolutionary History*, edited by Susan R. Schrepfer and Philip Scranton, 167–89. New York: Routledge, 2003.

Otter, Chris. *The Victorian Eye: A Political History of Light and Vision in Britain, 1800–1910*. Chicago and London: University of Chicago Press, 2008.

Otterloo, Anneke H. van. *Eten en eetlust in Nederland (1840–1990): een historisch-sociologische studie*. Amsterdam: Bert Bakker, 1990.

———. "Fast Food and Slow Food: The Fastening Food Chain and Recurrent Countertrends in Europe and the Netherlands (1890–1990)." In *Land, Shops and Kitchens: Technology and the Food Chain in Twentieth-Century Europe*, edited by Carmen Sarasúa, Peter Scholliers, and Leen Van Molle, 255–77. Turnhout: Brepols, 2005.

"Oude school werd jeugdgebouw: Vele jeugdverenigingen vonden eigen tehuis in de Violenstraat." *Nieuwsblad van het Noorden*, December 6, 1952.

Oudshoorn, Nelly, and Trevor J. Pinch, eds. *How Users Matter: The Co-Construction of Users and Technology*. Cambridge, MA: MIT Press, 2003.

Overbeeke, Peter van. *Kachels, geisers en fornuizen: Keuzeprocessen en energieverbruik in Nederlandse huishoudens, 1920–1975*. Hilversum: Verloren, 2001.

Öxler, Florian Karl. *Vom tragbaren Labor zum Chemiebaukasten: Zur Geschichte des Chemieexperimentierkastens unter besonderer Berücksichtigung des deutschsprachigen Raums*. Stuttgart: Wissenschaftliche Verlagsgesellschaft, 2010.

Panke-Kochinke, Birgit. *Göttinger Professorenfamilien: Strukturmerkmale weiblichen Lebenszusammenhangs im 18. und 19. Jahrhundert*. Pfaffenweiler: Centaurus, 1993.

Papon, Francis. "The Evolution of Bicycle Mobility in France." Paper presented at the 22nd International Cycling History Conference, Paris, 25–28 May, 2011.

Paris, Ivan. *Oggetti cuciti: l'abbigliamento pronto in Italia dal primo dopoguerra agli anni Settanta*. Milan: Franco Angeli, 2006.

Parusheva, Dobrinka. "Orient Express, or About European Influences on Everyday Life in the Nineteenth Century Balkans." *New Europe College Yearbook*, no. 9 (2001–2002): 139–67.

Patterson, Matthew. *Automobile Politics: Ecology and Cultural Political Economy*. Cambridge: Cambridge University Press, 2007.

Patterson, Walt. "A Decade of Friendship: The First Ten Years." In *The Environmental Crisis: A Handbook for all Friends of the Earth*, edited by Des Wilson and Friends of the Earth, 140–54. London: Heinemann Educational Books, 1984.

Peers, Juliette. *The Fashion Doll: From Bébé Jumeau to Barbie*. Oxford and New York: Berg, 2004.

Peiss, Kathy. *Cheap Amusements: Working Women and Leisure in Turn-of-the-Century New York*. Philadelphia: Temple University Press, 1986.

Pelka, Anna. *Jugendmode und Politik in der DDR und Polen: Eine vergleichende Analyse 1968–1989*. Osnabrück: Fibre, 2008.

Pells, Richard. *Not Like US: How Europeans Have Loved, Hated, and Transformed American Culture Since World War II*. New York: Basic Books, 1997.

Pennell, Elizabeth Robins. *Over the Alps on a Bicycle*. London: T. Fisher Unwin, 1898.

———, and Joseph Pennell. "Over the Alps on a Bicycle." *The Century Illustrated Monthly Magazine*, April 1898, 837–51.

Pennell, J. and E[lizabeth]. R[owe]. "The Most Picturesque Place in the World." *The Century Illustrated Monthly Magazine* LXVI (n.s. XXIV) (1893): 345–51.

Pennell, Joseph. "How to Cycle in Europe." *Harper's New Monthly Magazine*, April 1898, 680–92.

Perl, Louis de. "Exposé de la question des relations internationals (art. XIX du Questionnaire de la troisième session du congrès). Moyens de faciliter les relations internationals en ce qui concerne les transports des voyageurs et de leurs bagages." 1543–59. Paris: Congrès international des chemins de fer. Troisième session, 1889.

Perrot, Michelle. "Femmes et Machines au XIXe siècle." *Romantisme* 41 (1983): 5–18.

Peters, Peter F. *Time, Innovation and Mobilities: Travel in Technological Cultures*. London: Routledge, 2006.

Petrick, Gabriella M. "'Purity as Life': H.J. Heinz, Religious Sentiment, and the Beginning of the Industrial Diet." *History and Technology* 27, no. 1 (2011): 37–64.

Petzina, Dieter. *Autarkiepolitik im Dritten Reich: Der nationalsozialistische Vierjahresplan*. Stuttgart: Deutsche Verlags-Anstalt, 1968.

Petzold, Rudolf. *Die Altstoffe und die Rohstoffversorgung Deutschlands*. Leipzig: Noske, 1941.

Pfaffenberger, Bryan. "Technological Dramas." *Science, Technology, & Human Values* 17, no. 3 (1992): 282–312.

Pfister, Christian. "The '1950s Syndrome' and the Transition from a Slow-Going to a Rapid Loss of Global Sustainability." In *The Turning Points of Environmental History*, edited by Frank Uekötter, 90–118. Pittsburgh: University of Pittsburgh Press, 2010.

Phillips, R. Randal. *The Servantless House*. London: Country Life, 1920.

Piccinato, Giorgio. "Zum italienischen Volkswohnungsgesetz vom 31. Mai 1903: Entstehung des sozialen Wohnungsbaus in Italien 1896–1914." In *Die Kleinwohnungsfrage: Zu den Ursprüngen des sozialen Wohnungsbaus in Europa*, edited by Juan Rodríguez-Lores and Gerhard Fehl, 391–408. Hamburg: Hans Christians Verlag, 1988.

Pinch, Trevor J., and Wiebe E. Bijker. "The Social Construction of Facts and Artifacts: Or How the Sociology of Science and the Sociology of Technology Might Benefit Each Other." In *The Social Construction of Technological Systems: New Directions in the Sociology and History of Technology*, edited by Wiebe E. Bijker, Thomas P. Hughes, and Trevor J. Pinch, 17–50. Cambridge, MA: MIT Press, 1987.

Pinsolle, Dominique. "Gas Consumers' Strikes and the Rise of an Independent Consumers Movement in France (1892–1914)." Paper presented at NIAS, Wassenaar, 2011.

Pivato, Stefano. "The Bicycle as a Political Symbol: Italy, 1885–1955." *The International Journal of the History of Sport* 7, no. 2 (1990): 173–87.

Plans Bosch, Joan, "Bodas de oro de la fiesta del Pedal." *El Mundo Deportivo* (6 October, 1974): 25.

Polhemus, Ted. *Streetstyle: from Sidewalk to Catwalk*. New York: Thames and Hudson, 1994.

Popplow, Marcus. "Vom Lastenrad zum Cargobike—zum Neustart einer vergessenen Technologie." *Kultur & Technik: Das Magazin aus dem Deutschen Museum* 37, no. 2 (2013 in press).

Pouillard, Véronique. "In the Shadow of Paris? French Haute Couture and Belgian Fashion between the Wars." In *Producing Fashion: Commerce, Culture and Consumers*, edited by Regina L. Blaszczyk, 62–81. Philadelphia: University of Pennsylvania Press, 2008.

Power, Anne. *Hovels to High Rise: State Housing in Europe since 1850*. London and New York: Routledge, 1993.

Pratt, Mary Louise. *Imperial Eyes: Travel Writing and Transculturation*. New York: Routledge, 2008 [1992].

Prynn, David. "The Clarion Clubs, Rambling and the Holiday Associations in Britain since the 1890s." *Journal of Contemporary History* 11, no. 2/3 (1976): 65–77.

Pudney, John. *The Thomas Cook Story*. Stuttgart: Bernhard Tauchnitz, 1953.

Pursell, Carroll. "Domesticating Modernity: The Electrical Assocation for Women, 1924–1986." *British Journal for the History of Science* 32, no. 1 (1999): 47–67.

Putnam, Tim. "The Sewing Machine Comes Home." In *The Culture of Sewing: Gender, Consumption, and Home Dressmaking*, edited by Barbara Burman, 269–84. Oxford: Berg, 1999.

Pye, Denis. *Fellowship is Life: The National Clarion Cycling Club 1895–1995*. Bolton: Clarion Publishing, 1995.

Rabenstein, Ruediger. *Radsport und Gesellschaft: ihre sozial-geschichtlichen Zusammenhaenge in der Zeit van 1867 bis 1914*. Hildesheim: Weidmann, 1991.

Raetsch, Mechthilde. *Köstliche Küche: Anleitungen zum elektrischen Kochen*. 2nd ed. Berlin: Frauendienst-Verlag, 1933.

Rautio, Anna-Maria, and Lars Östlund. "'Starvation Strings' and the Public Good: Development of a Swedish Bike Trail Network in the Early Twentieth Century." *The Journal of Transport History* 33, no. 1 (2012): 42–63.

Reagin, Nancy Ruth. *Sweeping the German Nation: Domesticity and National Identity in Germany, 1870–1945*. Cambridge and New York: Cambridge University Press, 2007.

Redationeel. "Rijwielen en automobielen op de Wereldtentoonstelling te Parijs." *De Kampioen*, August 17, 1900, 687–91.

Redationeel. "Waarom?" *De Kampioen*, July 16, 1897, 757–79.

Reid, Susan E. "'Our Kitchen Is Just as Good': Soviet Responses to the American Kitchen." In *Cold War Kitchen: Americanization, Technology, and European Users*, edited by Ruth Oldenziel and Karin Zachmann, 83–112. Cambridge, MA: MIT Press, 2009.

Reitsma, S.A. "De rijwielbelasting als bestemmingsheffing voor het motorsnelverkeer." *De Opbouw* 2 (June, 1938).

Reynolds, Helen. "'Your Clothes are Materials of War': The British Government Promotion of Home Sewing during the Second World War." In *The Culture of Sewing: Gender, Consumption, and Home Dressmaking*, edited by Barbara Burman, 327–39. Oxford: Berg, 1999.

Richter, Amy G. *Home on the Rails: Women, the Railroad, and the Rise of Public Domesticity*. Chapel Hill, NC: University of North Carolina Press, 2005.

"Rijwielmishandeling op Duitsche spoorwegen." *De Kampioen*, June 17, 1898, 657.

Ring, Jim. *How the English Made the Alps*. London: Murray, 2001.

Rischbieter, Laura. "Globalisierungsprozesse vor Ort: Die Interdependenz von Produktion, Handel und Konsum am Beispiel 'Kaffee' zur Zeit des Kaiserreichs." *Comparativ* 17, no. 3 (2007): 28–45.

Robinson, Lisa Mae. "Safeguarded by Your Refrigerator: Mary Engle Pennington's Struggle with the National Association of Ice Industries." In *Rethinking Home Economics: Women and the History of a Profession*, edited by Sarah Stage and Virginia B. Vincenti, 253–70. Ithaca, NY: Cornell University Press, 1997.

Rodenstein, Marianne. *'Mehr Licht, mehr Luft': Gesundheitskonzepte im Städtebau seit 1750*. Frankfurt a.Main: Campus, 1988.

Rodgers, Daniel T. *Atlantic Crossings: Social Politics in a Progressive Age*. Cambridge, MA: Belknap Press of Harvard University Press, 1998.

Rodríguez-Lores, Juan. *Sozialer Wohnungsbau in Europa: Die Ursprünge bis 1918: Ideen, Programme, Gesetze*. Basel: Birkhäuser, 1994.

Rome, Adam. "The Genius of Earth Day." *Environmental History* 15, no. 2 (2010): 194–205.

———. "'Give Earth a Chance': The Environmental Movement and the Sixties." *The Journal of American History* 90, no. 2 (2003): 525–54.

Rommes, Els. "Gender Scripts and the Internet: The Design and Use of Amsterdam's Digital City." PhD diss., University of Twente, 2002.

Rosenberg, Emily S. "Consuming Women: Images of Americanization in the 'American Century'." *Diplomatic History* 23, no. 3 (1999): 479–97.

Ross, Robert. *Clothing: A Global History: Or, the Emperor's New Clothes*. London: Polity Press, 2008.

Rossem, Vincent van. *Het Algemene Uitbreidingsplan van Amsterdam: Geschiedenis en ontwerp*. Rotterdam: NAi Uitgevers, 1994.

Rossfeld, Roman. *Schweizer Schokolade: Industrielle Produktion und kulturelle Konstruktion eines nationalen Symbols 1860–1920*. Baden: hier + jetzt, 2007.

Ruane, Christine. "Clothes Shopping in Imperial Russia: The Development of a Consumer Culture." *Journal of Social History* 28, no. 4 (1995): 765–82.

———. *The Empire's New Clothes: A History of the Russian Fashion Industry, 1700–1917*. New Haven: Yale University Press, 2009.

Rubinstein, David. "Cycling in the 1890s." *Victorian Studies* 21, no. 1 (1977): 47–71.

Rudberg, Eva. *Stockholmsutställningen 1930: Modernismens genombrott i svensk arkitektur*. Stockholm: Stockholmia, 1999.

Rudolph, Nicole. "Domestic Politics: The Cité Expérimentale at Noisy-Le-Sec in Greater Paris." *Modern & Contemporary France* 12, no. 4 (2004): 483–95.

———. "'Who Should Be the Author of a Dwelling?' Architects versus Housewives in 1950s France." *Gender & History* 21, no. 3 (2009): 541–59.

Ruether, Kirsten. "Heated Debates over Crinolines: European Clothing on Nineteenth-Century Lutheran Mission Stations in the Transvaal." *Journal of Southern African Studies* 28, no. 2 (2002): 359–78.

Ryan, Nora. "The Apartment Question: The Avant-garde and the Problem of the Domestic Interior in 1920s Russia." PhD diss., University of California, 2008.

Saarikangas, Kirsi. "What's New? Women Pioneers and the Finnish State Meet the American Kitchen." In *Cold War Kitchen: Americanization, Technology, and European Users*, edited by Ruth Oldenziel and Karin Zachmann, 285–311. Cambridge, MA: MIT Press, 2009.

Saarikoski, Petri, and Jaakko Suominen. "Computer Hobbyists and the Gaming Industry in Finland." *IEEE Annals of the History of Computing* 31, no. 3 (2009): 20–33.

Sachs, Wolfgang. *For Love of the Automobile: Looking Back into the History of Our Desires*. Berkeley: University of California Press, 1992.

Saint-Simon, Henri. *De la réorganisation de la société européenne ou De la nécessité et des moyens de rassembler les peuples de l'Europe en un seul corps politique, en conservant à chacun son indépendance nationale*. Paris: A. Égron, 1814.

Samuel, Henry. "Women Banned from Wearing Trousers in Paris." *Daily Telegraph*, 17 November, 2009.

Sármány-Parsons, Ilona. "Villa und Einfamilienhaus: Urbanisierung und Veränderungen der Wohnkultur im internationalen Kontext." In *Bürgerliche Wohnkultur des Fin de siècle in Ungarn*, edited by Péter Hanák, 243–306. Vienna: Böhlau, 1994.

Scarpellini, Emanuela. "Shopping American-Style: The Arrival of the Supermarket in Postwar Italy." *Enterprise & Society* 5, no. 4 (2004): 625–68.

Schacht, Hans Joachim. *Der Radwegebau in Deutschland*. Halle: Akademischer Verlag, 1937.

Schein, Ernst. *Organisation und Technik des deutschen Rohproduktenhandels*. Ohlau: Hermann Eschenhagen, 1931.

Scheuring, Dirk. "Heavy Duty Denim: 'Quality Never Dates'." In *Zoot Suits and Second-Hand Dress: An Anthology of Fashion and Music*, edited by Angela McRobbie, 225–36. London: Macmillan, 1989.

Schipper, Frank. *Driving Europe: Building Europe on Roads in the Twentieth Century*. Amsterdam: Aksant, 2008.

Schivelbusch, Wolfgang. "Railroad Space and Railroad Time." *New German Critique*, no. 14 (1978): 31–40.

———. *The Railway Journey: The Industrialization of Time and Space in the Nineteenth Century*. Berkeley: University of California Press, 1986 [1977].

Schleinzer, Anika. "'Die Freude der Kinder': Spurensuche nach frühen Modellen von Baukästen als geschlechtsneutralen Artefakten technischen Spielzeugs." In *Gender schafft Wissen – Wissen*schaft *Gender: Geschlechtsspezifische Unterscheidungen und Rollenzuschreibungen im Wandel der Zeit*, edited by Dominik Groß, 61–78. Kassel: Kassel University Press, 2009.

Scholliers, Peter. "From Elite Consumption to Mass Consumption: The Case of Chocolate in Belgium." In *Food Technology, Science and Marketing: European Diet in the Twentieth Century*, edited by Adel P. den Hartog, 127–38. East Linton: Tuckwell Press 1995.

Schot, Johan, and Adri Albert de la Bruhèze. "The Mediated Design of Products, Consumption, and Consumers in the Twentieth Century." In *How Users Matter: The Co-construction of Users and Technology*, edited by Nelly Oudshoorn and Trevor Pinch, 229–45. Cambridge, MA: MIT Press, 2003.

Schot, Johan W., and Vincent Lagendijk. "Technocratic Internationalism in the Interwar Years: Building Europe on Motorways and Electricity Networks." *Journal of Modern European History* 6, no. 2 (2008): 196–217.

Schueler, Judith. *Materialising Identity: The Co-Construction of the Gotthard Railway and Swiss National Identity*. Amsterdam: Aksant, 2008.

Schulz, Thorsten. "Transatlantic Environmental Security in the 1970s? NATO's 'Third Dimension' as an Early Environmental and *Human Security* Approach." *Historical Social Research* 35, no. 4 (2010): 309–28.

Schweitzer, Marlis. "American Fashions for American Women: The Rise and Fall of Fashion Nationalism." In *Producing Fashion: Commerce, Culture and Consumers*, edited by Regina L. Blaszczyk, 130–49. Philadelphia: University of Pennsylvania Press, 2008.

Scott Smith, Giles. *Networks of Empire: The US State Department's Foreign Leader Program in the Netherlands, France, and Britain, 1950–70*. Brussels: Peter Lang, 2008.

Scott, William R. "California Casual: Lifestyle Marketing and Men's Leisurewear, 1930–1960." In *Producing Fashion: Commerce, Culture and Consumers*, edited by Regina L. Blaszczyk, 169–86. Philadelphia: University of Pennsylvania Press, 2008.

"Security for Railway Travellers." *The Irish Times*, December 14, 1860, 3.

Seligman, Kevin L. "Dressmakers' Patterns: The English Commercial Paper Pattern Industry, 1878–1950." *Costume* 37 (2003): 95–113.

Selin, Henrik, and Björn-Ola Linnér. "The Quest for Global Sustainability: International Efforts on Linking Environment and Development: CID Graduate Student and Postdoctoral Fellow Working Paper No. 5." Cambridge, MA: Harvard University, 2005.

Shearer, David. "Elements Near and Alien: Passportization, Policing, and Identity in the Stalinist State, 1932–1952." *Journal of Modern History* 76, no. 4 (2004): 835–81.

Sherwood, Shirley. *Venice Simplon Orient-Express: The Return of the World's Most Celebrated Train*. London: Weidenfeld & Nicolson, 1983.

Simmons, Jack. "Railways, Hotels, and Tourism in Great Britain 1839–1914." *Journal of Contemporary History* 19, no. 2 (1984): 201–22.

Simpson, Claire S. "Respectable Identities: New Zealand Nineteenth-Century 'New Women'—on Bicycles!" *The International Journal of the History of Sport* 18, no. 2 (2001): 54–77.

"Sjonge, sjonge." *De Kampioen*, December 1911, 945.

Sklar, Katherine Kish. *Catharine Beecher: A Study in American Domesticity*. New Haven: Yale University Press, 1973.

Sklar, Kathryn Kish. "The Consumers' White Label Campaign of the National Consumers' League, 1898-1918." In *Getting and Spending: European and American Consumer Society in the Twentieth Century*, edited by Susan Strasser, Charles McGovern, and Matthias Judt, 17–35. Cambridge: Cambridge University Press, 1998.

Sluijter, Babette. "Kijken is grijpen: Zelfbedieningswinkels, technische dynamiek en boodschappen doen in Nederland na 1945." PhD diss., Eindhoven University of Technology, 2007.

Söderberg, Johan. "Users in the Dark: Development of a User-Controlled Technology in the Czech Wireless Community." In *Hacking Europe: From Computer Cultures to Demoscenes*, edited by Gerard Alberts and Ruth Oldenziel. New York: Springer, 2013.

Somer, Kees. *The Functional City: The CIAM and Cornelis van Eesteren, 1928–1960*. Rotterdam: NAi Uitgevers, 2007.

Sørensen, Knut H. "Domestication: The Enactment of Technology." In *Domestication of Media and Technology*, edited by Thomas Berker, Maren Hartmann, Yves Punie, and Katie J. Ward, 40–61. Maidenhead: Open University Press, 2006.

Sörlin, Sverker. "Utopin i verkligheten: Ludvid Nordström och det moderna Sverige." In *I framtidens tjänst: Ur folkhemmets idéhistoria*, 166–95. Malmö: Gidlunds, 1986.

Spiekermann, Uwe. "Twentieth-Century Product Innovations in the German Food Industry." *Business History Review* 83, no. 2 (2009): 291–315.

Spigel, Lynn. *Welcome to the Dreamhouse: Popular Media and Postwar Suburbs*. Durham, NC: Duke University Press, 2001.

"Sport en politiek." *Tilburgsche courant*, March 1898.

Stadt- und Landesplanung Bremen, 1926–1930. Bremen: H.M. Hauschild, 1931.

Stage, Sarah, and Virginia B. Vincenti, eds. *Rethinking Home Economics: Women and the History of a Profession*. Ithaca, NY: Cornell University Press, 1997.

"Staten-Generaal. Tweede Kamer. Vergadering van Donderdag 16 October... Tariefwet." *Nieuwe Rotterdamsche Courant*, October 17, 1924.

"Staten Generaal. Eerste Kamer. Vergadering 23 dec....Belastingonderwerpen." *Nieuwe Rotterdamsche Courant*, December 23, 1926.

Steele, Valerie. *Paris Fashion: A Cultural History*. Oxford: Berg, 1999 [1988].

———, and Fashion Institute of Technology Museum. *Fashion, Italian Style*. New Haven, CT: Yale University Press, 2003.

Steendijk-Kuypers, Jacoba. "Freedom on the Bicycle: Women's Choice." *Cycle History* 10 (1999): 127–32.

Steiner, Edward Alfred. *On the Trail of the Immigrant*. New York: Revell, 1906.

Stevens, Thomas. *Around the World on a Bicycle: From San Francisco to Teheran*. Vol. I. London: Century, 1888.

———. *Around the World on a Bicycle: From Teheran to Yokohama*. Vol. II. New York: Charles Scribner's Sons, 1888.

Stewart, Mary Lynn. "Copying and Copyrighting Haute Couture: Democratizing Fashion, 1900–1930s." *French Historical Studies* 28, no. 1 (2005): 103–30.

Stieber, Nancy. *Housing Design and Society in Amsterdam: Reconfiguring Urban Order and Identitiy, 1900–1920*. Chicago and London: University of Chicago Press, 1998.

Stilgoe, John R. *Metropolitan Corridor: Railroads and the American Scene*. New Haven, CT: Yale University Press, 1985.

Stitziel, Judd. *Fashioning Socialism: Clothing, Politics and the Consumer Culture in East Germany*. Oxford: Berg, 2005.

———. "Shopping, Sewing, Networking, Complaining: Consumer Culture and the Relationship between State and Society in the GDR." In *Socialist Modern: East German Everyday Culture and Politics*, edited by Katherine Pence and Paul Betts, 253–86. Ann Arbor: University of Michigan Press, 2008.

Stoffers, Manuel, Harry Oosterhuis, and Peter Cox. "Bicycle History as Transport History: The Cultural Turn." In *Mobility in History: Themes in Transport*, edited by Gijs Mom, Peter Norton, Georgine Clarsen, and Gordon Pirie, 265–74. Neuchatel: Editions Alphil-Presses Universitaires Suisses, 2010.

Stoilova, Elitsa. "Bulgarian Yoghurt: Manufacturing and Exporting Authenticity." PhD diss., Eindhoven University of Technology, forthcoming.

"Strange Murder in Paris." *The Irish Times*, December 10, 1860, 3.

Strasser, Susan. *Waste and Want: A Social History of Trash*. New York: Henry Holt, 1999.

Sudrow, Anne. *Der Schuh im Nationalsozialismus: Eine Produktgeschichte im deutsch-britisch-ameranischen Vergleich*. Göttingen: Wallstein Verlag, 2010.

Tanner, Jakob. *Fabrikmahlzeit: Ernährungswissenschaft, Industriearbeit und Volksernährung in der Schweiz 1890–1950*. Zurich: Chronos, 1999.

Tascon, Ignazio Gonzales. "The Coffee Innovation Culture versus the Chocolate Tradition during the Eighteenth and Nineteenth Centuries." In *Kaffee im Spiegel europäischer Trinksitten: Coffee in the Context of European Drinking Habits*, edited by Daniela U. Ball, 145–52. Zurich: Johann Jacobs Museum, 1991.

Taylor, Karin. "From Trips to Modernity to Holidays in Nostalgia: Tourism History in Eastern and Southeastern Europe." *Tensions of Europe Working Paper*, no. 1, 2011.

Taylor, Michael. "The Bicycle Boom and the Bicycle Bloc: Cycling and Politics in the 1890s." *Indiana Magazine of History* 104, no. 3 (2008): 213–40.

Terhoeven, Petra. *Liebespfand fürs Vaterland: Krieg, Geschlecht und faschistische Nation in der italienischen Gold- und Eheringsammlung 1935/36*. Tübingen: Max Niemeyer, 2003.

Terlinden, Ulla, and Susanna von Oertzen. *Die Wohnungsfrage ist Frauensache! Frauenbewegung und Wohnreform 1870 bis 1933*. Berlin: Dietrich Reimer Verlag, 2006.

Teuteberg, Hans J., and Günter Weigelmann. *Unsere tägliche Kost: Geschichte und regionale Prägung*. Münster: F. Coppenrath, 1986.

Theien, Iselin. "Shopping for the 'People's Home': Consumer Planning in Norway and Sweden after Second World War." In *The Expert Consumer: Associations and Professionals in Consumer Society*, edited by Alain Chatriot, Marie-Emmanuelle Chessel, and Matthew Hilton, 137–50. Aldershot: Ashgate, 2006.

Thompson, Christopher S. "Bicycling, Class, and the Politics of Leisure in Belle Epoque France." In *Histories of Leisure*, edited by Rudy Koshar, 131–46. London: Berg, 2002.

Thoms, Ulrike. "Changes in the Kitchen Range and Changes in Food Preparation Techniques in Germany, 1850–1950." In *Food and Material Culture: Proceedings of the Fourth Symposium of the International Commission for Research into European Food History*, edited by Martin R. Schärer and Alexander Fenton, 48–76. East Linton: Tuckwell Press, 1998.

Thorsheim, Peter. "Salvage and Destruction: The Recycling of Books and Manuscripts in Great Britain during the Second World War." *Contemporary European History* 22, no. 3 (2013): 431–52.

Tobin, Gary Allan. "The Bicycle Boom of the 1890's: The Development of Private Transportation and the Birth of the Modern Tourist." *Journal of Popular Culture* 7, no. 4 (1974): 838–49.

Tolstoy, Leo. *Anna Karenina*. Translated by Joel Carmichael. New York: Bantam Dell, 2006 [1877].

Tomes, Nancy J. *The Gospel of Germs: Men, Women, and the Microbe in American Life*. Cambridge, MA: Harvard University Press, 1998.

Torpey, John. *The Invention of the Passport: Surveillance, Citizenship and the State*, Cambridge Studies in Law and History. New York: Cambridge University Press, 2000.

Trentmann, Frank. *Free Trade Nation: Commerce, Consumption, and Civil Society in Modern Britain*. Oxford: Oxford University Press, 2008.

Trischler, Helmuth, and Martin Kohlrausch. *Building Europe on Expertise: Innovators, Organizers, Networkers*. Basingstoke: Palgrave Macmillan, forthcoming 2013.

Troy, Nancy J. *Couture Culture: A Study in Modern Art and Fashion*. Cambridge, MA: MIT Press, 2003.

———. "The Theatre of Fashion: Staging Haute Couture in Early 20th-Century France." *Theatre Journal* 53, no. 1 (2001): 1–32.

Turner, Fred. *From Counterculture to Cyberculture: Stewart Brand, the Whole Earth Network, and the Rise of Digital Utopianism*. Chicago: University of Chicago Press, 2006.

Tympas, Aristotle, Fotini Tsaglioti, and Theodore Lekkas. "Universal Machines vs. National Languages: Computerization as Production of New Localities." Paper presented at the International Conference "Technologies of Globalization," Darmstadt, 30–31 October, 2008.

Uhlig, Günther. *Kollektivmodell 'Einküchenhaus': Wohnreform und Architekturdebatte zwischen Frauenbewegung und Funktionalismus 1900–1933*, Werkbund Archiv. Giessen: Anabas Verlag, 1981.

Urry, John. *The Tourist Gaze: Leisure and Travel in Contemorary Societies*. London: Sage, [1990] 2002.

Van Laak, Dirk. *Imperiale Infrastrukturen: Deutsche Planungen für eine Erschliessung Afrikas 1880 bis 1960*. Paderborn: Schöningh, 2004.

van Lente, Dick. "The Romance of Technology in the Cold War Era: Leonard de Vries and His Hobby Clubs, 1947–1966." Paper presented at the ICOHTEC, TICCIH & Worklab joint conference "Reusing the Industrial Past" Tampere, Finland, 10–15 August, 2010.

Veblen, Thorstein. *Conspicuous Consumption*. London: Penguin, 2005 [1899].

Veenis, Milena. "Dromen van dingen: Oost Duitse fantasieën over de westerse consumptiemaatschappij." PhD diss., University of Amsterdam, 2008.

Veraart, Frank. "Transnational (Dis)connection in Localizing Personal Computing in the Netherlands, 1975–1990." In *Hacking Europe: From Computer Cultures to Demoscenes*, edited by Gerard Alberts and Ruth Oldenziel. New York: Springer, 2013.

———. "Vormgevers van persoonlijk computergebruik: De ontwikkeling van computers voor kleingebruikers in Nederland 1970–1990." PhD diss., TU Eindhoven, 2008.

Veraart, Frank C.A. "Basicode: Co-Producing a Microcomputer Esperanto." *History of Technoloy* 28 (2008): 129–47.

———. "Geschiedenis van de fiets in Nederland, 1870–1940." MA Thesis, TU Eindhoven, 1995.

———. "Reis in de tijd: fietspaden, van gerieflijke paden tot fietsbeleid." *Verkeerskunde* 60, no. 5 (2009): 21–6.

"Verband- en reparatiekisten." *Algemeen Handelsblad*, November 3, 1896.

Vickery, Michael R. "Conservative Politics and the Politics of Conservation: Richard Nixon and the Environmental Protection Agency." In *Green Talk in the White House: The Rhetorical Presidency Ecounters Ecology*, edited by Tarla Rai Peterson, 113–33. College Station, TX: Texas A&M University Press, 2004.

"Vlijtige handen kan een succes worden: Ruim driehonderd inzendingen uit geheel Friesland." *Friesche Koerier*, 12 March 1953.

Vogel, David. *Trading Up: Consumer and Environmental Regulation in a Global Economy*. Cambridge, MA: Harvard University Press, 1995.

von Saldern, Adelheid. *Häuserleben: Zur Geschichte städtischen Arbeiterwohnens vom Kaiserreich bis heute*. Bonn: J.H.W. Dietz, 1995.

———. "Im Hause, zu Hause: Wohnen im Spannungsfeld von Gegebenheiten und Aneignungen." In *Geschichte des Wohnens, Vol. 3: 1800-1918: Das bürgerliche*

Zeitalter, edited by Jürgen Reulecke, 145–332. Stuttgart: Deutsche Verlags-Anstalt, 1997.

Vossen, Joachim. *Bukarest—Die Entwicklung des Stadtraums: Von den Anfängen bis zur Gegenwart*. Berlin: Dietrich Reimer Verlag, 2004.

Wagenaar, Michiel. *Stedebouw en burgerlijke vrijheid: De contrasterende carriers van zes Europese hoofdsteden*. Bussum: Uitgeverij THOTH, 2001.

Wagner, Curt. *Konserven und Konservenindustrie in Deutschland: Mit einem Anhang Upton Sinclairs 'The Jungle'*. Jena: Gustav Fischer, 1907.

Waldén, Louise. *Genom symaskinens nålsöga: Teknik och social förändring i kvinnokultur och manskultur*. Stockholm: Carlssons, 1990.

Walsh, Margaret. "The Democratization of Fashion: The Emergence of the Women's Dress Pattern Industry." *The Journal of American History* 66, no. 2 (1979): 299–313.

Ward, Barbara. *Spaceship Earth*. New York: Columbia University Press, 1966.

Wasiak, Patryk. "Playing and Copying: Social Practices of Home Computer Users in Poland during the 1980s." In *Hacking Europe: From Computer Cultures to Democenes*, edited by Gerard Alberts and Ruth Oldenziel. New York: Springer, 2013.

Weber, Heike. "Towards 'Total' Recycling: Women, Waste and Food-Waste Recovery in Germany, 1914–1939." *Contemporary European History* 22, no. 3 (2013): 371–97.

Wedgwood, Ralph L., and J.E. Wheeler. *International Rail Transport*. London: Oxford University Press, 1946.

"A Week's Paris Gossip." *The New York Times*, April 24, 1899.

Weindling, Paul Julian. *Epidemics and Genocide in Eastern Europe, 1890–1945*. Oxford: Oxford University Press, 2000.

Welke, Barbara. *Recasting American Liberty: Gender, Race, Law, and the Railroad Revolution, 1865–1920*. New York: Cambridge University Press, 2001.

Wessely, Anna. "Travelling People, Travelling Objects." *Cultural Studies* 16, no. 1 (2002): 3–15.

Westermann, Andrea. "When Consumer Citizens Spoke Up: West Germany's Early Dealings with Plastic Waste." *Contemporary European History* 22, no. 3 (2013): 477–98.

"Wet op de veiligheid van het verkeer." *De Kampioen*, December 1, 1899, 1143–44.

Wezel, Ruud van. "Doe-het-zelf: een klus die nooit geklaard is." In *Schoon genoeg: huisvrouwen en huishoudtechnologie in Nederland 1898–1998*, edited by Ruth Oldenziel and Carolien Bouw, 231–52. Nijmegen: SUN, 1998.

White Jr., John H. *The American Railroad Passenger Car*. Baltimore, MD: Johns Hopkins University Press, 1978.

Whitfield, Stephen J. *The Culture of the Cold War*. Baltimore, MD: Johns Hopkins University Press, 1996 [1991].

Whyte, William Hollingsworth. *The Organization Man*. New York: Simon and Schuster, 1956.

Wiedemann, Inga. *Herrin im Hause: Durch Koch- und Haushaltsbücher zur bürgerlichen Hausfrau*. Pfaffenweiler: Centaurus, 1993.

Wiesen, S. Jonathan. "Miracles for Sale: Consumer Displays and Advertising in Postwar West Germany." In *Consuming Germany in the Cold War*, edited by David F. Crew, 151–78. Oxford: Berg, 2003.

Wilkerson, Isabel. *The Warmth of Other Suns: The Epic Story of America's Great Migration*. New York: Random House, 2010.

Wilson, Elizabeth. *Adorned in Dreams: Fashion and Modernity*. Revised and updated ed. London: I.B. Taurus, 2007 [1985].

Winkelman, Hélene J.M. *Hier is Barbie en de rest van de Mattel-familie (1964–2003): Veertig jaar barbiepoppen in Nederland*. Amsterdam: Aksant, 2003.

Winner, Langdon. *The Whale and the Reactor: A Search for Limits in an Age of High Technology*. Chicago: University of Chicago Press, 1986.

Wright, Lawrence. *Home Fires Burning: The History of Domestic Heating and Cooking*. London: Routledge & Kegan Paul, 1964.

Zachmann, Karin. "Managing Choice: Constructing the Socialist Consumption Junction in the German Democratic Republic." In *Cold War Kitchen: Americanization, Technology, and European Users*, edited by Ruth Oldenziel and Karin Zachmann, 259–84. Cambridge, MA: MIT Press, 2009.

"Zadania Polskiej Ligi Nabiałowej." *Codzienna Gazeta Handlowa*, September 30, 1934, 3.

Zakim, Michael. "A Ready-Made Business: The Birth of the Clothing Industry in America." *The Business History Review* 73, no. 1 (1999): 61–90.

Żerkowski, J. "Spółdzielczość Spożywców w Polsce w latach 1869–1944." In *100 lat polskiej Spółdzielczości Spożywców*, 13–28. Warsaw: Zakład Wydawnictw CRS, 1969.

Zevenbergen, Cees. *Toen zij uit Rotterdam vertrokken: Emigratie via Rotterdam door de eeuwen heen*. Zwolle: Waanders, 1990.

Zezina, Maria R. "The Introduction of Motor Vehicles on a Mass Scale in the USSR: From Idea to Implementation." In *Towards Mobility: Varieties of Automobilism in East and West*, edited by Corinna Kuhr-Korolev and Dirk Schlinkert, 43–54. Wolfsburg: Volkswagen AG, 2009.

Zieger, Robert H. "The Paradox of Plenty: The Advertising Council and the Post-Sputnik Crisis." *Advertising & Society Review* 4, no. 1 (2003).

Zimmerman, M.[ax] M[endell]. *Los supermercados*. Madrid: Riald, 1959.

Zimmerman, Max Mandell. *The Super Market: A Revolution in Distribution*. New York: McGraw-Hill, 1955.

Zola, Émile. *The Beast Within*. Translated by Roger Whitehouse, Penguin Classics. London and New York: Penguin, 2007.

Zutphen, Nan van. "Sociale geschiedenis van het fietsen te Leuven, 1880–1900." In *Jaarboek 1979 Vrienden Stedelijke Musea: Fiets en film rond 1900: moderne uitvindingen in Leuvense samenleving*, edited by Nan van Zutphen and Guido Convents, 11–256. Leuven: Crab, 1981.

Zweiniger-Bargielowska, Ina. *Austerity in Britain: Rationing, Controls, and Consumption, 1939–1955*. Oxford and New York: Oxford University Press, 2000.

Illustration Credits

Permissions to reproduce the illustrations and photographs in this book have generously been granted by the institutions and collections named here, and are gratefully acknowledged by the editors, the authors, the Foundation for the History of Technology, and Palgrave Macmillan.

The Foundation for the History of Technology has carefully tried to locate all rights holders. Parties who despite this feel that they are entitled to certain rights are requested to contact the Foundation for the History of Technology (www. histech.nl).

Cover Original caption: "Deutsche StadtimfünftenKriegsjahr. Flugs radelt sich so ein Schnit aus, mit dessen Hilfe man dann die schönsten Kinderkleidchen oder auch anderes aus noch verwandbaren gebrauchten Altkleidern herstellen kann": Germany 1944–5. Orbis Photo. Permission by Centre for Historical Research and Documentation on War and Contemporary Society (CEGESOMA), Brussels (Belgium). Ref. no. 200804.

Making Europe:
Series Acknowledgements

Making Europe is the result/product of an unusual collaboration among a host of individuals and organizations. The *Making Europe* authors and series editors feel extremely fortunate to be working with them. We list here individuals and organizations who contributed to the entire series. Each volume in the series also has its own separate acknowledgements.

Making Europe was initiated by:

- Foundation for the History of Technology (www.histech.nl)

Making Europe is sponsored by:

- Eindhoven University of Technology (www.tue.nl)
- SNS REAAL Fonds (www.snsreaalfonds.nl)
- Next Generations Infrastructures (www.nextgenerationsinfrastructures.eu)
- Foundation for the History of Technology Corporate Program (www.histech.nl) that includes:
  - DSM (www.dsm.com)
  - EBN (www.ebn.nl)
  - FrieslandCampina(www.frieslandcampina.com)
  - Philips (www.philips.nl)
  - SIDN (www.sidn.nl)
  - TNO (www.tno.nl)

Making Europe has been made possible thanks to:

- Tensions of Europe Network (www.tensionsofeurope.eu)
- European Science Foundation EUROCORES Programme Inventing Europe – Technology and the Making of Europe, 1850 to the Present (www.esf.org)
- Research Theme Group Grant of the Netherlands Institute of Advanced Studies (NIAS) in 2010–11 (www.nias.nl)
- European University Institute in Florence, Italy for providing the support in developing the series and for the founding workshop (3–6 July 2008) (www.eui.eu)

Making Europe benefited from the feedback of a community of scholars who have been involved in the series from the start:

Håkon With Andersen, Alec Badenoch, Robert Bud, David Burigana, Cornelis Disco, Paul Edwards, Valentina Fava, Karen Johnson Freeze, Andrea Guintini, Gabrielle Hecht, Rüdiger Klein, Eda Kranakis, John Krige, Leonard Laborie, Vincent Lagendijk, Suzanne Lommers, Slawomir Lotysz, Dagmara Jajeśniak-Quast, Karl-Erik Michelsen, Matthias Middell, Thomas J. Misa, Dobrinka Parusheva, Kiran Patel, Pierre-Yves Saunier, Emanuela Scarpellini, Frank Schipper, Michael Strang, Ivan Tchalakov, Frank Trentmann, Aristotle Tympas, Hans Weinberger

Making Europe relied on the unflagging support of:

Picture editors
- Katherine Kay-Mouat
- Giel van Hooff
- Jan Korsten (Management)

Text editors

- Lisa Friedman
- James Morrison

PalgraveMacmillan

- Jenny McCall (Publisher)
- Holly Tyler (Editorial Assistant)
- Philip Hillyer (Copy-Editor and Editorial Services Consultant)
- Susan Boobis (Indexer)

Office Foundation for the History of Technology:

- Sonja Beekers (Secretarial Support)
- Jan Korsten (Business Director)
- Loek Stoks (Bookkeeping)
- Henk Treur (Volunteer)

Board Foundation for the History of Technology:

- Hans de Wit (chair)
- Martin Schuurmans (vice-chair/treasurer)
- Saskia Blom
- Herman de Boon
- Wim van Gelder
- Jacques Joosten
- Michiel Westermann
- Harry Lintsen (advisor)

Index

Page numbers in *italics* refer to endnotes

Printed and bound by CPI Group (UK) Ltd, Croydon, CR0 4YY